Praise for *Now You See It*

"As Ms. Davidson puts it: 'Pundits may be asking if the Internet is bad for our children's mental development, but the better question is whether the form of learning and knowledge-making we are instilling in our children is useful to their future.' In her galvanic new book, *Now You See It*, Ms. Davidson asks, and ingeniously answers, that question. One of the nation's great digital minds, she has written an immensely enjoyable omni-manifesto that's officially about the brain science of attention. But the book also challenges nearly every assumption about American education. . . . Rooted in field experience, as well as rigorous history, philosophy, and science—this book about education happens to double as an optimistic, even thrilling, summer read. It supplies reasons for hope about the future. Take it to the beach. That much hope, plus that much scholarship, amounts to a distinctly unguilty pleasure." —Virginia Heffernan, *The New York Times*

"A remarkable new book, *Now You See It* . . . offers a fresh and reassuring perspective on how to manage anxieties about the bewildering pace of technological change. . . . Her work is the most powerful yet to insist that we can and should manage the impact of these changes in our lives." —*Fast Company*

"In a chatty, enthusiastic style, the author takes us on a journey through contemporary classrooms and offices to describe how they are changing—or, according to her, should change. Among much else, we need to build schools and workplaces that match the demands of our multitasking brains. That means emphasizing 'nonlinear thinking,' 'social networks,' and 'crowdsourcing.' . . . *Now You See It* is filled with instructive anecdotes and genuine insights." —*The Wall Street Journal*

"Her book *Now You See It* celebrates the brain as a lean, mean, adaptive multitasking machine that—with proper care and feeding—can do much more than our hidebound institutions demand of it. The first step is transforming schools, which are out of touch with the radical new realities of the Internet era. . . . Davidson is such a good storyteller." —*The New York Times*

"Davidson has produced an exceptional and critically important book, one that is all-but-impossible to put down and likely to shape discussions for years to come." —*Publishers Weekly*

"*Now You See It* is humorous, poignant, entertaining, endearing, touching, and challenging. It is a book I would happily recommend to anyone engaged in teaching at any level, because it aims both to comfort and to disrupt; it is devised to convince readers that the human mind is ready for the next quantum advance into our collective future, whatever that may be. It is certainly all-embracing in its scope, demonstrating how a sound knowledge of the many ways we can learn in new, media-rich environments might provide a better understanding of how individuals can attain their optimum potential." —*Times Higher Education*

"Practice Collaboration by Difference: This idea is stolen directly from Cathy N. Davidson's marvelous book *Now You See It*. (Note: If you are looking for someone to visit your campus to both spark ideas and bring people together you would have a hard time doing better than Cathy.) Davidson's work suggests that homogenous teams will have a difficult time completing innovative work. If innovation is our goal then we must pay careful attention to the diversity of the people around our project tables."
—*Inside Higher Ed*

"A preview of the future from an educational innovator . . . It is becoming clear that our minds are capable of multitasking to a degree far beyond what the twentieth-century assembly-line worker or middle manager was trained to do. . . . [Davidson's] points are worth pondering."
—*Kirkus Reviews*

"There is an emerging consensus that higher education has to change significantly, and Davidson makes a compelling case for the ways in which digital technology, allied with neuroscience, will play a leading role in that change."
—*The Chronicle of Higher Education*

"[Davidson] makes a provocative case for radical educational and business reforms. She presents vivid examples of schools and workplaces unleashing learning and achievement. . . . Davidson's call to experiment with digital schemes that turn students and workers into motivated problem solvers rings as clear as a bell atop a little red schoolhouse."
—*Science News*

"The book's purpose and strength are in detailing the important lessons we can glean from the online world. If Davidson is right, twenty-first-century society will move away from categorizing people based on standardized tests, which are crude measures of intelligence at best. Instead we will define new metrics, ones that are better aligned with the skills needed to succeed in the shifting global marketplace. And those who cannot embrace this multidisciplinary world will simply be left behind."
—*Scientific American*

"Davidson takes the technological bull by the horns and argues that concentration hasn't gone downhill with the Internet: we're just operating with an outdated notion of attention, in the workplace and at home. . . . Davidson's claim that mono-tasking (the idea that a person can focus on one single task at hand) is an unrealistic model of how the brain works, seems strikingly persuasive. Davidson also calls for a reform in education, suggesting ways in which technology can be incorporated into the classroom and help kids become multitasking, problem-solving thinkers."
—*LA Weekly*

"The technological changes around us are of unprecedented proportions. What effects this has on us and what it tells us about human nature more generally is a central question for society and for all of us personally. In this book Cathy Davidson integrates findings from psychology, attention, neuroscience, and learning theory to help us get a glimpse of the future and more importantly a better understanding of our own individual potential."
    —Dan Ariely, James B. Duke Professor of Psychology and Behavioral
        Economics, Duke University, and author of *Predictably Irrational:
        The Hidden Forces That Shape Our Decisions*

"*Now You See It* is simply fantastic. Only Cathy Davidson could pull off such a sweeping book. It is about so much more than just education or even learning. It is about a way of being. Her book and stories are incredibly important for the true arc of life learning and for constantly becoming!"
    —John Seely Brown, former chief scientist, Xerox Corporation, current director
        of Palo Alto Research Center (PARC), and coauthor of *The Social Life of
        Information* and *A New Culture of Learning: Cultivating the Imagination
        for a World of Constant Change*

"Cathy Davidson has one of the most interesting and wide-ranging minds in contemporary scholarship, a mind that ranges comfortably over literary arts, literacy, psychology, and brain science. I've been stimulated by her writings. Her ambitious and timely book is certain to attract a lot of attention and to catalyze many discussions."   —Howard Gardner, Hobbs Professor of Cognition and
                Education, Harvard University

"The self-reprogramming capacity of the human mind, together with the way our communication technologies influence our thinking, are combining to reprogram our attention. One cutting edge of educational practice is participatory learning—giving students a more active, exploratory role based on critical inquiry—and one frontier of brain research is what is happening to our attention in the always-on era. Cathy Davidson is a natural to bring together these neuroscientific and educational themes."   —Howard Rheingold, lecturer at Berkeley and Stanford, and
                author of *Smart Mobs* and *Net Smart*

"*Now You See It* starts where Malcolm Gladwell leaves off, showing how digital information will change our brains. Think Alvin Toffler meets Ray Kurzweil on Francis Crick's front porch. We need this book."
    —Daniel Levitin, James McGill Professor of Neuroscience, McGill
        University, and author of the *New York Times* bestsellers
        This Is Your Brain on Music and *The World in Six Songs*

Cathy Davidson teaches at Duke University, where she codirects the Ph.D. Lab in Digital Knowledge and holds distinguished chairs in English and interdisciplinary studies. She served as Duke's first vice provost for interdisciplinary studies and helped to create the Program in Information Science + Information Studies and the Center for Cognitive Neuroscience. She is a cofounder of the global learning network HASTAC, which administers the annual two $2 million HASTAC/MacArthur Foundation Digital Media and Learning Competitions, and she was recently appointed by President Obama to the National Council on the Humanities. Her more than a dozen books include *Thirty-Six Views of Mt Fuji*, *Revolution and the Word*, and *The Future of Thinking* (with HASTAC cofounder David Theo Goldberg). A frequent speaker and consultant on institutional change at universities, corporations, and nonprofits around the world, she writes for *Harvard Business Review*, *The Wall Street Journal*, *Fast Company*, *The Chronicle of Higher Education*, *The Washington Post*, *Times Higher Ed*, and many other publications in the United States and abroad.

NOW YOU SEE IT

How Technology and Brain Science Will
Transform Schools and Business for
the 21st Century

Cathy N. Davidson

PENGUIN BOOKS

PENGUIN BOOKS

Published by the Penguin Group
Penguin Group (USA) Inc., 375 Hudson Street,
New York, New York 10014, U.S.A.
Penguin Group (Canada), 90 Eglinton Avenue East, Suite 700,
Toronto, Ontario, Canada M4P 2Y3
(a division of Pearson Penguin Canada Inc.)
Penguin Books Ltd, 80 Strand, London WC2R 0RL, England
Penguin Ireland, 25 St. Stephen's Green, Dublin 2, Ireland
(a division of Penguin Books Ltd)
Penguin Books Australia Ltd, 250 Camberwell Road, Camberwell,
Victoria 3124, Australia
(a division of Pearson Australia Group Pty Ltd)
Penguin Books India Pvt Ltd, 11 Community Centre, Panchsheel Park,
New Delhi – 110 017, India
Penguin Group (NZ), 67 Apollo Drive, Rosedale, Auckland 0632,
New Zealand (a division of Pearson New Zealand Ltd)
Penguin Books (South Africa) (Pty) Ltd, 24 Sturdee Avenue,
Rosebank, Johannesburg 2196, South Africa

Penguin Books Ltd, Registered Offices:
80 Strand, London WC2R 0RL, England

First published in the United States of America by Viking Penguin,
a member of Penguin Group (USA) Inc. 2011
Published in Penguin Books 2012

10 9 8 7 6 5 4 3 2 1

THE LIBRARY OF CONGRESS HAS CATALOGED THE HARDCOVER EDITION AS FOLLOWS:
Davidson, Cathy N., ———.
    Now you see it : how the brain science of attention will transform the way we live,
work and learn / Cathy N. Davidson.
        p. cm.
    Includes bibliographical references and index.
    ISBN 978-0-670-02282-3 (hc.)
    ISBN 978-0-14-312126-8 (pbk.)
1. Attention.   2. Distraction (Psychology)   3. Neuroscience.   I. Title.
BF321.D38 2011
153.7'33—dc22   2011009307

Printed in the United States of America
Set in Adobe Garamond   Designed by Alissa Amell

For Ken

Contents

||||||||||||||||||||||||

Introduction

I'll Count—You Take Care of the Gorilla *1*

|||||||

Part One

Distraction and Difference: The Keys to Attention and the Changing Brain

1

Learning from the Distraction Experts *23*

2

Learning Ourselves *44*

|||||||

Part Two

The Kids Are All Right

3

Project Classroom Makeover *61*

4

How We Measure *105*

5

The Epic Win *132*

|||||||

x **Contents**

Part Three

Work in the Future

6

The Changing Workplace *165*

7

The Changing Worker *208*

||||||||

Part Four

The Brain You Change Yourself

8

You, Too, Can Program Your VCR (and Probably Should) *247*

||||||||

Conclusion

Now You See It *277*

Acknowledgments *293*

Appendix

Twenty-first-Century Literacies—a Checklist *297*

Notes *301*

Index *331*

Introduction

||||||||||||||||||||||||

I'll Count—You Take Care of the Gorilla

Five or six years ago, I attended a lecture on the science of attention that was part of a luncheon series designed to showcase cutting-edge research by the best and brightest at my university. A philosopher who conducts research over in the medical school was talking about attention blindness, the basic feature of the human brain that, when we concentrate intensely on one task, causes us to miss just about everything else.[1] Because we can't see what we can't see, our lecturer was determined to catch us in the act. He had us watch a video of six people tossing basketballs back and forth, three in white shirts and three in black, and our task was to keep track only of the tosses between the people in white. I hadn't seen the video back then, although it's now a classic, featured on punked-style TV shows or YouTube versions enacted at frat houses under less than lucid conditions. The tape rolled, and everyone began counting.

Everyone except me. I'm dyslexic, and the moment I saw that grainy tape with the confusing basketball tossers, I knew I wouldn't be able to keep track of their movements, so I let my mind wander. My curiosity was piqued, though, when about thirty seconds into the tape, a gorilla sauntered in among the players. She (we later learned a female student was in the gorilla suit) stared at the camera, thumped her chest, and then strode away while they continued passing the balls.

When the tape stopped, the philosopher asked how many people had counted at least a dozen basketball tosses. Hands went up all over. He then asked who had counted thirteen, fourteen even, and congratulated those who'd scored the perfect fifteen. Then he asked, "And who saw the gorilla?"

I raised my hand and was surprised to discover I was the only person at my table and one of only three or four others in the large room to do so. Around me, others were asking, "Gorilla? What gorilla?" Some people were getting annoyed. Several muttered that they'd been "tricked." Instead of answering them, the philosopher rewound the tape and had us watch again. This time everyone saw the gorilla.

He'd set us up, trapping us in our own attention blindness, priming us for his lecture. Yes, there had been a trick, but he wasn't the one who had played it on us. By concentrating so hard on the confusing counting task, we had managed to miss the main event: the gorilla in the midst. In a brief experiment that had required us simply to pay attention, it was our own minds that had deceived us.

Except I hadn't been deceived. I'd seen the gorilla, not because I'm better at this than anyone else—I've taken enough attention tests since that day to know I'm not—but precisely because I wasn't paying attention to counting basketballs. That's how the visual cortex is structured. We think we see the whole world, but we actually see a very particular part of it. For a lot of neuroscientists, that's the cautionary message of the gorilla experiment: We're not nearly as smart as we think we are.[2]

In a very real sense, this book began that day. Attention blindness is the fundamental structuring principle of the brain, and I believe that it presents us with a tremendous opportunity. My take is different from that of many neuroscientists: Where they perceive the shortcomings of the individual, I sense opportunity for collaboration. If we see selectively but we don't all select the same things to see, that also means we don't all miss the same things. If some of us can accurately count basketballs in a confusing situation and some can see the gorilla, we can pool our insights and together see the whole picture. That's significant. The gorilla experiment isn't just a lesson in brain biology but a plan for thriving in a complicated world.

Without focus, the world is chaos; there's simply too much to see, hear, and understand, and focus lets us drill down to the input we believe is most useful to us. Because focus means selection, though, it leaves us with blind spots, and we need methods for working around them. Fortunately, given the interactive nature of most of our lives in the digital age, we have the tools to harness our different forms of attention and take advantage of them.

But there's an important first step, and if we pass over it, we'll never be able to capitalize on the benefits of our interactive world. It's not easy to acknowledge that everything we've learned about how to pay attention *means* that we've been missing everything else—including the gorilla. It's not easy for us rational, competent, confident types to admit that the very key to our success—our ability to pinpoint a problem and solve it, an achievement honed in all those years in school and beyond—may be exactly what limits us. For over a hundred years, we've been training people to see in a particularly individual, deliberative way. No one ever told us that our way of seeing excluded everything else. It's hard for us to believe we're not seeing all there is to see.

But here's the kicker: Unless we're willing to take attention blindness personally, we're going to either flub the basketball count *or* miss the gorilla every single time. We can't even develop a method for solving the dilemma until we admit there's a gorilla in the room and we're too preoccupied counting basketballs to see it.

A great cognitive experiment is like a fantastic magic trick performed by an exceptionally skilled magician. It makes us see things about ourselves that we don't normally see and helps us to believe what might otherwise be impossible to accept about the world we live in. An experiment allows us to see the imperfect and idiosyncratic way our own brain works. That's a key difference between magic tricks and scientific experiments. Scientists don't contrive experiments to trick, surprise, embarrass, or entertain us. They devise experiments so they can learn more about what makes humans tick.

When they were just starting their careers, the young Harvard psychologists Christopher Chabris and Daniel Simons first performed the now-famous gorilla experiment, or what they've come to call the invisible gorilla.[3] It was 1999, and they were determined to come up with a convincing way to illustrate the cognitive principle of selective attention that had been identified way back in the 1970s but that people simply refused to believe.[4] A colleague down the hall was doing a study on fear and happened to have a gorilla suit handy. The rest is history.

Under normal testing conditions, over half of the participants miss the gorilla. Add peer pressure and that figure goes way up. In a live reenactment of this experiment performed in London, with four hundred very social college students packed into an auditorium, only 10 percent noticed the gorilla stride across the stage.[5] We didn't keep an exact count at our event, but our numbers must have rivaled those of the college kids' in London. In our case, the most likely reason so few saw the gorilla was that academics like to do well on tests.

And that's the annoying lesson of attention blindness. The more you concentrate, the more other things you miss.

In one amusing experiment you can view on YouTube, a man and a woman perform a familiar card trick. While the man fans a deck of cards out before her, the woman selects a card at random from a deck, shows it to the viewing audience, and slips it back into the pack. And as is to be expected, the man then "magically" pulls the right card out of the deck.[6] In this case, however, he reveals his secret to the audience: While we were looking at the card the woman chose, he switched to another deck of a different color so that when she replaced the card, he could spot it instantly. But it turns out that's only the beginning of what we've missed. While we've been keeping our eyes on the cards, the man and woman have changed into different clothes, the color of the backdrop has been altered, and even the tablecloth on which the cards were spread has been changed to one of a dramatically different color. Throughout all of this frantic activity a person in a gorilla suit sits onstage, in homage to the Chabris and Simons experiment.

The scariest attention blindness experiment I know is used to train apprentice airplane pilots. You can imagine where this is going. Trainees using a flight simulator are told they will be evaluated on how well they land their plane in the center of a very narrow runway. They have to factor in numerous atmospheric variables, such as wind gusts and the like. But just as the plane is about to land, after the pilot has navigated all the treacherous obstacles, the simulator reveals an enormous commercial airliner parked crossways in the middle of the runway. Pilot trainees are so focused on landing their planes accurately that only about half of them see the airliner parked where it shouldn't be. When they see a tape of the simulation, they are made aware that they have landed smack on top of another plane. It's a good lesson to learn in a simulator.[7]

ATTENTION BLINDNESS IS KEY TO everything we do as individuals, from how we work in groups to what we value in our institutions, in our classrooms, at work, and in ourselves. It plays a part in our interactions with inanimate objects like car keys or computer screens and in how we value—and often devalue—the intelligence of children, people with disabilities, those from other cultures, or even ourselves as we age. It plays a part in interpersonal relations at home and in the office, in cultural misunderstandings, and even in dangerous global political confrontations.

For the last decade, I've been exploring effective ways that we can make use of one another's blind spots so that, collectively, we have the best chance of success. Because of attention blindness, we often arrive at a standstill when it comes to tackling important issues, not because the other side is wrong but because both sides are precisely right in what they see but neither can see what the other does. Each side becomes more and more urgent in one direction, oblivious to what is causing such consternation in another. In normal conditions, neither knows the other perspective exists. We saw this in the summer of 2010 when an explosion on the BP Deepwater Horizon drilling rig sent nearly 5 million barrels of crude oil gushing into the Gulf of Mexico. Some people reacted to the environmental disaster by wanting all offshore oil drilling banned forever. Others protested about the loss of jobs for oil workers in the area when the president of the United States declared a six-month moratorium on oil drilling to investigate what had gone so disastrously wrong. It was as if neither side could see the other.

But it doesn't have to be that way. If we can learn how to share our perspectives, we can see the whole picture. That may sound easy, but as a practical matter, it involves figuring a way out of our own minds, which as the gorilla experiment so perfectly demonstrates, is a pretty powerful thing to have standing in the way. Yet with practice and the right methods, we can learn to see the way in which attention limits our perspectives. After all, we learned how to pay attention in the first place. We learned the patterns that convinced us to see in a certain way. That means we can also unlearn those patterns. Once we do, we'll have the freedom to learn new, collective ways that serve us and lead to our success.

What does it mean to say that we learn to pay attention? It means no one is born with innate knowledge of how to focus or what to focus on. Infants track just about anything and everything and have no idea that one thing counts as more worthy of attention than another. They eventually learn because we teach them, from the day they are born, what we consider to be important enough to focus on. That baby rattle that captivates their attention in the first weeks after they're born isn't particularly interesting to them when they're two or twenty or fifty because they've learned that rattles aren't that important to anyone but a baby. Everything works like that. Learning is the constant disruption of an old pattern, a breakthrough that substitutes something new for something old. And then the process starts again.

THIS BOOK OFFERS A POSITIVE, practical, and even hopeful story about attention in our digital age. It uses research in brain science, education, and workplace psychology to find the best ways to learn and change in challenging times. It showcases inventive educators who are using gaming strategy and other collaborative methods to transform how kids learn in the digital age, and it highlights a number of successful innovators who have discarded worn-out business practices in order to make the most of the possibilities difference and disruption afford in a new, interconnected world.

We need these lessons now more than ever. Because of attention blindness, the practices of our educational institutions and workplace are what we see as "school" and "work," and many of the tensions we feel about kids in the digital age and our own attention at work are the result of a mismatch between the age we live in and the institutions we have built for the last 120 years. The twentieth century has taught us that completing one task before starting another one is the route to success. Everything about twentieth-century education and the workplace is designed to reinforce our attention to regular, systematic tasks that we take to completion. Attention to task is at the heart of industrial labor management, from the assembly line to the modern office, and of educational philosophy, from grade school to graduate school. Setting clear goals is key. But having clear goals means that we're constantly missing gorillas.

In this book, I want to suggest a different way of seeing, one that's based on multitasking our attention—not by seeing it all alone but by distributing various parts of the task among others dedicated to the same end. For most of us, this is a new pattern of attention. Multitasking is the ideal mode of the twenty-first century, not just because of our information overload but because our digital age was structured without anything like a central node broadcasting one stream of information that we pay attention to at a given moment. On the Internet, everything links to everything and all of it is available all the time, at any time. The Internet is an interconnected network of networks, billions of computers and cables that provide the infrastructure of our online communication. The World Wide Web lies on top of the Internet and is, in effect, all the information conveyed on the Internet. It is the brilliant invention largely of one person, Sir Tim Berners-Lee, who developed a way that the documents, videos, and sound files—all the information uploaded to the Internet—would have addresses (URLs) that allowed them to be instantaneously transferred anywhere around the billions of computers and networks worldwide without everything going through one, central switching point and without requiring management by one centralized broadcast system.[8]

Neither the Internet nor the World Wide Web has a center, an authority, a hierarchy, or even much of a filter on the largest structural level. That allows for tremendous freedom and also, in some circumstances, risk. Instead of reinforcing an idea of sustained attention the way, by comparison, television programming might, with the Internet we have no schedule to keep us on track from the beginning to the ending of a sixty-minute show. If I'm reading along and decide to click on a link, I can suddenly be in an entirely different universe of content. There's no guidebook. There are very few partitions. Everything is linked to everything, each network is a node on another network, and it's all part of one vast web. We blink and what seemed peripheral or even invisible a minute ago suddenly looms central. Gorillas everywhere!

Internet and *web* are great metaphors for the world we live in, too. The domino-like collapsing of markets around the world has brought home a truth we should have seen coming long ago: Like it or not, we *are* connected. We can no longer be living in an "us versus them" world because our fate and theirs (whoever "we" and "they" are) depend on each other. We are all inextricably interwoven. The flip side is that we also have infinite opportunities for making our interconnections as productive as possible. The Internet offers us the communication means that we need to thrive in a diverse and interdependent world.

By one recent accounting, in the last decade we've gone from 12 billion e-mails sent each day to 247 billion e-mails, from 400,000 text messages to 4.5 billion, from 2.7 hours a week spent online to 18 hours a week online. That's an incredible change in the amount and extent of the information taking over our time.[9] If life once seemed calmer and more certain (and I'm not sure it ever really did), that wasn't reality but a feature of a tunnel vision carefully crafted and cultivated for a twentieth-century view of the world. If we're frustrated at the information overload, at not being able to manage it all, it may well be that we have begun to see the problems around us in a twenty-first-century multifaceted way, but we're still acting with the individualistic, product-oriented, task-specific rules of the twentieth. No wonder we're so obsessed with multitasking and attention! You can't take on twenty-first-century tasks with twentieth-century tools and hope to get the job done.

Here's the real-life benefit of the gorilla story: If attention blindness is a structural fact of human biology equivalent to, say, our inability to fly, then that means we are faced with the creative challenge of overcoming it. It might take some thinking; it might require humility about our rationality and our vaunted self-reliance; it may require rearranging our twentieth-century training and habits; and it certainly will require lining up the right tools and the right partners. But once we acknowledge the limitations we've been living with, we can come up with a workaround.

||||||||||||||||||||||||||||||||||

I wasn't always dyslexic. I'm old enough that "learning disabilities" didn't exist as a category when I was a kid. Back then, there wasn't any particular diagnosis for my unusual way of seeing the world. I was already a twenty-seven-year-old professor at Michigan State University when I first heard myself described as "a severe dyslexic with tendencies to attention deficit disorder." Before that, the diagnosis was simpler: I was simply "obstinate."

One evening, a friend invited me to dinner. When I walked into her living room, I saw her sharp-as-a-tack daughter, who was six or seven at the time, lying on her side reading, with a dark blue cover pulled over her head.

"That's how I read too!" I told her mother.

When my friend explained that she was about to take her daughter for some experimental testing because she was exceptionally bright but was falling behind in school, I decided to go along. I was fascinated by what I saw at the testing office and arranged to come back and take the tests for dyslexia myself. It was the first standardized exam on which I'd ever made a perfect score.

In school, I had managed to frustrate just about every teacher I had ever had. According to family lore, I could solve equations with two unknowns in my head before I started school, but I could barely count. One summer, I won a competition that earned me a scholarship to math camp. Two or three of my fellow campers there were going straight from grammar school to MIT without benefit of first passing through puberty. I loved calculus but had never been able to add and spent practically every day of fourth grade (or so I recall) in after-school detention writing endless multiplication tables on the blackboard.

I didn't fare so well at reading and writing either. I loved perusing thick books where the small print ran from margin to margin, but I found picture books confusing and couldn't read anything out loud. I still can't. I happen to be in a profession where it's common to write out an entire talk in long, complex sentences and then stand up and read it word for word to your audience. Though I try to avoid these occasions, when I must, I practice reading the text maybe fifteen or twenty times, until familiarity with the rhythm of the words and phrases somehow helps me stay on track.

Because my learning disability was diagnosed a few years after I'd already earned a PhD, it was pretty easy to feel the relief of the diagnosis ("So *that's* what was going on all along!") without being burdened with a lifelong label of inferiority, disability, disorder, or deficiency that I see being used to describe many of my students today, including some of my smartest. There's no way of knowing if school would have been better or worse had I worn the label "dyslexic" instead of "obstinate," and no way of knowing if my career path would have taken a different turn. All I know is that succeeding against odds and not being afraid to veer off in a direction that no one else seems to be taking have become second nature to me. Tell a kid enough times that she's obstinate and she begins to believe you. Where would I be if the world had been whispering, all those years, that I was disabled?

I don't know, but I do know that I'm not alone. If you are a successful entrepreneur in the United States, you are three times more likely than the general population to have been diagnosed with a learning or attention disorder.[10]

Incidentally, my friend's young daughter has done quite well for herself. She, too, has a PhD now. I'm convinced the key to her success wasn't the label or the special classes or the drugs for her attention deficit disorder, but a mom who fought tirelessly for her daughter's unique and important way of being in the world.

I'll never forget the day my high school principal called me into his office, this time with something good to tell me. He had received a lengthy, typed letter from whoever is at the other end of ACT scores, telling him that he needed to sit down with me and explain that this was a multiple-choice test and I was doing myself a disservice with the long essays I had handwritten on the reverse. I had addressed all the questions that were either ambiguously worded or where all of the answers offered were incorrect. Those essays had wasted valuable time, which partly accounted, this person insisted, for my low test score. My principal had had me in his office more than once before, but I'll be eternally grateful to him for reading me that letter. For all its exhortatory tone and cautions, it ended by saying that the principal should tell me that the ACT committee had gone over all of my comments—there were fifteen or maybe twenty of them—and they wanted me to know that I'd been right in every case.

All these years later, this book is my way of paying back that anonymous ACT grader. I suspect he or she must have been a very fine teacher who knew this would be a good exercise not only for me but also for my principal.

Who pays attention to the reverse side of a test? Not many people, but sometimes the right answers are there, where we don't see them, on the other side. I'm convinced that's the case now. For all the punditry about the "dumbest generation" and so forth, I believe that many kids today are doing a better job preparing themselves for their futures than we have done providing them with the institutions to help them. We're more likely to label them with a disability when they can't be categorized by our present system, but how we think about disability is actually a window onto how attention blindness keeps us tethered to a system that isn't working. When we assess others against a standard without acknowledging that a standard is hardly objective but rather a shifting construct based on a local set of values, we risk missing people's strengths. We don't see them, even when they're in plain view. The implications for this are broader than you might expect. It means more than discouraging a few potential great novelists or professors or artists (though even that can be a huge loss; imagine a world without Warhol or Einstein); it means that we risk losing broad contributions from the people who are supposed to take us into the future.

||||||||||||||||||||||||||||||||

This book is designed as a field guide and a survival manual for the digital age. I have focused on the science of attention because it offers us clear principles that, once we see them, can be useful in thinking about why we do things the way we do them and in figuring out what we can do differently and better. The age we live in presents us with unique challenges to our attention. It requires a new form of attention and a different style of focus that necessitates both a new approach to learning and a redesign of the classroom and the workplace.

Many of these changes are already well under way. In our individual lives, we've gone through astonishing transformations in a little over a decade. A recent survey found that 84 percent of those polled said they could not accomplish their day's work if the computers were down at their office.[11] That's pretty remarkable given that the Internet has been prevalent in the modern office only since around 1995.

Because we're in a transitional moment, most of us aren't aware of how structurally different our life has become because of the Internet. We don't see how radical the changes of the last decade or so have been. It's a bit like growing up poor. If everyone is, you don't know you are. But to step back and look at the digital age from the long view of human history is to see that this is one of those rare times when change concatenates: A change in one place makes a series of other changes in others. With the Internet, we have seen dramatic rearrangements, in a little over a decade, in the most basic aspects of how we communicate, interact, gather knowledge of the world, develop and recognize our social networks and our communities, do business and exchange goods, understand what is true, and know what counts and is worthy of attention.

The eminent historian Robert Darnton puts our information age into perspective for us. He argues that, in all human history, there have been only four times when the very terms of human interaction and communication have been switched so fundamentally that there was no going back. When he says our digital shake-up is major, he's comparing it to *all* of human history. For the first great information age of humanity, Darnton goes back to ancient Mesopotamia, around 4000 B.C., and the invention of writing. He counts movable type as responsible for the dawning of a second information age. That happened in tenth-century China and in Europe in the fifteenth century with Gutenberg. The third information age, he says, came with the invention of mass printing and machine-produced paper and ink that made cheap books and newspapers and all other forms of print available to the middle and lower classes for the first time in history. That began at the end of the eighteenth century, in the Enlightenment. And then there's now, our very own information age, the fastest and most global of all the four great epochs in the history of human communication.[12] It's a bit startling and perhaps humbling to consider that one of the greatest transformations in human interaction is playing out across our everyday lives.

Yet as we've gone through enormous changes in our modes of social interaction and communication, in our attention and in the tasks we now set ourselves, our most important institutions of school and work haven't changed much at all. That's to be expected, perhaps. As Internet analyst Clay Shirky notes succinctly, "Institutions will try to preserve the problem to which they are the solution."[13] We rival the Cradle of Civilization (remember that?) in momentousness, and many of our institutions still look as if there's been no digital revolution at all.

Think about our kids' schools. My grandmother came to this country in steerage, by steamship, but when I look at the photograph of her standing tall and proud in her eighth-grade class in Chicago, surrounded by immigrants from other places, the schoolroom itself looks entirely familiar. Her classroom could be plopped down almost unchanged in any large, urban public school today. What's more, many features of that classroom and what she learned there were structured to help her adjust to the new industrial, manufacturing-based economy she was entering. That economy, as we all know, has been transformed irrevocably by globalization and the changes wrought by the information age. If kids must face the challenges of this new, global, distributed information economy, what are we doing to structure the classroom of the twenty-first century to help them? In this time of massive change, we're giving our kids the tests and lesson plans designed for their great-great-grandparents.

The workplace isn't much different. Unless you happen to be employed at the famous Googleplex, the Day-Glo Google campus in Mountain View, California, your office might still look like something out of a Charles Dickens novel—or a *Dilbert* cartoon, which is to say the same thing. Small cubicles or offices all in a row are another feature of the industrial age and the workplace of the late nineteenth and twentieth centuries. Is this really the most effective way to work in the twenty-first?

Is it possible for a whole society to have attention blindness? I think it is. We seem to be absorbed right now in counting the equivalent of digital basketballs: fretting about multitasking, worrying over distraction, barking about all the things our kids don't know. We're missing the gorilla in the room. We are missing the significance of the information age that is standing right in the midst of our lives, defiantly thumping her chest. It's not that we haven't noticed the change. Actually, we're pretty obsessed with it. What we haven't done yet is rethink how we need to be organizing our institutions—our schools, our offices—to maximize the opportunities of our digital era.

We're so busy attending to multitasking, information overload, privacy, our children's security online, or just learning the new software program and trying to figure out if we can really live without Twitter or Four Square, that we haven't rethought the institutions that should be preparing us for more changes ahead. Politically, on the right and on the left, we've got a lot to say about whether the globalized workforce of the twenty-first century is good or bad, but in some ways the politics of globalization are beside the point. The digital age is not going anywhere. It's not going to end and it's not going away. So it's long past due that we turn our attention to the institutions of school and work to see how we can remake them so they help us, rather than hold us back.

IN OTHER GREAT MOMENTS OF technological change, we've used education as the way to meet new challenges. After the Russians launched *Sputnik 1* on October 4, 1957, at the height of the Cold War, Americans got busy and devoted enormous energy, resources, and innovation to improving schools so our kids could compete in the future. Not every educational experiment from the time worked, but educators were determined to try to find new ways relevant to the moment. Yet the information age has so far failed to produce the kind of whole-cloth rethinking of policy and institution building necessary to meet the challenge, so as we sally forth into the fourth great information revolution in the history of humanity, we're armed with No Child Left Behind, a national "standards-based" educational policy based on standardized tests and standardized thinking, which trumpets tradition and looks steadfastly backward more than a hundred years for its vision of the future. We haven't done much better with many of our workplaces. A cubicle in a modern, global office is rather like the proverbial fish with a bicycle. It's hard to imagine anything less relevant.

Keep in mind that we had over a hundred years to perfect our institutions of school and work for the industrial age. The chief purpose of those institutions was to make the divisions of labor central to industrialization seem natural to twentieth-century workers. We had to be *trained* to inhabit the twentieth century comfortably and productively. Everything about school and work in the twentieth century was designed to create and reinforce separate subjects, separate cultures, separate grades, separate functions, separate spaces for personal life, work, private life, public life, and all the other divisions.

Then the Internet came along. Now work increasingly means the desktop computer. Fifteen years into the digital revolution, one machine has reconnected the very things—personal life, social life, work life, and even sexual life—that we'd spent the last hundred years putting into neatly separated categories, cordoned off in their separate spaces, with as little overlap as possible, except maybe at the annual company picnic.

Home and work? One click of the mouse and I've moved from the office memo due in half an hour to Aunt Tilly's latest banana bread recipe. Labor and leisure? The same batch of e-mail brings the imperious note from my boss and another wacky YouTube video from Cousin Ernie. One minute I'm checking corporate sales figures and the next I'm checking how my auction's going on eBay, while down the hall, I'm pretty sure the sales assistant has slipped yet again from spreadsheets to bedsheets (over 50 percent of Internet porn viewing takes place on company time). Whatever training the twentieth century gave us in separating the different facets of our life and work is scuttled by a gizmo such as the iPhone, which puts the whole, dazzling World Wide Web in the palm of our hand.

Even our brain seems to have changed because of new computational capacities, and those capacities have also supplied a nifty new set of metaphors to help us comprehend the complicated neural circuitry stirring between our ears. We're less inclined to think of the brain as a lump of gray matter lolling inside our cranium than to imagine it as an excited network of a hundred billion neurons, each connected to several thousand other neurons, and all firing several times a second, in constant weblike communication with one another all the time, even when we're asleep or resting. If multitasking is the required mode of the twenty-first century, thank goodness we now have a hyperactive, interactive brain that's up to the (multi)task!

The way we think about the brain has a lot to do with the technology of the era. Is the brain a machine, linear and orderly? Is the brain a mainframe computer, hardwired with certain fixed properties and abilities? Is the brain a network like the Internet, always changing and inherently interactive? Not surprisingly the metaphors for the brain have grown in complexity along with the evolution of ever more complicated technologies of interconnection.

From contemporary neuroscience we now know the brain is a lot like an iPhone. It comes with certain basic communication functions bundled within it, and it has apps for just about everything. Those apps can be downloaded or deleted and are always and constantly in need of a software update. These iPhone apps represent the things we pay attention to, what counts for us, what we are interested in. Our interests shape what apps our personal iPhone has, but our interests aren't isolated. If my best friend says, "Find me on Gowalla," I then add the GPS-based Gowalla app so we can follow one another's comings and goings around L.A., and before I know it, I've added a dozen more apps to my phone from things I've learned through our game play and social networking as we travel the city, separate but connected.

The brain is similar. How we use our brain (what we pay attention to) changes our brain. Those things that most capture our attention—our learning and our work, our passions and our activities—change our actual brain biology. In this way the iPhone brain also corresponds nicely with recent advances in what we know about neural plasticity, the theory that the brain adapts physically to the sensory stimuli it receives, or what psychiatrist Norman Doidge calls "the brain that changes itself."[14] This model was given real biological heft in the 1990s, when stem cells were discovered in different parts of the adult human brain. Stem cells can regenerate. They can also take on new functions—new apps—that have been lost due to damage to other parts of the brain.

This is exciting news. The first great era of brain science, the late nineteenth century, coincided with the industrial age, and it's not surprising that brains back then had distinctive parts and hardwired functions. Early brain scientists surveyed our lobes and nodes like explorers on the frontier, creating topographical maps distinguishing this region from that and describing what natural resources could be found there. Like gold rush mining towns, parts of the brain were named after the pioneers who planted a flag there first, so to speak, such as Broca's area, after Paul Pierre Broca, who first labeled the inferior frontal gyrus as the center of speech production. German neurologist Korbinian Brodmann identified fifty-two of these distinctive functional regions of the brain, still known as the Brodmann areas. In this view, each area has a specific function within a hierarchy, from the higher-level intellectual ordering and "executive functions" of the prefrontal cortex down to the base, emotional, "reptilian" amygdala.

The brain's development was also thought to be linear too. It was thought that the brain's capacities grew until about age twenty-five, plateaued for five or ten years, and then began to decline, first slowly, then more rapidly. It was all a bit mechanical, and that is no accident, for the image of the linear, orderly, machinelike brain hardwired with fixed capacities changing in a clear developmental pattern came into being about the same time as the assembly line and mass production.

Contemporary neuroscience insists that nothing about our brain is quite so fixed or static, including its progress and its decline. Rather, we're constantly learning, and our mental software is being updated all the time. As we get older, we can become obsessed with what we think we may have lost, but new brain science reveals that in healthy individuals the loss is far less than what was once believed. We stay smarter longer and our capacities expand in more interesting ways than was previously thought possible. There are even productive ways (including culture shock) to force this late-learning process, in much the same way that tulips can be forced to bloom indoors in winter. As we will see, a major factor reshaping the brain is technology.

How we perceive the world, what we pay attention to, and whether we pay attention with delight or alarm are often a function of the tools that extend our capabilities or intensify our interactions with the world. That expansion of human capacities can be scary. Albert Einstein famously observed that technological change is like an ax in the hands of a pathological criminal. It can wreak a lot of havoc. If you read newspapers published at the time of Broca and Brodmann, you find anxiety about new technologies of the day. A chief concern is speed. Trains, bicycles, and especially the automobile (the "horseless carriage") all seemed to push humans beyond their natural, God-given, biological limits. Early critics of the car, for example, simply refused to believe they could be safe because, after all, human attention and reflexes were not created to handle so much information flying past the windshield. That debate reached a crescendo in 1904, when the Hollywood film director Harry Myers received the world's first speeding ticket, when he was clocked rushing down the streets of Dayton, Ohio, at the death-defying speed of twelve miles per hour.[15]

It's commonplace in the history of technology for people to insist that "human biology" or the "human brain" simply cannot cope with the new technology or that our minds and bodies have been put at risk by that technology. Probably people said that about the wheel. What this line of argument overlooks is that the brain is not static. It is built for learning and is changed by what it encounters and what operations it performs. Retooled by the tools we use, our brain adjusts and adapts.

Right now, we're in a transitional moment. We are both adopting new information technologies all the time and being alarmed by them, even wondering if they are causing us harm, exceeding our human capacities. Fifteen years is a very brief time in the history of a major new technology. Basically, the Internet is still in its adolescence and so are we as users. We've grown up fast, but we still have much to learn. There's a lot of room for improvement. We are experiencing growing pains.

Because we learn to pay attention differently depending on the world we see, when the world changes, there is a lot we're suddenly seeing for the first time and even more we suspect we're missing. So that's a key question: How can we focus on what we do best without missing new opportunities to do better?

The starting place is recognizing the brain's habits, really taking in what these habits mean, and then working (by ourselves and with others) to find ways to break the old patterns that no longer serve us. A friend of mine has cultivated the habit of always putting her watch on the wrong wrist when she wants to remind herself later to remember something. She says when she gets home from the office and goes to take off the watch, that moment of awareness that the watch is on the wrong wrist forces her to think, "What was it that was so important today that I was sure I would forget it later and changed my watch to remind me?" She then takes an inventory and inevitably remembers. It's her device, her way of interrupting her habits to refocus her attention. The trick works and has application, as we will see, in individual and group situations in both everyday life and the workplace.

||||||||||||||||||||||||||||||||

If attention suddenly has our attention, it's because we live in a time when everything is changing so radically and so quickly that our mental software is in constant need of updating. We have heard many times that the contemporary era's distractions are bad for us, but are they? All we really know is that our digital age demands a different form of attention than we've needed before. If a certain kind of attention made sense in a world where all the news came through one of three television channels, then what form of attention do you need when your primary information source is Google, where a search for information about "attention" turns up 371 million entries, and there's no librarian in sight?

When Tim Berners-Lee invented the World Wide Web, he anticipated a new form of thinking based on process, not product: synthesizing the vast and diverse forms of information, contributing and commenting, customizing, and remixing. Do we even know how to assess this form of interactive, collaborative intelligence? Is it possible we're still measuring our new forms of associational, interactive digital thinking with an analog stopwatch? What if kids' test scores are declining because the tests they take were devised for an industrial world and are irrelevant to the forms of learning and knowing more vital to their own world?

By one estimate, 65 percent of children entering grade school this year will end up working in careers that haven't even been invented yet.[16] Take one of the "top ten" career opportunities of 2010: genetic counseling. Every hospital now needs this hybrid medical professional–social worker to advise on family planning, prenatal testing, and treatment protocols. The shortage of genetic counselors is considered a matter of national urgency. Before 2000, when the human genome was sequenced, such a career would have seemed the stuff of science fiction.

Nor is it just kids who face uncertain changes in their careers. My friend Sim Sitkin recently invited me to have lunch with some of his students in the Weekend Executive MBA Program at Duke's Fuqua School of Business. This is a program for executives who have been in the workforce at least five years and are seeking to retool. They commit to a nineteen-month intensive program that meets every other Friday and Saturday. Tuition is over $100,000. You have to need something badly to make that kind of professional and financial commitment. About seventy students at a time are admitted to this top-ranked program. Sim says they come from every imaginable profession.

We were joined at our table that Saturday by five executives: a twenty-eight-year-old middle manager for an international pharmaceutical firm, a forty-year-old sales rep at an international business equipment manufacturer, a software developer originally from China, a financial analyst now managing accounts handled offshore, and a distinguished physician whose job increasingly relies on telemedicine. All told me about different ways their occupations have changed in the last five years. The words *context, global, cross-cultural, multidisciplinary,* and *distributed* came up over and over. One person noted that his firm was bought out by a multinational corporation headquartered in India and that's where his boss lives and works. Learning how to communicate across cultural, linguistic, and geographical barriers by Skype is no small challenge. The physician compared his medical practice to the hub system in the airlines, with general practitioners across the region referring their most serious problems to his staff at a large research hospital, where they were counseled virtually in preparation for a possible actual visit. He was getting his MBA because medical school hadn't prepared him to be a traffic controller, bringing in patients from all over, some with major problems and some minor, some referred by highly skilled generalists and others by doctors or nurse practitioners with only rudimentary knowledge. His job was to land all patients safely within an enormous hospital system, making sure that they were not only well cared for, but that their procedures were carefully synchronized across the testing, diagnostic, and financial sectors of the hospital. He hoped the executive MBA would help him navigate this bewildering new world, one that bore little relationship to the medical specialization he'd mastered fifteen years before.

Futurist Alvin Toffler insists that, if you scratch beneath the surface of anyone's life in the twenty-first century, you will find the kinds of enormous change these executive MBA candidates experienced. Because change is our generation's byword, he believes we need to add new literacy skills to the old three *R*s of reading, writing, and arithmetic. He insists that the key literacy skill of the twenty-first century is the ability to learn, unlearn, and relearn.[17] *Unlearning* is required when the world or your circumstances in that world have changed so completely that your old habits now hold you back. You can't just resolve to change. You need to break a pattern, to free yourself from old ways before you can adopt the new. That means admitting the gorilla is there, even if you're the only person in the room who does (or doesn't) see it. It means putting the watch on your other arm. It means becoming a student again because your training doesn't comprehend the task before you. You have to, first, see your present patterns, then, second, you have to learn how to break them. Only then do you have a chance of seeing what you're missing.

As the attention blindness experiments suggest, learning, unlearning, and relearning require cultivated distraction, because as long as we focus on the object we know, we will miss the new one we need to see. The process of unlearning in order to relearn demands a new concept of knowledge not as a thing but as a process, not as a noun but as a verb, not as a grade-point average or a test score but as a continuum. It requires refreshing your mental browser. And it means, always, relying on others to help in a process that is almost impossible to accomplish on your own.

That's where this book comes in. It offers a systematic way of looking at our old twentieth-century habits so that we can break them. It proposes simple methods for seeing what we're missing and for finding the strategies that work best for the digital age. This book starts from some core questions:

Where do our patterns of attention come from?
How can what we know about attention help us change how we teach and learn?

How can the science of attention alter our ideas about how we test and what we measure?

How can we work better with others with different skills and expertise in order to see what we're missing in a complicated and interdependent world?

How does attention change as we age, and how can understanding the science of attention actually help us along the way?

These are the central questions this book will address, and I hope that by its end we'll have something even better than answers: We'll have new skills and new insights that will help us address problems as they arise in our everyday life in this digital age.

As we will see, the brain is designed to learn, unlearn, and relearn, and nothing limits our capabilities more profoundly than our own attitudes toward them. It's time to rethink everything, from our approach to school and work to how we measure progress, if we are going to meet the challenges and reap the benefits of the digital world.

We can do this, but we can't do it by ourselves. Whether prepping for a college interview, walking into a board meeting when a hostile takeover is in the air, interviewing a new manager for a key position in the firm, or, on a personal level, going for a follow-up test after a cancer diagnosis, the best thing we can do to ensure the most successful outcome is to have a partner accompany us. But as we'll see, it's not enough to just have someone be there. We need a strategy for working in tandem—a method I call "collaboration by difference." We have to talk through the situation in advance and delegate how each of us watches for certain things or keeps an eye on certain people. If we can trust our partner to focus in one direction, we can keep our attention in another, and together we can have more options than we ever imagined before: "I'll count—you take care of that gorilla!"

Our ultimate metric is our success in the world. And that world has changed—but we have not changed our schools, our ways of training for that world, in anything like a suitable way. By working together, by rethinking how we structure the ways we live, work, and learn, we can rise to meet the exciting opportunities presented by the information age.

That's the challenge this book addresses. Let's get started.

Part One

||||||||||||||||||||||||

Distraction and Difference

The Keys to Attention and
the Changing Brain

1

||||||||||||||||||||||||||

Learning from the Distraction Experts

The slim, attractive father in the T-shirt and the loose fitting "dad jeans" twirls his eight-year-old daughter around their living room as an infant slumbers nearby. His home is neat, modest, comfortable. Here's a good provider, a loving and reliable family man. Leafy afternoon light casts a golden glow as soothing classical music plays gently in the background. Well-being and being well infuse the scene. I know what will happen next; I've witnessed this scene a hundred times before. The music swells, we're building to the happy-ever-after ending. This story will turn out fine.

Especially if I can convince my doctor to order me some Cymbalta.

That's the move that drug manufacturer Eli Lilly and Company wants me to make, extrapolating from the contented man in the ad to myself, moving from my couch in front of the television to my doctor's office where I will request Cymbalta. We've seen this man at the beginning of the commercial leaning his head back against a wall, his face etched in despair in a bleak hallway. By the end of the commercial, there he is, dancing with his daughter before the drug's logo appears and the words "depression hurts" fade in and the voice-over reassures us, "Cymbalta can help." The space between "before" and "after" is where the magic happens.[1]

Not by coincidence, that same space is occupied by the familiar voice-over required by the Food and Drug Administration warning of possible side effects. Thanks to FDA regulations, Eli Lilly can't advertise on television without listing for me the possible negative consequences of taking their drug, which most of us would probably agree is a good thing, at least for consumers. For Eli Lilly, who's in the business of selling as many pills as possible, those regulations are more of a nuisance, which is why presenting the negative side effects of a drug in the least damaging manner falls to the advertising companies who help the drug makers design their commercials. Draftfcb Healthcare, the advertising firm that holds the Cymbalta account, has the job of fulfilling the FDA requirements for stating side effects clearly while somehow distracting my attention from the mandatory but scary warnings about liver damage, suicidal thoughts, and fatality.

It turns out that Draftfcb performs this task marvelously. Its ads for Cymbalta comprise the single most successful prescription drug campaign on television, the gold standard in the competitive direct-to-consumer marketing industry. After this ad campaign began, Cymbalta's first-quarter sales rose 88 percent, with Draftfcb credited with securing that dominance of market share.[2] So, then, they are the distraction experts.

It should come as no surprise that advertisers spend millions of dollars each year studying the science of attention, both through empirical research and testing with target audiences. Draftfcb's motto is "6.5 seconds." That motto is based on their research indicating that a typical viewer really pays attention to only 6.5 seconds of any TV ad. Their goal is to "capture the consumer's attention and motivate them to act" in that brief time.[3] They want to ensure that we spend our 6.5 seconds of attention on "Cymbalta can help" and not on "liver damage."

How do they do that? How do they know what will motivate us to disregard the frightening side effects and believe in the twirling promises of happiness ever after? We are rational human beings, and, more than this, we're seasoned consumers, growing ever more skeptical of advertising in general. Yet these ads work. They convince us, even against our better judgment. How is that possible? What do the makers of the Cymbalta ad know about us that we don't know about ourselves?

You can find lots of parodies of the Cymbalta ad on YouTube. Comedians, too, poke fun at happy ads that minimize the potent side effects, but it turns out that even this knowing attitude doesn't make us as wary as one might expect. A study of direct-to-consumer pharmaceutical ads presented as part of testimony to Congress found that viewers were more likely to remember hearing positive rather than negative claims in the ads by a margin of nearly two to one. Among those viewers who recalled hearing any recitation of the side effects at all, twice as many could recall the content and substance (the message) of the positive information over the negative.[4]

This is an alarming figure, given that the FDA maintains an entire division dedicated to monitoring and screening these ads for balance and accuracy.[5] We have to wonder what we're doing with the side-effect information on the screen. Are we tuning it out? Are we laughing at the obviousness and silliness of the positive parts of the ads *and* still being convinced by them? Or do the makers of drugs like Cymbalta understand our own patterns of attention and distraction so well that it's easy for them to hide the negative right there, in plain sight?

We'll be looking, in this chapter, at how distraction is used to persuade us to choose differently, how attention is shaped by our emotions, and how what we value focuses what we notice in our world. By looking closely at how Draftfcb assesses the components of our attention and distraction, we can begin to see ourselves better. If they know enough about our cognitive patterns to fit them to their own goals, surely we can learn how to manipulate those processes for our own success.

|||||||||||||||||||||||||||||||||||

The Cymbalta ad runs for 75 seconds, not 6.5 seconds. Draftfcb's job is to carefully craft the ad so our attention is hooked by the right 10 percent.[6] Entitled "No One," the commercial begins with the viewer placed in the position of the lonely outsider to his or her own life. The first scenes make us feel as if we're watching home movies taken during some happy time in the recent past. The home video tape has a date on it: 10/17/06, roughly three years before the release of the commercial. In this narrative, happiness existed not so long ago; the proof is in the pudding of this home video with its jumpy, amateur footage, intimate and fun, its brightly colored scenes of couples and families romping together on a camping trip, acting goofy for the camera.

But all is not well. The edges of the home-video images are seared in ominous black that threatens to engulf the happy scene. The voice-over begins almost immediately, with a concerned female asking in a maternal alto rich with sympathy, "When you're depressed, where do you want to go?" She speaks the answer—"nowhere"—as the word appears on screen, typed in white lowercase letters on a black card, as in an old silent movie. The same soothing female voice next asks, "Who do you feel like seeing?" and "no one" appears in white letters against the black background.

It's at this point that the images begin to change. As the voice begins to spell out the symptoms of depression, we see its victims. A middle-class African American man (a subgroup statistically underrepresented in commercials but more likely to suffer from depression than white men), probably in his thirties, leans his head against a wall at a workplace, perhaps a hospital or a modern office. He looks very much a victim of depression, rooted nowhere, attached to no one. Next we see an attractive (but not too attractive) young woman in her twenties with flat black eyes and olive-toned skin. She doesn't meet the camera's gaze. All around her is deep black. Then a pale white woman, this one in her fifties or sixties, somewhat dowdy, sits nervous and alone at a kitchen table, fingers to her lips, her eyes wandering and anxious. It's a broad enough spectrum of age, race, and gender to invite in almost any potential sufferer.

"Depression hurts . . . in so many ways," that sympathetic female voice-over is saying, with pauses between the hard words: "Sadness. Loss of interest. Anxiety." And then, with a hopeful rising inflection, the punch line: "Cymbalta can help." Suddenly, the images of the sad and anxious individuals come to a halt. In their place is the vibrant Cymbalta logo, the antithesis and antidote to depression. It's a whirl of innocent, exuberant energy, an almost childlike doodle of turquoise and lime green, a bit of stylistic whimsy to relieve the grimness.

If the narrative spun by the commercial seems obvious, that's part of the point. We already know the story of this ad from a lifetime of happy tales, beginning with bedtime stories, children's books, fairy tales, myths, and adventure stories, all the optimistic fables we were raised on. Cymbalta to the rescue! This ad is carefully crafted as a transformation story, with scenes of peril and punishment rewarded by a happy ending. We know Cymbalta will fulfill our expectations for the psychological version of the rags-to-riches plot.

And it is just then, with the happy logo promising redemption, that, with no break in the cadence of her speech, no pause or even shift in the sympathetic tone of her mother-voice, that the voice-over actress starts to list the side effects. Why here? Because this is the best place to take advantage of our mind's training in following the very type of narrative this commercial is plying. The advertiser understands the power of associations built up over a lifetime, the cues that will help us make a leap to the way this story ends. The side-effects roll call comes at the place in the commercial where we are least likely to pay attention because our mind already knows the end to the story.

That doesn't mean that Draftfcb isn't going to do everything it can to lead us along, even if we already know the path. Every aspect of this Cymbalta ad is rooted in the science of attention. Not much is left to chance in a multi-billion-dollar industry.[7]

Music helps keep us on our way. The Cymbalta ad has a sound track, an adaptation of a poignant piece by Robert Schumann, *Kinderszenen*, opus 15, no. 1, the same music played in a dramatic moment of memory at the piano in the movie *Sophie's Choice*, starring Meryl Streep. In this ad, the evocative music plays as a woman's voice murmurs the side effects of Cymbalta sotto voce.

The maternal-sounding voice is carefully planned too. Like narrative, characteristics of voice come with cultural expectations and preconceptions that help to persuade or dissuade us during our 6.5 seconds of attention. We think we listen to what people are saying, but it turns out we're a little like dogs in that we sometimes hear the tone of voice and don't even pay attention to what that voice is actually saying. How we hear, how we listen, and how we comprehend (three very different topics, it turns out) are all relevant to the side-effect voice-over. In ordinary conversation, Americans speak about 165 words per minute. For New Yorkers, it's closer to 200. News announcer Walter Cronkite famously trained himself to speak more slowly, at around 125 words per minute, to encourage people to pay more attention to the news. Since Cronkite, it is the convention of American newscasters to articulate slowly, especially for the most important part of a story. In general, in the American English speech pattern, our voice drops to a deeper pitch the more slowly we speak. Test audiences associate both slowness and a deeper sound with truthfulness, importance, and calm.

In the voice-over acting business, drug commercials are their own specialty niche. They require special training. For side effects, you need to speak clearly enough to pass FDA muster but you want to make sure you avoid anything that will actually capture listeners' attention. The most common method involves using a nonconventional speaking pattern in which the voice drops, not rises, as you speed up your delivery. Standard acting classes teach you how to hit the beginnings and endings of words (think of just about anyone trained in traditional methods of British stage acting doing the "To be or not to be" speech in *Hamlet*). By contrast, voice-over classes help you master selective mumbling. When you deliver the news about potential liver damage, you soften consonants and end sounds, so you are speaking in a murmur as soothing and inconsequential as a parent's babble to a baby. When you introduce the product at the beginning of the commercial and when you deliver the punch line at the end, your diction is much slower and your words are carefully e-nun-ci-a-ted.

If viewers remember the positive but not the negative content of the ad, it is because highly skilled professionals have worked hard to make the conditions for distraction possible, even probable. They have researched other technical matters, such as camera angles and speeds, cutting techniques, the science of color, and on and on. Things that we nonprofessionals think of as "incidental" have typically been tried on test audiences and are anything but. I know someone who participated in a study in which two ads were shown, one with the side effects recited with a little cartoon of a bee flying across the TV screen, another identical but for the bee. The audience was half as likely to remember the negative side effects if they watched the version with the distracting bee.[8]

This speaks to the power of the images in focusing our attention, and the Cymbalta ad proves a fine example of this power. As the soothing alto voice recites potential side effects, the images transform from tragic to happy: The attractive young woman isolated against the black background now smiles sweetly from the window, despite the rain; there's more camping with couples and kids and dogs; the anxious older woman has left the kitchen table and now tenderly holds hands with an older man, presumably her loving husband, in another window, this one full of sunlight. And there's that man, the one who had leaned his head against the wall in despair, now twirling his young daughter, nuzzling his baby. Liver damage? Suicidal thoughts? Not here! It's all happening, the happy ending is nigh. With Cymbalta, relief is on the way.

Of course none of these techniques designed to manipulate our attention would matter if the message weren't the one we already wanted to believe in. From infancy on, our sense of ourselves has gone hand-in-hand with our need to think of ourselves as in control (even when we're not). Growing up in Western culture is almost synonymous with being in control, and everything about the ad plays upon that message. It is Draftfcb Healthcare's challenge to play into these deep-seated cultural values of self-worth. The Cymbalta ad moves seamlessly across all the realms of work, home, friends, loved ones, and family. Failure at one is failure at all; the man who is in despair at work is playing with his daughter after help comes his way. Success at one is key to success at all, which means that, without Cymbalta, everything is in jeopardy.

It is a brilliant commercial, no matter what you think about the ethics of direct-to-consumer pharmaceutical advertising. In seventy-five seconds, the panoply of our culture's values is condensed into a powerful and compelling narrative, all supported by the most sophisticated tricks of the trade. For the commercial to work, we have to be able to identify so closely with the people—actors, of course—that we see their pain as our pain, their solution as ours. If they can get past all the shameful feelings required in admitting one's weakness and getting help, we can too. Surely you've seen the commercial in which the guy's reflection in a store window speaks to him and tells him he shouldn't be embarrassed about talking to his doctor about his erectile dysfunction. The guy listens to his reflection and walks out of his doctor's office a taller, more potent man.

BY DECONSTRUCTING THE SEVENTY-FIVE SECOND CYMBALTA ad, we can make ourselves aware of how unnatural persuasion is. All the components in this ad have been researched, planned, constructed, tested, edited, tried out in front of studio audiences, discussed in focus groups, re-edited, and tested yet again. Those seventy-five seconds distill hundreds of research studies, experiments, insights, acting lessons, film production classes, and more.

But what is it that all these tests are reaching for? In this case, advertisers are seeking out the very fabric of our attention, the patterns and values that we have come to see as so natural that we really don't even see them anymore. We know what the series of images in this narrative means, so we don't really have to think about it. We have internalized the voice cues, the swelling music, so deeply that we hear the meaning without having to hear the words. By plugging into the unconscious patterns of our attention, the only thing that really needs to disrupt our attention—the added element to the narrative we're meant to see anew—is the drug itself. Everything else is constructed so as to be as unobtrusive as possible. And it works.

At least this ad does, here in the United States. Part of the trick is that we can be wooed by the logic of the admen only if we are part of the cultural script it writes for us and writes us into. I learned this for myself when I went to teach in Japan for the first time in the 1980s. I hadn't intended to go, and it would be hard to be less prepared than I was for the trip. I'd been to Canada but not much of anywhere else in the world, and in a matter of weeks, I suddenly found myself living in a tiny apartment outside of Osaka, in a lovely town with only one or two other *gaijin,* "foreigners." Having had barely enough time to sublet my apartment back in Michigan and pack my bags, I arrived with no Japanese and quickly enrolled in a local language school. My teacher was sure that, if I watched television commercials, I'd pick up the language quickly, because commercials often use simple language and repetition to make their point.

I admit I'm terrible at learning languages, so that may have had something to do with my confusion. But I also found that commercials were a perfect shorthand to cultural values—and upon first arriving in Japan, I had no idea what those were. So much in the commercial was unspoken, or wrapped up in euphemism, or so steeped in cultural references I didn't know that I often missed the sales pitch entirely. I could reduce my bilingual Japanese friends and students to tears with my guesses, which more often than not were miles off the mark. Was that gray-haired man in the traditional men's *haori* robe pitching ramen, funeral services, or a cure for constipation? You'd think you'd be able to tell which was which, but I was so decisively wrong so often that it became almost a shtick, a cross-cultural icebreaker, during my first stint in Japan for me to narrate what I thought was going on, which invariably wasn't at all.

The experience of finding TV ads in Japan indecipherable made me see how attention is rooted in cultural values embedded so deeply that we can barely see them.

As such, not only is attention learned behavior, but it is shaped by what we value, and values are a key part of cultural transmission, one generation to another. The absorption of those values into our habitual behavior is also biological. We change brain pathways, and we make neural efficiencies when we learn. That is the *physiology* of attention blindness. Of course those values change with time and circumstances. We are constantly disrupting previous patterns and forming new ones, which puts ad agencies like Draftfcb in a position not just to reinforce those patterns, but to change them.

In ways less commercially driven, parents and teachers are crafting attention too. We're all researching, planning, constructing, testing, editing, trying out, editing again, shaping, reshaping, and, in other ways, typically without any conscious thought, crafting our values and expectations into a narrative as compelling and as acceptable as the Cymbalta ad. Our attention in the rest of our lives is no more natural, innate, genetic, or universal than is the crafted, produced version of attention in the Cymbalta ad.

But then the questions remain: Why? Why are we prisoners of attention? How did we get this way? Why do we see certain things so naturally and miss others that seem so obvious once others point them out to us? Where do our patterns of attention come from, and do we have any say in how they are shaped?

To answer those questions, we need to go back, all the way back to the nursery, where our own brains were formed and where we learned what to value. To say that another way, in the nursery we began to learn what is important to pay attention to—and what isn't. It's also in the nursery that we learned that the world is far, far too vast to be mastered one bit at a time. We need to organize all the stuff of the world around us. We need priorities and categories that make navigating through life easier and more efficient. And that's exactly where we get into trouble.

|||||||||||||||||||||||||||||||

Years ago, I visited friends with their four-month-old baby. It was a large family gathering, with all the attention of parents, grandparents, nephews, cousins, aunts, uncles, plus a gaggle of friends all trained on little Andrew. Baby Andrew, on the other hand, had no interest in us. He was lying in his crib in the nursery, totally and happily absorbed by whirring fan blades making shadows and swooshing sounds overhead. When his mother picked him up and began rocking him, he began staring in another direction, oblivious to all the preening adult attention.

"What's he looking at?" Grandfather asked peevishly as Andrew stared in rapt attention elsewhere.

A few of us squatted down beside the rocker to see in the direction that Baby Andrew was looking. That's when we saw the light shine through the slits in a Venetian blind as the rocker moved back, then disappear as the rocker moved forward, a dramatic chiaroscuro of light and shadow that was far more fascinating to Baby Andrew than his grandfather's attentions. So what did we do to divert him from that fascinating display and convince him to look at us instead? We did what any self-respecting adult would do. We shook a rattle at him. We distracted him from the absorbing light by an unexpected sound.

In a nutshell, all attention works that way, even for adults. Even multitasking is really multidistraction, with attention not supplemented but displaced from one point of focus to another. That's basic Brain Biology 101. But that's not the whole story. Sometimes there is continuity as we move from one distraction to another, and sometimes there's not very much. Many distractions happen and are simply forgotten. Others we take with us. We apply information from one experience when we need it in another. What makes the difference between the forgettable and the important is what I call learning.

From birth on, Baby Andrew is being taught, by those smiling, rattling grandparents and everyone else, what is and what is not worth paying attention to. Light through Venetian blinds is pretty, but not worth paying attention to in most situations. We count it as a distraction. The rattle isn't worthy of attention either, except as a mechanism for getting Andrew's attention and trying to make him focus on what *is* important in his world: all the relatives and friends gathered to focus love, affection, and what anthropologists would call "kinship bonds" on him. In shaking the rattle, we didn't make the light go away. We just put it in a proper hierarchy of unimportant to very important things, of things that don't matter and other things that are worthy of attention—that is, the things we adults consider to be more important. We diverted Baby Andrew's attention from the light in the same way Draftfcb stows the side effects.

That day in the nursery with Baby Andy happened decades ago, and grown-up Andy now has a healthy law practice and kids of his own. How did he go from fascinated preoccupation with the motions of fan blades and sunlight through Venetian blinds to passing the bar exam and raising a family? There was no rule book. And yet, if we want to understand how our attention works— and how we can be brought along by the makers of Cymbalta or stumped in times of tremendous change such as our own—we need to uncover the patterns that shaped Andy. And just as these are revealed in the Cymbalta commercial to be diverse and multifaceted, covering everything from sense inputs to values, so are the patterns that shape each of us woven together beginning in our earliest moments.

Infants are not born paying attention. When babies come into the world, they think everything is worthy of attention. They see the whole world unsorted. They use those around them—their attitudes, gestures, emotions, language— as the scaffolding on which to build their own sense of what they should be paying attention to.[9] Particularly from their main caregivers (typically, their mothers), they learn what they should be paying attention to, what counts, what is rewarded, and how to categorize all that *does* count into language, the single best organizer of what does or doesn't count in a society. Each language sorts and categorizes the world in a unique way, marking differences between approved and disapproved, good and bad, and other distinctions that an infant being raised into that society needs to learn.

Adults looking at infants are in awe of how "quickly" kids learn language, but, when you realize that every contact they make reinforces language in the most positive and powerful ways possible, with affection and reward being the fulfillment of the baby's basic needs, it becomes clear that language learning doesn't actually happen quickly at all. The rule of thumb is that it takes 10,000 hours of practice to become a virtuoso at just about anything worth doing. There are approximately 8,765 hours in a year, and babies are awake and getting feedback from the world a lot of that time. The process of learning the complexities of vocabulary, syntax, and grammar can extend into adolescence and beyond.[10] Even the full and clear articulation of all of the complex sounds of a language (including the difficult consonant blends in English) is a long process, taking up to seven or eight years. Adult learning builds on the efficiencies of past learning—which is why it is so difficult to unravel the process in order to see how much we've learned.

Infant psychologists, cognitive scientists, neurologists, and cultural anthropologists have studied how infant learning happens, all in an effort to understand the boundaries between culture and biology. They have studied our developmental patterns of attention as carefully as Draftfcb has. This is important because, as we have seen, we can't really see ourselves very clearly. Many times, we assume we are acting "naturally" when we're in fact enacting a cultural script that has been repeated for us many times and that we are beginning to imitate even as early as four months into life.

Andy came into a very particular world with a very specific genetic inheritance. The interaction between those two things, nature and nurture, is where attention begins. We're looking at a few minutes in the life of Baby Andrew. But look in the mirror! Baby Andy R Us. Our attention and selectivity begins here.

WHAT'S ANDREW'S BACKSTORY? BEFORE WE can even focus on seventy-five seconds in his crib life, we need to situate him, because these details, though selected randomly, will come together in all sorts of different ways to create the norms, values, and patterns that he sees as the world. Those details are analogous to the particulars in the home movie at the beginning of the Cymbalta ad, the one that makes clear we're looking at a middle-class, nuclear family, with enough leisure to go on a vacation in a camper. In our own little scene, Andrew is lying in a crib in his very own nursery in suburban Philadelphia. Let's say his mother is a professional woman but, as far as we can tell, she's now devoting herself to child care. We could say, for example, that she is on leave from her city law firm to take care of him for the first six months of his life. Mother and Daddy like to believe they are liberated co-parents; they both talked for months about "being pregnant." But, like most men in his world, Daddy doesn't take time off from his firm for parental leave. How can he? He's part owner of a small advertising firm. Maybe his biggest accounts are direct-to-consumer television ads, and like everyone in television, he's a bundle of anxieties about what the move to all-digital television format, with shows on demand anytime you want them, will do to his bottom line. He can't be taking time off to play with Andy. Someone has to be the family breadwinner.

Andy has a big sister, Ashley, who is seven. Ashley's best friend is named Sharon. The two of them visit Baby Andy in the nursery from time to time. Andy's parents would call themselves Protestants, maybe Episcopalian, if asked, although there was at least one great-grandparent back there reputed to be Jewish and another who was German Catholic, and they aren't a particularly observant family in any religious tradition.

Why are these categories important? One reason is that once everything is located in a proper category, the category itself (for better or worse) answers a host of unaskable questions. A category is a shorthand. "He's a nerd" is a categorical statement that does and doesn't tell you much about the whole person but says a lot about a society's attitude toward people who are focused on intellectual matters. In Andy's case, even if no one specified the various sociological and demographic labels, there are telltale signs of what category Andy fits into, just from the seemingly neutral fact that Andy sleeps alone in a crib in his own room (that he even has his own room!), and he isn't even weaned yet.

Andy's too little to know it yet, but in many societies, if his parents had isolated him alone in a crib in a separate room, they'd be considered cruel and abusive. If Andy had been born in that kind of society, he'd be getting more sympathy when he whimpered over being left alone again. Not in Philadelphia, though. Be tough, little Andy!

How well Andy fares alone in his crib represents a system of values that Andy learns by example, by the responses of those around him to his behavior, or by the concern or happiness of those around him, rarely by actual instruction. At four months, he's already aware that people pay more attention to him when he cries about some things than about others. He is already aware that crying when he's hurt brings succor fastest. Even at four months, he's starting to realize crying has certain powers to command attention—but also that this tactic has to be used selectively or people aren't as nice to him. They hold him, pat him, soothe him, all of which signals crying should stop.

Four-month-old Andy is also becoming aware that not everyone cries. He's learning that crying is a kind of alarm when he is around other infants, and he's becoming attuned to the crying of those around him. He's beginning to make a "how crying is valued" map of the world. Big Sister cries less than Andy, Mama cries less than Big Sister. Daddy never cries.

If we share Baby Andrew's backstory, we might not think twice about anything in our little scene (unless we happen to be one of those psychologists, neuroscientists, or anthropologists). In situations where we share the same cultural script, we proceed as if that script were natural. In most circumstances, it is invisible. That's what "natural" means. In certain situations, though, we are jolted from our sense of the natural and then are suddenly forced to think about how and why we have certain habits. This is culture shock. When we're plunked down into a very different society, especially if we do not know the language and don't have a guide from our own culture, we suddenly start seeing what is "strange" about other customs and behaviors; we also see others treating our way of doing things as strange. "Natural" is always relative to our world.

Andrew learns the rules from family members who provide scaffolding for the values he is developing. Andy's family may not be aware they are teaching Andy what to pay attention to and what to ignore. He'd be happy watching the fan blades twirl all day, so long as other basic needs for food, warmth, protection, and affection were being met. But the rest of the family, even his older sister, has already figured out that there isn't much reward in this world for staring at the shadows. If Andy is going to live with these bigger people, he has to learn their strange ways. That's what you do in a new culture, and if you are Andy, you have begun to enter the culture of your parents long before you've even begun to speak, simply by shifting your attention from fans and Venetian blinds to faces and wordlike sounds.

So what can seventy-five seconds of Baby Andy's crib life tell us about the specifics of his developing attention? Let's see. Start with Mama bending over his crib. "Is that a cute little nose, Baby Andrew?" she asks. Andrew is too young to realize it and Mama doesn't either, but in those few words she is helping to form many complex patterns of attention all at once. Attention is being shaped around values of "cuteness" (an idea of beauty), "little" (more aesthetics but also, when applied to a nose, with a whiff of ethnic values), and "nose" (a distinctive body part with functions that define it and that also need to be controlled). Americans, it turns out, are particularly obsessed about noses, not only their shape but their function. Americans stigmatize people of different national origins, class, ethnicity, and race based on scent, suggesting those who are not white and middle class "smell bad," a vague value advertisers exploit extremely well. As a white, middle-class American, I probably wouldn't have noticed our obsession with smell before I lived in Japan. I sometimes encountered Japanese, especially older ones, who would physically stand away from me and other *gaijin*. I asked a friend about it once and she informed me that, to Japanese, all Americans smell like wet dogs.

Mama's question is also very American linguistically. It exhibits the habit of asking Andrew a question even though he is far too little to understand it or to answer, in a high, lilting voice that she doesn't use with anyone else. Linguists call that voice, with its special sounds, inflections, vocabulary, and syntax, Motherese. It turns out that American mothers use Motherese statistically more often than just about anyone else on the planet. Asking babies questions and then answering them also turns out to be how you build a specifically Western baby. That question-and-answer structure is foundational for a lot of Western linguistic, philosophical, and even personality structures, extending, basically, from Socrates to *The Paper Chase* to Hogwarts. We want to build an Andy who asks questions and demands answers.

When his mother coos at him, we can call it love. Or we can call it being an American. Love may be natural, but how we express it is entirely learned and culturally distinctive.

Let's say Mama turns on a bright Mozart flute concerto. "Pretty," Mama says, pointing to the music player. She's building values again, in this case class values, an aesthetic appreciation of a particular kind of music. That's what Mama wants for her baby. In fact, Andrew might even recognize the music, because it was the same music she played while she was pregnant. Somewhere in the late 1990s, new audio technologies confirmed that babies can hear from the sixth month of gestation, and a smart advertising campaign was mounted trying to convince anxious pregnant mothers that, if they played Mozart, it would give their infants an appreciation for classical music that somehow would make them smarter. It is true that Andrew would be born recognizing Mozart as well as the sounds of American English and that he would be startled into attention if suddenly his nursery filled with the sounds of African music and people speaking Somali. It is not true he would be any smarter for being born used to hearing Mozart.

If Andrew's parents were listening to him as carefully as he was listening to the languages in his world, they might also have recognized that their baby cries with an accent. This is actually quite a new discovery. A study conducted in 2009 revealed that newborns cry in the language patterns they hear in utero. French newborns cry with an upward inflection, German babies cry with a falling inflection, mimicking the speech patterns of their parents.[11] Andy cries in American.

Even at four months, a number of Andy's behavior patterns are already distinctively American. For example, he is already smiling more than his Mexican or Kenyan or Japanese equivalents would. His American parents interact with him with a lot of talking, looking, and smiling—more than parents would in most other countries, in fact.[12] Smiling and crying are also Andy's two best gambits for eliciting the much-desired touch from his American parents. American parents rank almost at the bottom on the international scale of parent-infant physical affection, meaning that we just don't touch and hold our babies as much as people do in most other cultures. Some people believe that is why American babies are more verbally boisterous—crying, babbling, claiming attention by making noise, using their voices to claim some of the touch they crave.[13] Because he's an American baby *boy*, Andy gets even less touching than Big Sister did when she was his age. That disparate treatment of male and female infants is more extreme in America than in most other countries. Andy doesn't realize that he is being touched (or not touched) American-style.[14]

The cognitive psychologist Richard Nisbett and his international team of researchers have now conducted dozens of experiments with mothers and children that show how radically different child rearing is in different cultures. Even how we go about communicating our social building blocks varies, it turns out. For example, Americans love to name things—like "nose" or "Dada"—for their infants. We do this far more than a Japanese, Korean, Hong Kong, or Taiwanese mother would. We're "noun-obsessed." We *like* organizing the world into categories. We like nouns and we like giving the things of the world names and labels. We might think this is natural. How else would anyone group the world efficiently? That turns out not to be a rhetorical question but one with myriad answers, varying from culture to culture.

In one experiment, in which mothers were supposed to present their children with toys—a stuffed dog, a pig, a car, and a truck—the American mothers used twice as many object labels as Japanese mothers, even with their very youngest children. The American mother would say something like: "That's a car. See the car? You like it? It's got nice wheels." The Japanese mother would say, "Here! It goes *vroom vroom*. I give it to you. Now give this to me. Yes! Thank you." The Japanese mothers performed for their kids twice as many social routines that introduced the concept of polite giving and receiving, and they spent twice as much time emphasizing relationships, community, process, and exchange without describing the objects themselves. They used far more verbs and fewer nouns. As Nisbett notes wryly, "Strange as it may seem to Westerners, Asians don't seem to regard object naming as part of the job description for a parent."[15]

This is how babies learn their world. They are not simply learning difference. They are beginning to chart out the concepts and the distinctions made by those whose world they mirror; they are learning the value of those distinctions and the values of those who make them.[16] People depend on those distinctions and teach them to their children. But distinctions are normative, sensory, behavioral, social, cognitive, and affective *all at once*. Learning happens in categories, with values clumped together in our words, concepts, and actions. And this is where attention and its concomitant attention blindness come from.

WE'VE ANALYZED ONLY THE TIP of the iceberg in a scene that, so far, has probably lasted only ten seconds or so. What else might go on in this little scene? Let's say little Andrew cries and, using Motherese, his mama asks, "Is my little Andrew hungry?" She starts to pick him up but then puts him back down, wrinkling her nose. "Stinky! Let's get this diaper changed." American again! Attitudes toward defecation rank way up there in the taboos and practices that anthropologists chart in every culture.

Let's add another person to our scene. "Daddy's here!" the mother calls out. Mama's announcing Daddy's arrival is another way of signaling to her baby that this is an occasion to pay attention to, a personage commanding attention. "Look! Daddy's put away his cell phone!" Mama might add, words of praise that give weight and importance to any attention Andrew receives from Daddy. On the other hand, because attention is based on difference, remarking Daddy's attention also suggests it's not something that is routine. If Andrew wants to get his needs met quickly, as a general rule, he better holler first for Mama.

Again, these gestures happening in an instant do not come with lesson plans or Post-it notes or annotations. It's how culture is learned, how attention is learned. But it is learned so early, reinforced so often, that we do not stop to think of it as anything other than reflexive, automatic, "natural" behavior. Mama reports glowingly to Daddy about how Baby Andrew stopped crying when she put on the Baby Mozart tape and says that he might even have said "Dada" today. Here we have another lesson in attention blindness in action, the sifting of the world into meaningful language. Amid all the other babbling of a four-month-old, those two special syllables—"da-da"—were plucked out by Mama, emphasized, rewarded, encouraged, and thereby given meaning along with affection. Different cultures reward early language acquisition in different ways. Perhaps not surprisingly, given our culture's high valuation of language, in the West we consider precocious language skills as a distinguishing mark of an intelligent child. We spend a lot of time discerning words in baby babble, praising what we think are words, and bragging about an infant's early word mastery to others.

If Andy hasn't been fed yet because of the interruptions of the stinky diaper and the appearance of Daddy, it's time for him to start fussing. If into our scene come big sister Ashley and her BFF Sharon and they want to "play with the baby," it may require an actual shriek from Andrew to train the room's attention back on his needs. He may not even notice when Ashley touches him but, if he's fussing, Mama may well react by cautioning Ashley to be careful around the baby. If it is best friend Sharon who did the touching, Mama's warning will be more insistent. (Yes, psychologists have studied all of these things.) "Not too hard," Mama might warn her, a great way of communicating kinship relationships and rules, even without the help of a local Margaret Mead to describe what those are. Every society defines kinship differently, something I certainly saw in Japan when my students were baffled that "aunt" and "cousin" had so many kinship possibilities in English. We use "aunt" for the sister or the sister-in-law of a parent; "cousin" is so vague in our kinship system it basically means sharing a common ancestor. Other cultures describe familial relationships with different degrees of specificity.

When little Andy reaches up and grips Sis's finger, everyone takes turns commenting on how strong he is. Infant psychologists inform us that strength and smartness are far more often assigned to boy infants and cuteness to girl infants, even for newborns and even when the boy is smaller than the girl in the nursery. Boys aren't touched as much, but they are praised more for being smart.

"Señor?" a stranger says from the doorway. Mama doesn't know she draws closer to Baby Andrew at that time, but she probably does, even as she spins around to see who said that word. She wasn't expecting anyone. Daddy turns too, and then walks over to the stranger in the door. He pats the man on the back and, without introducing him, speaking of him in the third person, says, "Here's our plumber," and then walks him out the door. Mama's shoulders relax but Andy has a lot here to figure out, and even at four months, he is starting to put together a world where some people belong, some move back and forth unexpectedly but freely, and others are really not welcome over the threshold of the nursery without special permission by those who usually are in the nursery. "Strangers" get attention. They are different; they distract.

By the normal ideas in developmental psychology, such as those advanced by the brilliant Swiss philosopher Jean Piaget, complex social ideas such as concepts of race do not really enter a child's mind until much later, even in the early teen years. Yet increasingly, studies of infants as young as Andrew are showing that, on the most basic level of paying attention—turning a head in the direction of someone who talks a new language, who has different food or bodily smells, or who elicits apprehension from a parent—race and ethnicity begin to be factored in long before the concepts themselves are defined.

Ethnographic studies of preschoolers using picture cards and other games show that they develop a remarkably accurate map of the subtle socioeconomic, racial, and gender hierarchies implicit (and often not explicitly articulated) in a society.[17] Even before they can talk, infants are able to understand the dynamics of fear in their society. Babies being walked down a city street can be seen to bristle more when they pass adults of a different race, although in America, tragically, black babies, like white babies, are more likely to be frightened by an encounter with a black male stranger than a white one. Similar dynamics around complex social constructs such as gender and taboo subjects such as sexuality also manifest themselves at a far earlier, and even preverbal, stage than was previously believed possible. The new work being done in the lab of French neuroscientist Olivier Pascalis suggests that, by six months of age, babies notice the difference between faces of their own race and those of other races and ethnicities.[18]

Cute, nose, Mozart, Daddy, pretty, hungry, stinky, strong, careful, and stranger are some of the labels being built into Andy's cognitive system for mapping the world. He's learning how to pay attention to these, and he's learning the values so thoroughly that they will be close to automatic by the time he starts school, and then formal education will complete the process. He won't know why certain things go with other things, but he will act as if there's no other way for this to be—because that's how he's built. He won't even see any differently unless he has to.

WE'VE ONLY SCRATCHED THE SURFACE of all there is to know in a mere seventy-five seconds in Andrew's world, but it is a start at understanding how simple events, attitudes, sensory experiences, and linguistic cues are encouraging Andrew's habits of paying attention to what counts. Andrew doesn't understand this in anything like a rational or systematic way, but he is certainly getting the message that he's getting a message. He doesn't know the word for it yet, but he is noticing, even at four months, that there are patterns in his world and that those patterns have something important to do with him. For example, he is already wondering what everyone is trying to tell him, what it could possibly mean, and why some things are repeated over and over in so many ways, as if people are afraid he's not going to understand them.

And one other thing: Why, Baby Andrew wonders, is just about everything in his room—the walls, the carpet, the blankets, the stuffed toys, the clothes—so darn blue?

|||||||||||||||||||||||||||||||

Attention begins in the nursery, but as we age, we also learn to shape and reshape the values we learn there. Learning is the cartography of cultural value, indistinguishable from the landscape of our attention—and our blindness. We map the world around us in our own behaviors. And we can be tricked or herded in ways we might not wish to be when we encounter features of the world that aren't anywhere on our map. When we encounter a mismatch between our values and some new experience, we have a choice to either hold on to our values against all the evidence, to insist they are right or natural no matter what; or we can rethink them and even reject them, a process that can be smooth or traumatic, partial or complete. In any case, this process is a key component of the science of attention.

Once you have a category, it's hard to see past it without some serious reconsideration. That's the unlearning necessary to break an old habit before we can adopt a new one. It's a constant process and a crucial one.

That is exactly why it's so useful to understand what is going on with Andy. This story is not just about child rearing. It is about our own blind spots, and how we came to have them. By understanding that process, we increase our chances of intercepting the process, of intervening, and of changing its outcome. If the makers of the Cymbalta ad can understand our expectations well enough to manipulate them, so can we. But the only way we have a chance of paying attention differently is by understanding what we pay attention to when we're not thinking about it and where our reflexes and habits of attention came from. We cannot disrupt them, we cannot change them, until we understand these basic facts.

ANDY'S STORY IS OUR STORY. We no longer know our own preverbal history. Because of the categories by which we bundle our world, we can see efficiently. But those same categories make us miss everything else. Andy's job is to mimic and to master his culture's categories, long before he has the words to describe them and long before he's developed sophisticated sociological terms to explain them away. Mostly, in infancy, he detects the *feeling* of the categories, as conveyed to him by the people who care for him most and on whom he depends for everything.

Andy's process is the one we all go through as human beings. This process teaches our brain to pay attention, which means it is how our brain learns *not* to pay attention to the things that aren't considered important to those around us. But those other things do exist, all the time, even if we are not noticing them. That is simply how the brain science of attention works, with the world around us focusing our attention on what counts. What we are counting makes the things that don't count invisible to us.

As adults, we are not as helpless as little Andy anymore. We have the power to make choices. When confronted with the new—with what seems odd or outrageous, annoying or nonsensical—we can deny it has any value. We can label it a worthless distraction. Or we can try, on our own or with the help of others, to redraw our maps to account for the new. We can actually use the new to reshape *how* we focus our attention.

We have the capacity to learn, which is to say we have the capacity to change. We were born with that capacity. Just ask Andy. This is why we had to return to the scene of the nursery, to discover how we came to know the world. By understanding how we learned our patterns of attention, we can also begin to change them.

2

||||||||||||||||||||||||

Learning Ourselves

The first thing we need to know if we're going to make sense of the patterns of attention that are normally invisible to us is that the changes we talked about in the last chapter transform not merely our behavior but the underlying neural networks that make attention possible. Every manifestation of attention in the real world begins in the brain, which means that we need to look at a few basic principles of brain biology and at the way neural networks develop.

Neurons are the most basic cells in the nervous system, which includes the brain and spinal cord. Neurons are excitable, meaning that they process the body's electrical and chemical signals. There are various kinds of neurons, each with a specialized function, all interconnected in astonishingly intricate ways. The adult brain has been estimated to contain over a hundred billion neurons, each of which fires several times a second and has several thousand connections to other neurons. There are over a million billion neural connections in your brain. That's a lot of zeros.[1]

Like so much else we believe we know, the basics of brain biology are often not what we think they are. In fact, for much of the twentieth century, it was believed that the number of neurons in the brain increased as we aged. It was thought that connections must expand in number in much the same way that we grow taller or gain more knowledge over time. That's a logical assumption, but a false one.[2] The way the brain actually works, then, is counterintuitive: An infant has *more* neurons, not fewer, than anyone old enough to be reading this book. Our little Baby Andrew has an excess of neurons. If his development unfolds as it should, he will lose 40 percent of his extra neurons before he grows up. If he does not, he will not be able to function independently in society and will be considered mentally handicapped or disabled.

On a structural level, the process by which neurons are shed is remarkably similar to, and in fact is prompted by, the processes of selecting what to pay attention to that we discussed in the previous chapter; an infant's brain matures by selection. As the infant selects from all the world's stimuli those that matter—that deserve attention—he is also "editing" neurons. As he begins to select, concentrate, and focus on some things and not others, his brain is shearing unused neural pathways. When the brain is *not* making connections, the unused linkages wither away. The excess is eliminated so only the relevant data and relevant neural pathways remain.

Canadian Donald O. Hebb is often called the father of neuropsychology because he was the first person to observe that learning occurs when neurons streamline into pathways and then streamline into other pathways, into efficient clusters that act in concert with one another. This is now called the Hebbian principle: *Neurons that fire together, wire together.* This means that the more we repeat certain patterns of behavior (that's the firing together), the more those behaviors become rapid, then reflexive, then automatic (that's the wiring). They become patterns, habits, groupings, categories, or concepts, all efficiencies that "wire together" sets of individual reflexes or responses. Reflexive behaviors combine into patterns so we don't have to think about the components of each series of reactions each time we call upon them. We don't have to think, every time it happens, "Now I see the red light, now I am lifting my right foot off the gas pedal and moving it to the brake, now I am pushing down on the brake in order that I can stop the car before this yellow crosswalk." I see the red light, I stop.

If it weren't for the Hebbian "wiring together" of actions we perform in tandem, we would feel like everything we do is taxing, everything requires extra energy, everything is inefficient—in a word, that everything requires multitasking. Seeing the light, releasing the gas pedal, moving to the brake, all the while keeping my eyes on the road, keeping control of the car, while the kids are chattering in the backseat and the radio is on; all of this is multitasking. Fortunately, the Hebbian principle of our wired-together, firing-together neurons makes most of life's multiple tasks easier to process and respond to at once.

We don't know exactly how many repetitions it takes to create the pathways that produce automatic responses in infants—clearly there is tremendous variation, depending on whether the behavior is learning to smile or beginning the long conceptual and biological process toward toilet training. But we know that babies who do not receive a lot of care in infancy have a difficult time catching up, and some never do. This is why early infant care and preschooling are so crucial to a productive adulthood. All the pathways that will shape attention are being laid and reinforced, over and over, long before the child goes to school—where patterns will be named, mapped, and systematized—and long before that grown-up child enters the workforce, where the efficiencies that are developed and built upon since infancy will come to productive use.

Neural pathways connect the different parts of the brain and nervous system, translating ideas into actions in patterns learned over and over, reinforced by repetition, until they seem to us to be automatic. Repetitions literally shape specific patterns in very particular and extremely complex ways, coordinating impulses and activities across parts of the brain and nervous system that might be quite distant from one another. For example, the desire to walk and the ability to walk may seem automatic or natural to an able-bodied adult, but they are a complex operation involving many different parts of the brain that, with repetition, become more and more connected via neural pathways. To an infant, connecting the parts is a mysterious process. To a toddler, the process is clearer although not necessarily smooth. In those early stages, there are still many extraneous movements bundled into "toddling." We are constantly correcting this process (as we correct everything little Andy does) by our reward system, applauding some behaviors, moderating others.

Once human babies have learned to walk without thinking about it, they've reached a new level, both in learning and in the efficiency of the neural pathways. That means that the basic task of walking seems automatic: There are few impediments in the way between thinking you want to walk and walking. When transmittal is this reflexive, you can then build in other activities. You can think and walk. You can carry things. You can perform certain movements. Of course, all these are forms of multitasking, but because the basic task—walking—has happened so often and the complex network of neurons is so efficiently created and reinforced, we don't perceive it as multitasking. That which seems automatic doesn't feel to us like a "task." Because the most fundamental of the tasks—walking—seems natural and automatic and requires no premeditation, we can do other things easily as we walk. When you get really good at it, you might even be able to walk and chew gum.

Perhaps the single most important and certainly the most striking example of how an infant's neurons are firing together and wiring together—selecting relevant features from the irrelevant ones to be ignored—is language learning. At four months, an infant can still hear all the different sounds in all the world's languages.[3] But barely. At four months, he's already beginning to shape pathways that allow him to hear in English. That means he is excluding—shearing away—potential language sounds that do not occur in English. On a neural level, that means he is losing an ability he was born with: to be able to recognize all linguistic sounds.[4] Most infants in the United States lose this ability in order to focus on the sounds required for learning English and not French or Swahili or Icelandic or Arabic or Sanskrit or Chinese, all of which possess certain sounds that the other languages do not.

In the little scene with Andy, among all the babbling sounds he might make, Mama thinks she hears him say "Dada." She praises him, remarks on it, offers him affection as a reward, reinforcing the significance of those two syllables. "Dada" *means* something. When two syllables mean something, they are reinforced. "Mada" is ignored.[5] "Dada" is a word; in English, "Mada" is noise, so it receives no reinforcement. He stops trying to say "Mada."

Andy's babbling includes many sounds that aren't used in English at all. Those become meaningless; surprisingly quickly, they become unhearable. If, later, Andy decides to take up another language where those sounds are crucial, he will have to studiously retrain his ear to hear the sounds he's taught himself to ignore.

So a Japanese infant can distinguish *r* from *l*. A Japanese toddler cannot. There is no distinction between *r* and *l* in Japanese. Westerners hear these as two sounds because they both are sounds in our language. They aren't in Japanese; there is no careful distinguishing of them to the Japanese Andy, no one is constantly speaking to him using the two sounds, correcting his use of them, and so the distinction simply goes away for him. A Japanese infant can't hear a difference between *r* and *l* once he starts being able to speak Japanese.[6]

The principle we can learn from Andy and apply to our own lives is that this process doesn't stop with infancy. What we learn is also something we unlearn. Learn Japanese, unlearn the difference between *r* and *l*. It's not nearly as easy to relearn that difference, but it is possible so long as we remember that it is. If we believe capacities are "natural," we're lost. We need to remember how Andy's process of learning categories and concepts makes him believe that he is seeing the whole world, even though he isn't. He isn't even seeing or hearing all of the world that was once available to him, before he got rid of that overabundance of neurons.

Even if Andy were raised from infancy to speak more than one language, there would still be innumerable sounds that would be lost to him. He would still be paring down countless potential neural connections to relatively few. By definition, making new neural connections means severing others—the yin and yang of attention is mapped in the yin and yang of neural development. That's another key principle of learning. It's so basic that it has been given an extremely dramatic and powerful name that sounds like science fiction: *programmed cell death*.[7] Programmed cell death means that unused cells must die. They are *use-less* and soon don't exist. Learning requires this selection and discarding. Learning makes speedy, efficient, seemingly automatic neural pathways out of a tangle of haphazard connections.

The world would be chaos if unused neurons didn't atrophy and die. Without strong and efficient neural pathways, we'd be overwhelmed by the constant overstimulation of everything. Perceptually, it would be like being in the woods, lost and without a path, versus being in the woods on a well-marked path. Emotionally, it would feel terrifying to be constantly startled by events that always felt new and random.

An unsorted world would be almost impossible to navigate with any efficiency or success at all. One recent theory of severe autism is that something like this absence of categories happens in the early neurological development, around the same time that language learning is coalescing. Instead of being able to understand and assimilate and use categories, the autistic child can't make the groupings. Bundling never happens. The world may well make sense to the autistic individual, but that "sense" is incomprehensible to those around him. Communication falters, and so does the autistic child, whose development, in many cases, takes a very different course.

THOSE OF US WHO DON'T suffer from an unusual neural condition rarely pay attention to the efficiency of our neural pathways until something gets in their way, as might happen if one were to experience a crippling disease like Parkinson's or an injury that affects the limbs or spinal cord. In the aftermath of some catastrophic disruption of the neural pathways, the relay between the desire to walk and the act of walking can once again become a conscious process, often an arduous, sometimes impossible one.

These instances, when we need to consciously "rehab" and relearn what once seemed automatic, reveal to us the complexity of the task made efficient by neural shearing. Because learning how to walk again as an adult is very different from learning it as an infant, there is a significant component of unlearning, on a physical and neurological level, for one first has to break one's old habits and expectations in order to learn how to effectively walk again. The end result may seem the same, but because the process is so different, you actually need to retrain your mind to a new concept of "walking." Neural shaping and shearing that occurred during childhood made walking easy. After the injury, one has to disrupt the old patterns in order to find a new way to learn to walk, thus forming new neural pathways that eventually will make the relearned skill of walking more automatic. Because the injury unbundles well-trodden neural pathways, each part of learning to walk again requires creating new pathways, new patterns that, with extensive rehabilitation, may become automatic once more.

On a biological level, attention blindness is located deep within the brain and nervous system. If things are habitual, we do not pay attention to them—until they become a problem. Attention is about difference. We pay attention to things that are *not* part of our automatic repertoire of responses, reflexes, concepts, preconceptions, behaviors, knowledge, and categories and other patterns both mental and physical (if we can make such a crude distinction) for which we have, over time, developed more and more efficient neural pathways. We are constantly developing efficient ways of processing information so that certain sequences become automatic, freeing up valuable strategic thinking for novel tasks that have not yet been incorporated into our brain's repertoire of automatic actions.

It's only when something major and disruptive happens—say a kitten steps into the road right before I get to a stoplight—that I might break my pattern. Instead of bringing the car to a calm halt, I might jam on the brakes. I might even swerve sharply to miss the kitty. That change in my behavior, stimulated by the kitten, shakes me up in ways that bringing my car to a halt at a red light does not. I've been more or less paying attention as I drive, but braking for the cat makes me aware of paying attention.

Being aware of where and when I'm paying attention marks the difference from the usual forms of attention in everyday life. Suddenly being aware of having to pay attention is stressful, in good ways (exhilaration, inspiration) and in bad ways (anxiety, anger). On a biological level and on a pedagogical level, we become conscious of things that we normally don't think about. As sociolinguist George Lakoff says, we can be "reflective about our reflexes."[8] Self-reflexiveness or self-awareness is not necessary in all situations, but it *is* a key aspect of all successful informal and formal learning.

In times of major, global changes such as our own, a lot of life's incidents leave an indelible mark in the same way as slamming on the brakes to avoid the kitty, and for the same reason: They disrupt patterns that were laid down long ago. They unbundle neurons that have been firing together for a while. They start a new process of bundling, but until that process is successful—until enough firing and rewiring occur to become habitual—we will feel the stresses of the new. We will be aware that we need to pay attention in order to learn.

With new experiences, neurons that were not wired together by previous experience now have to fire urgently and independently. In the example of the straying kitty, there is also an emotional jolt to the system that I won't forget for a long time. Those powerful moments of learning—ones that come accompanied by some trauma or thrill—are the ones that make a difference. This is one reason why, as we shall see, a number of educators are advocating game principles as a learning system. If learning is exciting and as instantaneously self-reinforcing as winning a new game challenge, which comes with its own emotional bells and whistles to signal our learning victory, we are much more likely to remember and to incorporate the experience of our own success into other aspects of our life. We not only learn the content but we learn the form of winning, making us more adaptable and receptive to change in the future.

THE UPSHOT OF ALL THIS is that disruption in all its forms has the same effect: It makes us realize that what we thought was natural is actually a learned behavior that has come to feel that way thanks to the biological consequences of repetition. And *natural* defines not merely behavior, but culture and environment as well. In the Cymbalta ad, many of the most skillful manipulations were playing to our cultural efficiencies, to the thoughts, feelings, and desires we take for granted. Thanks to how often these are reinforced, they of course have neurological underpinnings just as ingrained as those that allow us to walk without thinking about it. Any new experience disrupts them in small or large ways, from a first taste of Ethiopian food to learning to drive stick shift after a decade on automatic to learning how to navigate the Internet. You thought you had those neural pathways nicely sheared and shaped only to find them disrupted.

Our institutions—family, friends, churches, social organizations, schools, workplaces—reinforce biological patterns all the time, thereby *shaping* those biological processes on the most fundamental levels. Sometimes, in periods of great change, there is a mismatch between the patterns our institutions reinforce and the patterns we need to operate efficiently in the new situation we are facing. If we had been in a terrible car accident or were suffering from a degenerative disease, we'd be using physical rehabilitation and the expert advice of others to make our motor neurons (which connect the spinal cord to muscles) work as smoothly as possible, given impediments that no longer allow us to work as efficiently as we once did.

The same is true in times of tremendous change. *That* is when we need to unlearn the previous patterns because they are not serving us. That is when we need to unlearn old habits so we can begin to relearn how to learn again.

||||||||||||||||||||||||||||

At the nexus of culture and biology we find the catastrophic neurological condition called Williams syndrome, which has much to tell us about both the necessity of neural shearing and the relativity of cultural norms that we often take for granted.[9] It is now thought that this rare genetic disorder (technically, the absence of twenty-six genes from the seventh chromosome) results in a systemic aberration of the architecture of the cortex in which too few neural networks are cut away. A child with Williams syndrome is bombarded with too much information and has no efficient way to sort it all out.[10]

Children with Williams syndrome typically test very low on IQ tests, in the 40–50 range, and have very little working memory, lacking the ability to remember a sequence of simple operations required to perform such simple tasks as shoe tying or counting.[11] Yet despite these inabilities, the disorder's unique genetic makeup also bestows personality traits and aptitudes that we might find quite valuable. Preschoolers with Williams syndrome exhibit exceptional ability at facial recognition, a difficult cognitive task not mastered in most children with "normal" intelligence until the age of five or six. Williams syndrome children often also have a love of music, typically have perfect pitch, and tend to be oblivious to negative cultural cues, such as those for racial bias that adults passively (and often unknowingly) transmit to children. On tests for racism, Williams syndrome children often test virtually prejudice-free.

The single most immediately obvious characteristic of children with Williams syndrome is a tendency to have an abundant and precise vocabulary and exceptional storytelling abilities.[12] Asked to name ten animals in a timed test, an average child might name such animals as *cat* or *dog.* The child with Williams syndrome might say *ibex* or *newt* or *alpaca* but might not be able to understand a simple processing concept such as naming ten animals. The child will therefore keep going, naming more and more animals until the tester stops him.

Things get even more interesting when we look at perceptions of Williams syndrome across cultures. In the United States, the diagnostic literature on Williams syndrome invariably defines a variety of personality traits thought to be characteristic of the disease. The words for these are almost entirely positive. Children with Williams syndrome are considered remarkably affable, inquisitive, charming, smiling, laughing, cheerful, affectionate, loving, and gregarious. In the West, that string of attributes is valued. The pleasing personality of the child with Williams syndrome is often considered a saving grace or blessing, some compensation for all the disabilities, varying in degree, of those born with this neurodevelopmental disorder. Children with Williams syndrome are sometimes described as "elfin," both because of their characteristic appearance and their spritely personalities. American researchers are studying oxytocin levels of Williams children with the idea that perhaps there is something about those twenty-six deleted genes on chromosome 7 that contributes to this powerful neurotransmitter, which helps us regulate pleasure, maternal feelings, empathy, and other positive responses.

In Japan, however, gregariousness, intrusiveness into other people's business, effusiveness, and affection in public or to casual acquaintances are fundamental offenses against the social fabric. Rather than being positive, these emotional and social traits rank as disabilities, as significant as neurological disorders in the catalogue of problems inherited by children with Williams syndrome. In Japan, they are considered continuous with the other mental and physical disabilities, and Williams syndrome children are not held up as a model for positive personality traits. They are pitied because of those same characteristics of personality. They are not studied for potential positive characteristics that might be used someday to help genetically engineer better human beings. They are far more likely than Western children with Williams syndrome to be institutionalized because of (what the Japanese perceive as) their antisocial nature.[13]

||||||||||||||||||||||||||||||||||

There is one last feature of brain biology we need to understand before we move on to schools, that formal place where our categories, concepts, patterns, and all our other habits of learning become reinforced in the most rigid way possible: through grades, rankings, evaluations, and tests. Everything we've seen about attention will be enacted—for good or ill—in the schoolroom.

The final principle of learning—and unlearning and relearning—we need to understand is mirror neurons. They were discovered in the 1980s and 1990s, and some people consider their discovery to be as important for neuroscience as sequencing the genome has been for genetics.

It happened in the lab of Giacomo Rizzolatti and his colleagues at the University of Parma in Italy. His lab didn't set out to find mirror neurons, as no one really knew they existed. At the time, the Parma scientists were studying how neurons synchronize hand-eye coordination. They placed electrodes in the ventral premotor cortex of macaque monkeys to see how their neurons were firing when they were picking up food and then eating it.[14] By doing so, the neurophysiologists were able to record the activity of single neurons when the monkeys were feeding themselves. That, alone, was news.

Then, one day, something *really* interesting happened. The Parma scientists began to notice that some neurons were firing in exactly the same pattern whether the monkey was picking up a piece of food and eating it or was watching a human or another monkey pick up a piece of food to eat. It didn't matter whether the monkey was performing the activity or watching it: The neural response was the same.

Many years later, we are finding that humans, too, have mirror neurons. Mirror neurons respond in the exact same way when a person performs a certain action and when a person observes someone else performing that action. That is so startling and so counterintuitive that it bears restating: These specialized neurons mirror the person (or monkey) observed as if the observer himself were performing the action.

Not all neurons act this way, only a particular subset. But this discovery was far more revolutionary than anything the scientists had set out to find about the neural mechanism of hand-eye coordination. The Parma neuroscientists switched the hypothesis and the protocols of their experiments. Soon they were isolating and identifying a subset of neurons that they named mirror neurons. They argued that approximately 10 percent of the neurons in the monkey's frontal cortex had these mirroring properties.

Since those early experiments, fMRIs have been used to study the human brain, and more and more mirror neurons are being found in new areas. There are mirror neurons that register the sounds we hear others make, as well as visual mirror neurons too. Recently, mirror neurons have also been located in the somatosensory areas of the brain associated with empathy.[15] Throughout our life, our mirror neurons respond to and help us build upon what we see, hear, and experience from others around us. In turn, their mirror neurons respond in the same way to us.

Primatologist Frans de Waal likes to say that imitation doesn't begin to comprehend the complex, mutually reinforcing interactivity of human teaching and learning.[16] He notes that teaching requires that mirroring work both ways. The child watches the parent do something and tries it, and then the parent watches the child trying and reinforces what she's doing right and corrects what the child is doing wrong: an intricate, empathic dance. Mirror neurons help to make the correction.

De Waal is highly skeptical of the claim, through the ages, that one or another distinctive feature "makes us humans" or "separates us from the animals." We *are* animals after all. De Waal, in fact, believes that animals can do just about everything we think defines us as humans. However, he still thinks that there is something special, exceptional even, about the calibrated, interactive nature of teaching. The animals he studies feel, think, solve problems, and have all kinds of astonishing capabilities that humans don't even approach— animals fly on their own power, navigate on and beneath the oceans without vessels, and migrate thousands of miles without instruments, and so on and so forth.

But teaching is a very complicated, interactive process. And it is possible, he suggests, that humans are among the only animals that actually teach not just by modeling behavior, but by watching and correcting in a complex, interactive, and empathetic way. De Waal claims that there is only one other very clear example in the nonhuman animal kingdom of actual teaching in this sense: a pod of killer whales off Argentina. He says this particular pod of killer whales actually trains its young in the dangerous practice of pursuing seals onto shore to eat them, managing this feat with barely enough time to ride the surf back out to safety again without being beached there to die. The older whales evaluate which of the young ones are good enough to do this, and they encourage this behavior relative to the young whales' skills. They don't just model but actually calibrate how they teach to those gifted students capable of learning this death-defying feat. These whales school their young the way humans do.

This particular ability to teach individually, to the best skills and abilities of the students, in a way that's interactive and geared to the particular relationship of student and teacher, seems if not exclusive to humans then certainly one of our very special talents. De Waal doesn't know of other animals able to calibrate empathy with instruction in this way. It is how we humans learn our world.[17] In the classroom and at work, it is the optimal way to teach any new skill and to learn it. It is the way most suited to all aspects of our brain biology, the way most suited to what we know about the brain's ways of paying attention.

Mirror neurons allow us to see what others see. They also allow us to see what we're missing simply by mirroring people who see different things than we do. As we will find in the subsequent chapters on attention in the classroom and at work, that feature alone can be world-changing for any of us.

||||||||||||||||||||||||||||

To live is to be in a constant state of adjustment. We can change by accident—because we have to, because life throws us a curveball. But we can also train ourselves to be aware of our own neural processing—repetition, selection, mirroring—and arrange our lives so we have the tools and the partners we need to help us to see what we might miss on our own. Especially in historical moments such as our own rapidly changing digital age, working with others who experience the world differently than we do and developing techniques for maintaining that kind of teamwork can help to take the natural processes of repetition, selection, and mirroring, and turn them to our advantage.

One guide to keep in mind—almost a mnemonic or memory device—is that when we *feel* distracted, something's up. Distraction is really another word for saying that something is new, strange, or different. We should pay attention to that feeling. Distraction can help us pinpoint areas where we need to pay more attention, where there is a mismatch between our knee-jerk reactions and what is called for in the situation at hand. If we can think of distraction as an early warning signal, we can become aware of processes that are normally invisible to us. Becoming aware allows us to either attempt to change the situation or to change our behavior. In the end, distraction is one of the best tools for innovation we have at our disposal—for changing out of one pattern of attention and beginning the process of learning new patterns.

Without distraction, without being forced into an awareness of disruption and difference, we might not ever realize that we are paying attention in a certain way. We might think we're simply experiencing all the world there is. We learn our patterns of attention so efficiently that we don't even know they are patterns. We believe they *are* the world, not a limited pattern representing the part of the world that has been made meaningful to us at a given time. Only when we are disrupted by something different from our expectations do we become aware of the blind spots that we cannot see on our own.

MANY OF OUR ANXIETIES ABOUT how the new digital technologies of today are "damaging" our children are based on the old idea of neural development as fixed, or "hardwired," and on notions of distraction and disruption as hindrances instead of opportunities for learning. Our fears about multitasking and media stacking are grounded in the idea that the brain progresses in a linear fashion, so we are accumulating more and more knowledge as we go along. Most of us, as parents or teachers or educational policy makers, have not yet absorbed the lessons of contemporary neuroscience: that the most important feature of the brain is Hebbian, in the sense that the laying down of patterns causes efficiencies that serve us only while they really are useful and efficient. When something comes along to interrupt our efficiency, we can make new patterns. We don't grow or accumulate new patterns. In many cases, new ones replace the old. Slowly or rapidly, we make a new pattern when a situation requires it, and eventually it becomes automatic *because* the old pattern is superseded.

Pundits may be asking if the Internet is bad for our children's mental development, but the better question is whether the form of learning and knowledge making we are instilling in our children is useful to their future. The Internet is here to stay. Are we teaching them in a way that will prepare them for a world of learning and for human relationships in which they interweave their interests into the vast, decentralized, yet entirely interconnected content online?

As we will see, the answer more often than not is no. We currently have a national education policy based on a style of learning—the standardized, machine-readable multiple-choice test—that reinforces a type of thinking and form of attention well suited to the industrial worker—a role that increasingly fewer of our kids will ever fill. It's hard to imagine any pattern of learning less like what is required to search and browse credibly and creatively on the free-flowing Internet than this highly limited, constricted, standardized brand of knowledge.

If some pundits are convinced that kids today know nothing, it may well be because they know nothing about what kids today know. A young person born after 1985 came into a world organized by different principles of information gathering and knowledge searching than the one into which you were born if your birthday preceded that of the Internet. Their world is different from the one into which we were born, therefore they start shearing and shaping different neural pathways from the outset. We may not even be able to see their unique gifts and efficiencies because of our own.

When we say that we resent change, what we really mean is that we resent the changes that are difficult, that require hundreds or even thousands of repetitions before they feel automatic. Adults often feel nostalgic for the good ol' days when we knew what we knew, when learning came easily; we often forget how frustrated we felt in calculus class or Advanced French, or when we played a new sport for the first time, or had to walk into a social situation where we didn't know a soul, or interviewed for a job that, in our hearts, we knew was wrong for us. We also tend to forget that, if we did poorly in French, we stopped taking it, typically narrowing our world to those things where we had the greatest chance of success.

We humans tend to worry about the passing of what and who we once were, even though our memories, with distance, grow cloudy. When calculators were invented, people were concerned about the great mental losses that would occur because we no longer used slide rules. With programmable phones, people wonder if anyone will memorize phone numbers anymore. Both predictions have probably come true, but once we no longer think about the loss, the consequences stop seeming dire. But, yes, that's how it *does* work, on a practical level and also on a neural level. Unlearning and relearning, shearing and shaping.

All of us like to believe we are part of the 50 percent in any situation who see it all clearly, act rationally, and make choices by surveying all the options and rationally deciding on the best course of action. But for most of us, it takes something startling to convince us that we aren't seeing the whole picture. That is how attention works. Until we are distracted into seeing what we are missing, we literally *cannot* see it. We are, as my colleague the behavioral economist Dan Ariely has shown us, "*predictably* irrational." You take courses in psychology and business on your way to designing direct-to-consumer ads in order to understand the ways most of us, most of the time, think. We believe we are rational, but in quite predictable patterns, we are not.[18]

We do not have to be stuck in our patterns. Learning happens in everything we do. Very little knowledge comes by "instinct." By definition, instinct is that which is innate, invariable, unlearned, and fixed. Instinct applies to those things that cannot be changed even if we want to change them. So far as anyone can test, measure, or prove, instinct doesn't account for much in humans. Biologists unanimously define as "instinctive" only a few very basic reflexive responses to stimuli. One of these, known as the Babinski reflex, is an involuntary fanning of the newborn's toes when her foot is stroked, a primitive reflex that disappears by around twelve or eighteen months.[19]

Except for this very specific reflex, babies come into the world having learned nothing by instinct and eager to pay attention to everything. We focus their attention and, in that process, also focus their values and deepest ways of knowing the world. The world we want them to know and explore should be as expansive and creative as possible.

As we move from the world of the infant to the world of formal education, the big questions we will be asking are: What values does formal education regulate? What forms of attention does formal education systematize? How valuable are both in the contemporary world? And, most important, are the educational areas on which we're placing our attention a good match with the world for which we should be preparing our children?

Part Two

||||||||||||||||||||||||||

The Kids Are All Right

3

IIIIIIIIIIIIIIIIIIIIIIII

Project Classroom Makeover

The *Newsweek* cover story proclaimed, "iPod, Therefore I Am."

On MTV News, it was "Dude, I just got a free iPod!"

Peter Jennings smirked at the ABC-TV news audience, "Shakespeare on the iPod? Calculus on the iPod?"

The online academic publication *Inside Higher Ed* worried for our reputation. How would Duke University "deal with the perception that one of the country's finest institutions—with selective admissions, a robust enrollment, and a plush endowment—would stoop to a publicity ploy?"

And *The Duke Chronicle* was apoplectic: "The University seems intent on transforming the iPod into an academic device, when the simple fact of the matter is that iPods are made to listen to music. It is an unnecessarily expensive toy that does not become an academic tool simply because it is thrown into a classroom."[1]

What had these pundits so riled up? In 2003, we at Duke were approached by Apple about becoming one of six "Apple Digital Campuses." Each campus would choose a technology that Apple was then developing and would propose a campuswide use for it. It would be a partnership of business and education, exploratory in all ways. One university selected Apple PowerBooks loaded with iLife digital audio and video production software. Another chose e-portfolios, online workspaces where students could develop multimedia projects together and then archive them. Another selected audio software for creating audio archives and other infrastructure. What landed us in hot water was that, at Duke, instead of any of these, we chose a flashy new music-listening gadget that young people loved but that baffled most adults: iPods.

In 2003, the iPod did not have a single known educational app, nor did it seem to fall into that staid, stolid, overpriced, and top-down category known as IT, or instructional technology. Gigantic billboards had sprung up everywhere showing young people dancing, their silhouettes wild against brilliant bright backgrounds. What could possibly be educational about iPods? No one was thinking about their learning potential because they were so clearly about young users, not about IT administrators. That's why they intrigued us.

Our thinking was that educators had to begin taking seriously the fact that incoming students were born after the information age was in full swing. They were the last entering class who, as a group, would remember the before and after of the Internet. If they were born roughly in 1985 or so, they would have been entering grade school around the time that Tim Berners-Lee was inventing the protocols for the World Wide Web. These kids had grown up searching for information online. They had grown up socializing online, too, playing games with their friends online and, of course, sharing music files online. Categories and distinctions that an earlier generation of students would have observed in school and at home, between knowledge making and play, came bundled in a new way for this first generation of kids who, in their informal learning, were blurring that boundary. Their schools hadn't changed much, but at home, online, they were already information searchers. They had learned by googling. What if instead of telling them what they should know, we asked them? What if we continued the lesson of the Internet itself and let them lead us into a new, exploratory way of learning in order to see if this self-directed way might mean something when it came to education? What if we assumed that their experiences online had already patterned their brains to a different kind of intellectual experimentation—and what if we let them show us where the pedagogical results of such an experiment might lead?

From the way most schools operated in 2003—from preschool to graduate schools—you wouldn't have had much of an idea that the Internet had ever been invented. It was as if young people were still going to the library to look in the *Encyclopaedia Britannica* for knowledge, under the watchful eye of the friendly local librarian. Schools of education were still training teachers without regard for the opportunities and potential of a generation of kids who, from preschool on, had been transfixed by digital media.

The opportunity seemed to be staring us in the face. At home, five-year-olds were playing Pokémon every chance they could, exchanging the cards at preschool with their pals, and then designing tools online to customize their characters and even writing elementary code to streamline their game play. They were memorizing hundreds of character names and roles and mastering a nine-year-old reading level just to play, but teacher training on every level was still text-based. It was as if schools were based on a kind of "hunt-and-peck" literacy, whereas kids were learning through searching, surfing, and browsing the Web. They were playing games in 3-D multimedia, learning to read and write not through schoolbooks but as they played games online and then traded their Pokémon cards with their friends.

When Duke announced that we would be giving a free iPod to every member of the entering first-year class, there were no conditions. We simply asked students to dream up learning applications for this cool little white device with the adorable earbuds, and we invited them to pitch their ideas to the faculty. If one of their profs decided to use iPods in a course, the prof, too, would receive a free Duke-branded iPod and so would all the students in the class (whether they were first-years or not). We would not control the result. This was an educational experiment without a syllabus. No lesson plan. No assessment matrix rigged to show that our investment had been a wise one. No assignment to count the basketballs. After all, as we knew from the science of attention, to direct attention in one way precluded all the other ways. So we asked our basic questions in as broad and open-ended a way possible: *Are there interesting learning applications for this device that is taking over young America as a source of entertainment?* And then the most revolutionary question of all: *What do you students have to tell us about learning in a digital age?*

If it were a reality show, you might call it Project Classroom Makeover. It was a little wild, a little wicked, exactly what you have to do to create a calculated exercise in disruption, distraction, and difference: a lesson in institutional unlearning, in breaking our own patterns and trying to understand more of the intellectual habits of a new generation of students and providing a unique space where those new talents might flourish. Instead of teaching, we hoped to learn. We wanted to tap into a wellspring of knowledge young people brought to college from their own informal learning outside of school. We didn't know what would happen, but we had faith that the students would come up with something interesting. Or not. We couldn't deny that failure was also a possibility.

At the time, I was vice provost for interdisciplinary studies at Duke, a position equivalent to what in industry would be the R & D (research and development) person, and I was among those responsible for cooking up the iPod experiment and figuring out how it could work in the most interesting ways.[2] We wanted to stir up some of the assumptions in traditional higher education. We didn't count on causing the uproar that we did. We assumed some of our fellow educators would raise an eyebrow, but we didn't imagine an educational innovation would land us on the cover of *Newsweek*. Usually, if education is on the cover, it's another grim national report on how we are falling behind in the global brain race. Come to think of it, that *is* what the *Newsweek* cover story was about! Like Socrates before us, Duke was leading youth astray, tugging them down the slippery slope to perdition by thin, white vinyl iPod cords.

We were inverting the traditional roles of teacher and learner, the fundamental principle in education: hierarchy based on credentials. The authority principle, based on top-down expertise, is the foundation of formal education, from kindergarten playgroups to advanced graduate courses. At least since the GI Bill that followed World War II, and the rapid expansion at that time of the public university system, a college degree has been the entry card to middle-class, white-collar achievement. Not graduating from high school and lacking a college degree has constituted failure, and education has constructed its objectives backward from that (negative) goal, in some cities all the way down to competition for the right private nursery school.

What this means for young people who come to an elite private university is that they have taken one of a number of specific routes to get there. One way is to test to get into the best preschools so you can go to the best private grammar schools so you can be admitted to the most elite boarding schools so you can be competitive at the Ivies or an elite school outside the Ivies like Stanford or Duke. Another way is through public schools, a lifetime of determined and focused study, getting A's and even A+ grades in every class, always taking the most difficult courses, earning perfect scores on tests, and doing lots of extracurricular work, too. These students have been focused toward educational achievement their entire lives.[3] We wondered what these astonishing young overachievers would do if given the chance not to follow the rules but to make them.

IN THE WORLD OF TECHNOLOGY, *crowdsourcing* means inviting a group to collaborate on a solution to a problem, but that term didn't yet exist in 2003 when we conducted the iPod experiment. It was coined by Jeff Howe of *Wired* magazine in 2006 to refer to the widespread Internet practice of posting an open call requesting help in completing some task, ranging from writing code (that's how the open source code that powers the Mozilla browser was written) to creating a winning logo (such as the "Birdie" design of Twitter, which cost a total of six bucks).[4] Crowdsourcing is "outsourcing" to the "crowd," and it works best when you observe three nonhierarchical principles. First, the fundamental principle of all crowdsourcing is that difference and diversity—not expertise and uniformity—solves problems. Second, if you predict the result in any way, if you try to force a solution, you limit the participation and therefore the likelihood of success. And third, the community most served by the solution should be chiefly involved in the process of finding it.

In the iPod experiment, we were crowdsourcing educational innovation for a digital age to our incoming students. We were walking the walk. Crowdsourced thinking is very different from credentialing, or relying on top-down expertise. If anything, crowdsourcing is suspicious of expertise, because the more expert we are, the more likely we are to be limited in what we even conceive to be the problem, let alone the answer. While formal education typically teaches hierarchies of what's worth paying attention to, crowdsourcing works differently, in that it assumes that no one of us individually is smarter than all of us collectively. No matter how expert we are, no matter how brilliant, we can improve, we can learn, by sharing insights and working together collectively.

Once the pieces were in place, we decided to take our educational experiment one step further. By giving the iPods to the first-year students, we ended up with a lot of angry sophomores, juniors, and seniors. They'd paid hefty private-university tuitions too! So we relented and said *any* student could have a free iPod—just so long as she convinced a prof to require one for a course and came up with a learning app in that course.

Does that sound sneaky? Far be it from me to say that we *planned* this, but once the upperclassmen coveted the iPods, once they'd begun to protest enviously and vehemently, those iPods suddenly tripled and quadrupled in perceived value: Everyone wanted one.

If "Shakespeare on the iPod" is the smirking setup, here's the punch line: Within one year, we had distributed more free iPods to students in forty-eight separate "iPod courses" than we had given without strings to the 1,650 entering first-year students.

That was vindicating enough, but it wasn't all. The real treasure trove was to be found in the students' innovations. Working together, and often alongside their profs, they came up with far more learning apps for their iPods than anyone—even at Apple—had dreamed possible. No one has ever accused Steve Jobs of not being cagey, and Apple's Digital Campus strategy was an R & D winner. The company's flagship technology now had an active lab of students creating new apps for it. There was also plenty of publicity for the iPod as a potential learning tool—the teenagers of America should all thank us for making it easier to pitch the purchase to their parents. In the first year of the iPod experiment, Duke students came up with dozens of stunning new ways to learn. Most predictable were uses whereby students downloaded audio archives relevant to their courses—Nobel Prize acceptance speeches by physicists and poets, the McCarthy hearings, famous trials, congressional debates, or readings by T. S. Eliot or Toni Morrison, or Thomas Edison's famous recitation of "Mary Had a Little Lamb"—one of the first sound recordings ever made. Almost instantly, students figured out that they could also record lectures on their iPods and listen at their leisure. Classes from Spanish 101 to Introduction to Jazz to organic chemistry could be taped and listened to anywhere. You didn't have to go to the library or the language lab to study. You could listen to assignments on the bus, at the gym, while out on a run—and everyone did. Because everyone had the device, sound suddenly had a new educational role in our text- and visuals-dominated classroom culture.

Some version of this convenient form of listening was possible with that radical eighties technology, the Sony Walkman. But the Walkman connected to radio and to tapes, not to the World Wide Web, with its infinite amount of information ready for downloading.

Interconnection was the part the students grasped before any of us did. Students who had grown up connected digitally gravitated to ways that the iPod could be used for *collective* learning. They turned the iPods into social media and networked their learning in ways we did not anticipate. In the School of the Environment, with the encouragement of Professor Marie Lynn Miranda, one class interviewed families in a North Carolina community concerned with lead paint in their homes and schools. Each student would upload the day's interviews to a course Web site, and any other student could download and comment on the interviews. At the end of the course, they combined their interviews, edited them digitally, and created an audio documentary that aired on local and regional radio stations and all over the Web.[5]

Some med students realized that there was an audio library of all the possible heart arrhythmias, but no way to access it in a real-time health exam. They came up with a way to put a stethoscope in one ear, using very simple signal-tracking technology to match what they were hearing in the patient's chest to the cataloged conditions. The implications of this original use were obvious, and soon students studying to be doctors and nurses were "operationalizing" such techniques for the diagnostic use of doctors in rural North Carolina and Africa. Dr. Martha Adams, a senior administrator at the Duke School of Medicine, grasped how revolutionary it was to be able to make state-of-the-art medical research available to those far outside major research centers, and to also make it possible for doctors elsewhere to report on health problems and patterns they were observing in their own communities, thus advancing medical research in both directions. Soon she was working with the National Institutes of Health and leading a national outreach iPod initiative. Once again, attention was being focused in multiple directions at once, not just on outcomes but on process and on interaction, the mirroring happening (as it must, definitionally) in both directions.

In the music department, composing students uploaded compositions to their iPods so their fellow students could listen and critique. Music performance students inserted their voices or their instruments into duets or choruses or orchestras. You could listen to how you sounded as first chair in the flute section of a famous philharmonic orchestra. Students in Duke's engineering department had a field day mangling and dissecting their iPods to study (hack, some would say) everything from Apple's ultrasecret computer code to the physical properties of the famous white plastic exterior of the original iPods.

And they began exploring apps, developing applications that could be added to the iPod's repertoire of abilities without Apple having to give away its proprietary code. In other words, the iPod could still remain an iPod with its own distinctive characteristics, but it could change and morph as new features were added and new capabilities emerged, including some developed by users. To me, this was a conceptual breakthrough: that a commercial product might also be susceptible to consumer customization, a way of extending the infinitely changeable open-source properties of the Internet itself to a product with a far more fixed, finite identity. It was a hybrid of old and new thinking. If that isn't a metaphor for attention in the digital age, I don't know what is.

By the end of our first experimental year, Duke was part of a new movement to transform the iPod from a listening device into an interactive broadcasting device. We were proud to host the world's first-ever academic "podcasting" conference early in 2005. I recently found one of our announcements for the conference and was amused to see those quotation marks around *podcasting*. No one was quite sure even what to call this new phenomenon, in which you could record a lecture, upload it to a Web site, and then anyone anywhere in the world could download it. Shakespeare on an iPod? Absolutely. And that lecture on Shakespeare delivered in the Allen Building at Duke could later be listened to by a student riding a bus in Bangkok or Brasília. That may not seem revolutionary now. It is hard to remember way back then, in the distant past of the Internet, before iPhones and netbooks, before MySpace and Facebook, and a full two years before YouTube was invented with its motto to "Broadcast Yourself."

The first podcasting conference drew standing-room-only attendance. It was sponsored by one of the first programs I'd spearheaded at Duke, something (another hybrid) called Information Science + Information Studies, or ISIS for short—artists and computer scientists, social scientists and engineers, and everyone in between in a new configuration. Lots of news media crowded into the auditorium at the Center for Interdisciplinary Engineering, Medicine, and Applied Science to witness the event. In a short span, the message had changed from "How could anyone possibly think this device could be used for learning?" to "This device facilitates sophisticated academic research and has the potential to make that learning instantly available to anyone in the world—for free."

The conceptual breakthrough of podcasting was access. It was expensive buying all those iPods, but the result was a breakthrough in education far beyond Duke, one whose purpose was to make a world of information cheaper to access than it ever had been before. With very little outlay, you had the potential of transmitting anything you heard, anywhere: You could download anything you heard worldwide. Not prerecorded programs made by professionals but content created and uploaded by anyone, ready for downloading—and for remixing and uploading again. When we launched the iPod experiment, no one expected that someday there would be an iTunes U (formed in 2007) with over 350,000 lectures and other educational audio and video files compiled by universities, libraries, and museums all around the world and available for download.

Duke took a lot of heat for being a "rich, privileged institution" that could afford this frivolity, but a revolution in the democratization of knowledge is not frivolous, especially considering that, once customized, an individual mobile device is actually an inexpensive computer. Several years after the Duke experiment, in the fall of 2008, Culbreth Middle School, a public school in nearby Chapel Hill, North Carolina, created its own iPod program for an experimental group of staff and students. They chose the iPod instead of a more traditional laptop because of "the mobility of the device in one's pocket with instant access to information and apps."[6] In January 2010, seventh graders were encouraged to explore the different ways their iPods could be used to keep them informed in the wake of the disastrous earthquake that brought destruction to Haiti. They used iPods to gather measurements of earthquake magnitude and related information, including demographic data, humanitarian assistance updates, local Haitian news podcasts, and historical information on Haitian culture and politics. The device also performed Creole-language translation. Students were even able to maintain up-to-date communication with a local graduate student who was in Haiti at the time and was badly injured in the earthquake. They used their iPods to educate themselves about a terrible disaster far away and produced their own podcasts from the information they gleaned. The experiment left little doubt that in the event of an emergency closer to home, students would be able to contribute their new knowledge to disaster-relief and fund-raising efforts locally.

The iPod experiment was not an investment in technology. It was an investment in a new form of attention, one that didn't require the student to always face forward, learn from on high, memorize what was already a given, or accept knowledge as something predetermined and passively absorbed. It was also an investment in student-led curiosity, whose object was not a hunk of white plastic, but the very nature of interactivity, crowdsourcing, customizing, and inspired inquiry-driven problem solving. At our most ambitious, we hoped to change the one-directional model of attention that has formed the twentieth-century classroom.[7]

THIS IPOD EXPERIMENT WAS A start at finding a new learning paradigm of formal education for the digital era. As we have seen, an infant's neural pathways are being sheared and shaped along with his values and his behavior in constant interaction with the people around him who exert influence over his life. The iPod experiment was an acknowledgment that the brain is, above all, interactive, that it selects, repeats, and mirrors, always, constantly, in complex interactions with the world. The experiment was also an acknowledgment that the chief mode of informal learning for a new generation of students had been changed by the World Wide Web. It was an attempt to put the new science of attention together with the new digital technology that both demanded and, in some ways, helped produce it.

I'm not going to argue that the *interactive* task of surfing is better or worse than the reception model that dominated mass education in the twentieth century. "Better" and "worse" don't make a lot of sense to me. But there's a difference and, as we have seen, difference is what we pay attention to. Said another way, we concentrate in a different way when we are making the connections, when we are clicking and browsing, than when we are watching (as in a TV show or movie) or listening or even reading a book. Indisputably, the imagination is engaged in making connections in all of those forms, as it is in anything we experience. It is engaged in a different way when we ourselves are making the connections, when we're browsing from one to another link that interests us and draws our attention. We don't need a "better or worse" because we have both, and both are potentially rich and fascinating cognitive activities. But the relative newness of the surfing/searching experience drove our interest in the potential of the iPod experiment; in 2003, educators already knew how to mine traditional media, but we had not yet figured out how to harness the new forms of attention students who had grown up surfing the Web were mastering. The Web does not prescribe a clear, linear pathway through the content. There is no one way to move along a straight-and-narrow road from beginning to end.

The formal education most of us experienced—and which we now often think of when we picture a classroom—is based on giving premium value to expertise, specialization, and hierarchy. It prepared us for success in the twentieth century, when those things mattered above all. Yet what form of education is required in the information age, when what matters has grown very different? What form of education is required in a world of social networking, crowdsourcing, customizing, and user-generated content; a world of searching and browsing, where the largest-ever encyclopedia is created not by experts but by volunteers around the world—as is the world's second most popular Web browser (Mozilla's Firefox), the world's most massive online multiplayer game (World of Warcraft, with over 11 million subscribers a month), and all the social networking and communication sites, from MySpace and Facebook to Twitter? Another way of asking the question is: How do we make over the twentieth-century classroom to take advantage of all the remarkable digital benefits of the twenty-first century?

The iPod experiment was a start, but to get a sense of just how big a task we face, it's useful to have a sense of how schools came to be the way they are, shaped by the values of a very different world.

|||||||||||||||||||||||||||||||

Do you remember the classic story by Washington Irving, "The Legend of Sleepy Hollow"? It was written in 1820 and features a parody of the pompous schoolmaster in the form of Ichabod Crane, a homely, gawky, and self-satisfied pedant who is confident in his role as a dispenser of knowledge. He knows what does and does not constitute knowledge worth having and is equally sure that students must be drilled in that knowledge and tested to make sure they measure up. If you blindfolded Ichabod Crane, spun him around, and set him down in a twenty-first-century classroom, he would be baffled by electricity, dumbfounded by moving images, confused by the computers and cell phones, but he'd know exactly where to stand, and he'd know exactly where he stood.

It's shocking to think of how much the world has changed since the horse-and-buggy days of Sleepy Hollow and how little has changed within the traditional classroom in America. On March 10, 2010, the National Governors Association and the Council of Chief State School Officers even called for "sweeping new school standards that could lead to students across the country using the same math and English textbooks and taking the same tests, replacing a patchwork of state and local systems in an attempt to raise student achievement nationwide."[8] Ichabod Crane lives!

What in the world is going on? In the past in America, times of enormous innovation in the rest of society, including in technology and in industry, have also been times of tremendous innovation in education. What has happened to us? Rather than thinking of ways we can be preparing our students for their future, we seem determined to prepare them for our past.

Literally. The current passion for national standards is reminiscent of the conversations on education at our country's beginnings, back in 1787, the year the U.S. Constitution was adopted. Technology was changing the world then, too. At the time of the signing of the Constitution, the new invention of steam-powered presses, coupled with the invention of machine-made ink and paper, made for mass printing of cheap books and newspapers, putting print into the hands of middle-class readers for the first time in human history. The new institution of the circulating library made books available even to the working poor. Books proliferated; newspapers sprang up everywhere. And that's when a cry for standards and public education was born in America, in response to a new democratic government that needed informed citizens and new technologies of print that made books and newspapers widely available.

Thomas Jefferson himself advocated that America had to launch a "crusade against ignorance" if the nation was to survive as an independent representative democracy.[9] By 1791, when the Bill of Rights was added to the U.S. Constitution, seven states were making provisions for public education. There was not yet anything like an "educational system" in the United States, though. Education was attended to unevenly by local, regional, state, and private institutions, some secular, some sectarian, an inheritance that continues to this day in the form of state-controlled educational policy, local and regional school boards, and other decentralized means of oversight.

Horace Mann, whose name can be found over the entrance of many public schools in America, was the first great champion of national educational reform. The son of a farmer of limited means, Mann clawed his way to an education, earning money by braiding straw to pay the local tuitions for the elementary schools he attended for only six weeks at a time, leaving the rest of his time free to help with family farming operations. He enrolled at Brown University at age twenty, graduated in three years as valedictorian of the class of 1819, and dedicated himself to the creation of the "common schools," which after around 1840 became the predecessor of a free, publicly supported education system.

The common schools were scheduled around the agricultural year so farm kids could attend too. The schools were open to both boys and girls, regardless of class, although several states explicitly forbade the attendance of nonwhite children. The schools were locally controlled, with the kind of local politics governing curriculum and textbook assignments then that we see now in the state-by-state regulation of education, even after our "national educational policy" has been adopted.

Mandatory, compulsory public schooling developed over the course of the last half of the nineteenth century and got its full wind at the turn into the twentieth century as part of America's process of industrialization. Public education was seen as the most efficient way to train potential workers for labor in the newly

urbanized factories. Teaching them control, socializing them for the mecha-
nized, routinized labor of the factory was all part of the educational imperative
of the day. Whether meant to calm the supposedly unruly immigrant populace
coming to American shores or to urbanize farmers freshly arriving in the city,
education was designed to train unskilled workers to new tasks that required a
special, dedicated form of attention. School was thought to be the right training
ground for discipline and uniformity. Kids started attending school at the same
age, passed through a carefully graduated system, and were tested systematically
on a standardized curriculum, with subjects that were taught in time blocks
throughout the day. In ways large and small, the process mimicked the forms of
specialized labor on the assembly line, as well as the divisions of labor (from the
CEO down to the manual laborers) in the factory itself.

Many features now common in twenty-first-century public education
began as an accommodation to the new industrial model of the world ush-
ered in during the last part of the nineteenth century. With machines that
needed to run on schedule and an assembly line that required human precision
and efficiency, schools began to place a great emphasis on time and timeliness.
Curriculum, too, was directed toward focusing on a task, including the mastery of a specified syllabus of required learning. "Efficiency" was the byword
of the day, in the world of work and in the world of school. Learning to pay
attention as directed—through rote memorization and mastery of facts—was
important, and schools even developed forms of rapid-fire question-and-answer,
such as the spelling bee or the math bee. This was a new skill, different from
the elite models of question-and-answer based on the Socratic method; the
agrarian model of problem solving, in which one is responsible for spotting a
problem and solving it (whether a wilted corn tassel or an injured horse); and
the apprenticeship model of the guild wherein one learned a craft by imitating
the skills of a master. An assembly line is far more regular and regulated. One
person's tardiness, no matter how good the excuse, can destroy everyone else's
productivity on the line. Mandatory and compulsory schooling for children was
seen as a way of teaching basic knowledge—including the basic knowledge of
tasks, obedience, hierarchy, and schedules. The school bell became a symbol of
public education in the industrial era.[10]

So did specialization. With the advent of the assembly line, work became
segmented. A worker didn't perform a whole job but one discrete task and
then passed the job on to the next person and the next and so forth down the

assembly line. The ideal of labor efficiency displaced the ideal of artisanship, with increased attention paid to the speed and accuracy of one kind of contribution to a larger industrial process. Focused attention to a task became the ideal form of attention, so different from, for example, the farmer on his horse scanning his land for anything that might look out of place or simply in need of care.

By 1900, state and regional schools were becoming the norm, replacing locally managed ones, and by 1918, every state had passed laws mandating children to attend elementary school or more. A concession was made to Catholics in that they could create a separate, parochial school system that would also meet these state regulations, another legacy that comes down to the present in the form of "faith-based schools."

During the first six decades of the twentieth century, as America ascended to the position of a world power, the rhetoric of education followed suit, with an increasing emphasis on producing leaders. While the nineteenth-century common schools had focused on elementary education, the twentieth-century focus was increasingly on the institution of high school, including improving graduation rates. In 1900, approximately 15 percent of the U.S. population received a high school diploma, a number that increased to around 50 percent by 1950.

After World War II, there was a rapid expansion of both high schools and higher education, invigorated after 1957 when the Russians surprised the world by launching *Sputnik,* the first man-made object ever to orbit the earth. As America competed against Russian science in the Cold War, policy makers placed more and more emphasis on educational attainment. Many economists argue that America's economic growth through the 1960s was fueled by this educational expansion.

Robert Schwartz, dean of the Harvard Graduate School of Education, notes that since the last quarter of the twentieth century, the pattern of educational expansion that has characterized the United States from the Revolutionary War forward has changed.[11] Since 1975, American educational attainment has leveled off or even dropped while there has been a dramatic increase in the number of jobs requiring exploratory, creative problem solving typically encouraged by postsecondary education. We are seeing the first signs that our education system is slipping in comparison to our needs.

The current high school graduation rate is roughly the same as it was in 1975, approximately 75 percent. Our graduation rate from four-year colleges is 28 percent, also roughly the same as it was thirty-five years ago. That this has even remained consistent in the face of all that has changed in the last thirty five years is remarkable enough, a credit to both the drive and quality of American students and the patchwork, piecemeal reforms we've used to hold the system together. Yet while we've been holding steady, other countries have made rapid gains. Whereas in the 1960s we ranked first in the proportion of adults with high school degrees, we now rank thirteenth on the list of the thirty countries surveyed by the Organisation for Economic Co operation and Development (OECD), an organization that coordinates statistics from market-based democracies to promote growth. By contrast, South Korea has moved from twenty-seventh place on that list to our old number one spot.

Most troubling is what happened from 1995 to 2005. During this one decade, the United States dropped from second to fifteenth place in college completion rates among OECD nations. For the wealthiest and most powerful nation on earth to rank fifteenth is nothing short of a national disgrace. This is especially the case, given that our system of education *presumes* college preparation is the ideal, even in environments where most kids are not going on to college. By that standard, we are failing.

It's not that we cared about education before 1975 but don't today. Our heart is not the problem. Look at the numbers. The Swiss are the only people on earth who spend more per child on public education than Americans. According to OECD, we spend over $11,000 per year per child on public education. That's more than double the rate of South Korea. Education spending accounts for more than 7 percent of our GDP.[12] However, the OECD statistics show that our graduation rates now are roughly on a par with those of Turkey and Mexico, not nations to which we like to compare ourselves by other indicators of our power or success.[13]

It is little wonder that educators and parents are constantly reacting to the comparative, global numbers with ever more strident calls for standards. The problem, however, is the confusion of "high standards" with "standardization." Our national educational policy depends on standardized tests, but it is not at all clear that preparing students to achieve high test scores is equivalent to setting a high standard for what and how kids should know and learn.

The real issue isn't that our schools are too challenging. It's the opposite. Among the top quartile of high school students, the most frequent complaint and cause of disaffection from schooling is boredom and lack of rigor. That also happens to be true among the lowest group, for whom low expectations lead to low motivation.[14] Kids aren't failing because school is too hard but because it doesn't interest them. It doesn't capture their attention.

Relevance has been proved to be a crucial factor for keeping students in high school, especially mid- and lower-level students. Tie what kids learn in school to what they can use in their homes, their families, and their neighborhood—and vice versa—and not surprisingly, that relevance kicks their likelihood of staying in school up a few notches. Because in the United States (but not in many countries with higher college attendance), going to college requires money for tuition, our emphasis on college as the grail of secondary education only rubs in its inaccessibility (its irrelevance) to lower-income kids—a fact that contributes to high school dropouts. Finally, for all groups, and especially for students in the lowest-achieving group, relationships with teachers and counselors who believe in them and support them (often against peer, familial, or cultural pressure) is a determining factor in remaining in school. These key factors for educational success—rigor, relevance, and relationships—have been dubbed the new three *R*s, with student-teacher ratio being particularly important. Small class size has been proved to be one of the single most significant factors in kids' staying in and succeeding in school. Twenty seems to be the magic number.[15] Even on a neurological level, brain researchers have shown that kids improve with directed, special attention to their own skills and interests, the opposite of our move toward standardization.[16]

The biggest problem we face now is the increasing mismatch between traditional curricular standards of content-based instruction and the new forms of thinking required by our digital, distributed workplace. At any level—blue collar or white collar—those jobs requiring "routine thinking skills" are increasingly performed by machine or outsourced to nations with a lower standard of living than the United States. Yet virtually all of contemporary American education is still based on the outmoded model of college prep that prepares students for middle management and factory jobs that, because of global rearrangements in labor markets, in large part no longer exist.

We've all seen industrial jobs for manual laborers dwindle in the United States and other First World economies, either taken over by machines or outsourced abroad to workers who are unprotected by unions or fair labor laws. The same is now the case for routinized white-collar office jobs. In exploitative "digital sweatshops" all around the world, workers at minimal wages can do everything from preparing your tax return to playing your online strategy games for you, so your avatar can be staging raids while you are on the trading floor or in your executive office on Wall Street.

To be prepared for jobs that have a real future in the digital economy, one needs an emphasis on creative thinking, at all levels. By this I mean the kind of thinking that cannot be computerized and automated. This creative thinking requires attention to surprise, anomaly, difference, and disruption, and an ability to switch focus, depending on what individual, unpredictable problems might arise. Perhaps surprisingly, these noncomputational jobs, impervious to automation, occur at all levels across the blue-collar and white-collar spectrum. Many of these jobs require highly specialized and dexterous problem-solving abilities or interpersonal skills—but do not require a college degree.

We were criticized for the iPod experiment. Many treated it as if it were an extravagance, superfluous to real learning and real education. But the iPod experiment exemplifies a form of inquiry-based problem solving wherein solutions are not known in advance and cannot be more successfully outsourced to either a computer or to a Third World laborer who performs repetitive tasks over and over in horrific conditions at minimal wages. The new global economics of work (whatever one thinks about it politically) is not likely to change, and so we must. And that change begins with schooling. Learning to think in multiple ways, with multiple partners, with a dexterity that cannot be computerized or outsourced, is no longer a luxury but a necessity. Given the altered shape of global labor, the seemingly daring iPod experiment turns out actually to be, in the long run, a highly pragmatic educational model.

PART OF OUR FAILURE RATE in contemporary education can be blamed on the one-size-fits-all model of standards that evolved over the course of the twentieth century; as we narrow the spectrum of skills that we test in schools, more and more kids who have skills outside that spectrum will be labeled as failures. As what counts as learning is increasingly standardized and limited, increasing numbers of students learn in ways that are not measured by those standards. This is the lesson of attention blindness yet again: If you measure narrowly, you see results just as narrowly. In other words, the more standardized our assessment, the more kids fail. Their failure is then diagnosed as a learning disability or a disorder. But they are failing when assessed by a standard that has almost nothing to do with how they learn online or—more important—what skills they need in a digital age.

The mismatch is just wrong. It's as if we're still turning out assembly-line kids on an assembly-line model in an infinitely more varied and variable customizing, remixed, mashable, user-generated, crowdsourced world. As long as we define their success by a unified set of standards, we will continue to miss their gifts, we will not challenge their skills, and, in the end, we will lose them from our schools just as, implicitly, we have lost interest in them.

We need far more inquiry-based opportunities for our kids. It doesn't have to be as expensive or as radical as the iPod experiment. The world is full of problems to solve that cost little except imagination, relevant learning, and careful guidance by a teacher with the wisdom to *not control* every outcome or to think that the best way to measure is by keeping each kid on the same page of the same book at the same time.

I recently visited a middle school where one girl with green- and blue-striped hair, creatively and eccentrically dyed, sat against the wall, remote from the other kids, as drawn into herself as she could possibly be without disappearing, looking for all the world like the kid who will never hear anything. When I, a stranger, came into the classroom, some of the other kids fussed over my unexpected appearance there, and my *difference*: One girl admired my purple leather jacket, another asked if I was an artist, because I was wearing a black turtleneck and skinny black pants. One boy asked about the image of an electric fan on my long, illustrated scarf from South Africa, and a waggish little boy hummed, with perfect pitch, "Who's That Lady?" when he saw me in his classroom. When I asked how in the world a twelve-year-old knew the Isley Brothers, he said snappily, "The Swiffer commercial." There was a lot of buzz in the class about the visitor, in other words. The green-haired girl in the corner slowly shifted her gaze in my direction, gave the smallest upward movement at the corner of her lips before returning to her frown and letting her eyes move back to outer space again, away from the strange visitor, determinedly *not there*.

I thought of a blog post I'd read earlier that week by the prolific business writer Seth Godin, creator of the popular user-generated community Web site Squidoo. Godin's post was called "What You Can Learn from a Lousy Teacher," and his list of what you can learn from the teacher you cannot please included: "Grades are an illusion, your passion and insight are reality; your work is worth more than mere congruence to an answer key; persistence in the face of a skeptical authority figure is a powerful ability; fitting in is a short-term strategy, standing out pays off in the long run; and if you care enough about work to be criticized, you've learned enough for today."[17] This remote young woman against the wall didn't look defeated by school. There was something *resolute* in her. I could sense that she had, somehow, taken in intuitively the kinds of lessons Godin was preaching, even if it would take her another decade to fully realize what, in her body language, she was already showing she'd absorbed.

She stayed that way, barely making eye contact with the teacher or any other students, until drawings begun during the previous class were handed around to be worked on again. The transformation I witnessed then was so rapid and thorough that I thought of what Pygmalion must have seen the first time his statue came to life. She went from being still, glassy-eyed, self-contained, and entirely not-present to a concentrated, focused, dedicated bundle of intensity. She still didn't interact with the other kids, but all eyes were on her as she began her day's work on a highly detailed line drawing she was executing. Unlike the bustle at the other tables in the room, there was a silence around her, with the kids practically tiptoeing into the circle of her aura to watch, not speaking to her or to one another, then moving away. I was dying to see her drawing but even more interested in the compact energy she'd created around herself and the other kids, generated from the passion of her drawing. Rather than interrupt that magic, I moved away.

Later in the day, I saw her waiting for her ride home, in the homeroom period at the end of the day, and again the sullen remoteness of I am not here returned. After all the kids had gone, I asked her teacher to tell me about her. She'd been diagnosed as profoundly learning disabled, with attention deficit disorder. Her parents sent her to this magnet arts school after she had failed elsewhere. At home, the only thing she seemed to enjoy doing was doodling, but her elementary school had laid off its art teachers, sacrificed to educational cutbacks and the fact that, in our standardized forms of testing, there is no EOG (end-of-grade exam) for art. Art is therefore an add-on that many public schools cannot afford. It was only at the new school that her skills as an artist were recognized. She quickly went from the class failure to the class artist. She mastered the editing tools on the class computer and transferred her imagination and creativity there, too, much to the admiration of her classmates.

I think again about what learning disabilities signify for kids today. In the *Diagnostic and Statistical Manual of Mental Disorders* (the infamous DSM), attention deficit disorder is characterized by distractibility, frequent switching from one activity to another, boredom with a task after just a brief time trying to execute it, and difficulty organizing and completing a task or in focusing on a task in which one is not interested. That last phrase is key. ADD almost never applies to *all* activities, only those in which the child is not interested. This isn't a disability (a fixed biological or cognitive condition) but a disposition (susceptible to change depending on the environment). Keep the kids interested and ADD goes away.

The girl with the green hair has special skills that show up nowhere on her compulsory EOG state tests, on which she continues to score poorly. "Your work is worth more than mere congruence to an answer key." This girl's talents don't count on those tests, and yet she has a special and valued ability that cannot be replaced by a computer program. The problem is that her fate is to a large extent controlled by her performance on the EOG tests, and unless the adults in her life—teachers and parents—are resolute in shepherding her along a path where her talents are valued, they may ultimately wind up undeveloped.

I identified with this girl. When I was in school, my talents had been math and writing, and there was a teacher, Miss Schmidt, who saw those gifts, despite my abysmal test scores, despite the fact that we had to memorize the preamble to the Constitution and the Gettysburg Address to graduate from eighth grade and I just couldn't. I tried and failed over and over many times, staying after school, trying to say it out loud and failing again. Miss Schmidt was feeling certain that I wasn't really trying but just being "obstinate." Then, during one of our painful after-class sessions, she had a hunch. She offered me the opportunity to write an essay instead, one essay about each brief text I was failing to memorize. I stayed up all night working on this and returned the next day with my project. My scrawl filled every page in a small spiral binder—*two hundred* pages.

I remember her eyes widening as she took this smudgy, worn binder from my hand. She looked through it in disbelief and bewilderment as she turned over pages and pages of my barely readable handwriting. There were even footnotes. "You still don't have them memorized, do you?" Miss Schmidt asked. I shook my head. After a moment, she got out the special state certification and, beside my name, put the check mark I needed to graduate from middle school. I'd passed.

Sometimes you learn from the good teachers too. It's not easy to be a good teacher all the time. The girl with the green hair was lucky to find one who honored her talent. Before I left his classroom, I told him, if he thought it appropriate, to let the girl know that the visitor, that professor from Duke University, had noticed her beautiful artwork and admired it. I told him, if he thought it appropriate, to let the girl know that, when the visitor was her age, her hair was purple.

As MATTHEW B. CRAWFORD ARGUES in *Shop Class as Soulcraft,* his eloquent study of the importance of making and fixing things, our national standards for secondary schools miss kids like this young lady. Our schools are geared, implicitly and explicitly, to be college preparatory. They are weighted toward a twentieth-century, white-collar, middle-management economy.[18] Our standardized education not only bores kids but prepares them for jobs that no longer exist as they once did. Attention blindness again. Our one-size-fits-all educational model focuses steadily and intently on the past.

The space age was America's educational glory moment, when we occupied the number one spot. At that moment, there was a splendid flowering of many kinds of educational theory, including progressive theories, free schools, Montessori schools, science and technology schools, new math, old math, and on and on. It was thought that we were facing a brave new world of science for, after all, incomprehensibly, men had walked on the moon. Education took that fervor to its heart, blossoming, experimenting with innovations as bold and expansive as the solar system itself.

Now, in a digital age, we can communicate and learn in ways not even imagined by Neil Armstrong and Buzz Aldrin walking for the first time on the face of the moon: "One small step for man, one giant leap for mankind." Yet our educational vision has shrunk to the tiny bubble of a multiple-choice exam. Our dreams of new standards of knowledge have shrunk to the right or wrong answer, chosen not among infinitely variable possibilities but among A, B, C, D, and, perhaps, none of the above.

I hate this kind of education. I'm prejudiced, I admit it. So don't ask me. Ask any adult. Ask yourself. What was the most formative experience of your life in school? I suspect your answer isn't "I scored in the 94th percentile on my fifth-grade end-of-grade compulsory exam." Whenever I ask this question, invariably I hear a story of a great teacher who had faith, who inspired and challenged someone to do more than she ever thought was possible before.

Ichabod Crane may nod approval at our current national educational policy. It may seem familiar to him and right. For those of us who remember what inspired us, what commanded our attention, there was always something experimental and daring, there was always something more.

||||||||||||||||||||||||||||||||

Some of the best teaching I ever witnessed did not happen courtesy of the Internet. It was many years ago, in rural Alberta, Canada. Inez Davidson, the mother of my first husband, Ted, began her teaching career at eighteen, teaching in a one-room schoolhouse in Pincher Creek, a coal-mining town in the Canadian Rockies. She rode half-wild horses to school each morning, alone in the dark and the cold through grizzly country, "putting miles on them," as she would say, before they could be sold to local ranchers as well-trained working horses. Inez was maybe five feet tall, but no one messed with her, ever.

Ted's father was a rancher in the town of Mountain View when he married Inez. Mountain View is as pretty as its name, a gorgeous ranching community in the foothills of the Rockies. The Davidson homestead looked out upon Chief Mountain, the sacred mountain of the Blackfoot and Blood Indian tribes. The area was made up mostly of Native peoples and Mormons or adamant ex-Mormons like the Davidsons, plus elk, moose, black bear, grizzlies, mountain lions, and so forth. Two hundred people in two hundred square miles. The area was poor, too. Rural electrification only came to Alberta in the late 1950s. Even in the 1970s and 1980s, when I was there, the kitchen had a stove that also burned wood (for when the electricity went out), and the entire house had only two additional propane heaters. This was *Canada*, in the foothills of the Rocky Mountains. In winter, it was sometimes 20 or 40 degrees below zero. The hallways and bathrooms and bedrooms were unheated. There was an outhouse for when the plumbing was overtaxed. It was a lot for a gal from Chicago to take in.

My mother-in-law taught in the three-room schoolhouse in Mountain View. For many years, there were more former residents of Mountain View with PhDs, MDs, or law degrees than from any other town in Alberta besides the two major cities in the province, Edmonton and Calgary. By the time I came on the scene, graduate students earning their degrees in education departments in the province were coming to Mountain View to find out what was happening there, that such a backwater was sending students not only to college but well beyond.

Mrs. Davidson, as she was called, was a main reason for this educational success story. How did she do it? First, she got in a lot of trouble, every year, with the school superintendent because she refused, ever, to teach to a test. She covered what the province demanded of third, fourth, and fifth graders and far more, but she always did it her way, as a challenge, a game, an interactive and communal learning activity. She made learning fun—and she was tough.

She was very skeptical of the concept of learning disabilities. She taught long after the normal retirement age and began to see kids coming to school who were more hyperactive than she'd seen in her earlier career and was convinced that part of the reason was environmental and part behavioral. She was concerned about both food additives and contaminants in the air and drinking water even in cattle country, where the water supply was tainted by runoff from oil and smelting plants far away. She was also shocked at how parents were driving kids to school instead of having them walk or ride horses in; these were forms of physical exercise she believed were important to kids' health, well-being, and concentration in the classroom. She loathed it when she saw kids medicated as a remedy for their learning disabilities, and was much more interested in all the subtle and obvious ways kids learned and learned differently.

She insisted that everyone had a unique way of learning and believed that to be true of the smartest students as well as the academically weakest. She took pride in finding challenges that inspired kids who had clear cognitive restrictions or physical ones, and defied anyone to bring her a kid she couldn't teach. Rodney, a close friend to all the Davidson boys, was a brilliant athlete but very poor in school. Reading was hard and math unbearable. He couldn't conceptualize arithmetic on any level and used his fingers to count even the simplest arithmetic. Other teachers wrote him off as "slow." Mrs. Davidson bought him an abacus and had him research all the sophisticated things the ancient Egyptians could calculate on one. She then taught him to use the different parts of each of his ten fingers as if they were beads on an abacus. He could put his hands under the table and, without anyone seeing, do "rapid calculation" instantly. He did so as a kid, acing math tests, and he did so as an adult, gaining a reputation for uncanny speed and accuracy in the brawny, take-no-prisoners arena of the cattle auctions. Rodney was no one's fool.

Two of Inez Davidson's best teaching tricks involved games. One was that, on every Friday, she would divide up the kids, the fifth graders on one side of the room, the little kids third and fourth graders—on the other. Whatever they had learned that week would be the subject of the Friday contests. The first part of the day would be spent with the previous week's winners figuring out what kind of competition they would make. They would devise the rules and explain them to the other group. The competition would then require one group testing the other on the answers to various problems in all the different areas they had studied that week.

Mrs. Davidson would serve as the referee. The little kids worked hard to beat the older kids, who of course had their honor to defend, and each team tried to ask harder and harder questions to stump the other. Once Mrs. Davidson felt that the week's work had been covered, she would declare a winning team. The winning group could head out to the playground fifteen minutes early while the losers sat there an *interminable* fifteen minutes—nine hundred seconds!—getting a head start on next week's contest. Some weeks, though, they had so much fun scheming on the hard questions they'd ask the following week that Mrs. Davidson would have to point to the clock to remind them that study hall was over and that it was time to go outside to the schoolyard to play with their friends.

Sometimes they had an hour of these competitions on Friday, sometimes two hours. The other teachers grumbled that Mrs. Davidson's kids got away with murder. Ted, who had his mother as a teacher for all three years because there wasn't anyone else in Mountain View, ended up winning a full scholarship to the University of Chicago, where he tested out of virtually all of his first- and second-year college courses. Not bad for a kid from a cow town in the Canadian Rockies.

Mrs. Davidson taught her kids to dream. Every year, there would be some new, ambitious project with a theme that would unfold over the course of the year. One year, the project was for each child to find a pen pal in another town called Mountain View somewhere in the world. There was only one map in the school, so a first step was to take an enormous piece of newsprint, cover the map, and spend days and days tracing out all the countries on their own map. Then, each child went to the library, which held far fewer books for the entire school than my study holds today, and they each started reading about all the countries in the world, looking for other Mountain Views. They would then mark their particular Mountain View on the hand-drawn map tacked up in the schoolroom. They had to think of ways to communicate with kids in these other Mountain Views. Since this was a Mormon area, there were families who had been on missions to other parts of the world, so that was the obvious way to make contacts, but this was a contest, and ingenuity was rewarded.

One kid remembered that Hang Sang, the elderly Chinese man who ran the local general store, the only store in town, had come over to Canada to work on the railroad, as had so many Chinese immigrants. Taciturn, with a thick accent, Mr. Sang was amused and delighted when one child suddenly wanted help writing a letter—in Chinese. The kid had somehow found out about a town called Mountain View in China. That was the child who won the contest, and long after the contest was over, he spent time with Mr. Sang, talking to him in the back of his store.

But of course they all won. The globe became smaller through the connections they made, and their town became larger. They learned geography and anthropology and foreign languages too. The project lasted not just one full year but many years, and some kids visited those other Mountain Views when they grew up. To this day, I don't drive through a town named Mountain View (there are a lot of them in the world, actually) without wondering if one of Mrs. Davidson's kids sent a letter there and, through the connection made, was inspired to go on, later, to become a professor or a doctor or a veterinarian.

None of what happened in Mrs. Davidson's classroom in Mountain View, Alberta, Canada, depended upon an iPod. None of it required the Internet. But the Internet requires this kind of interactive, motivated, inspired, and curious form of learning. The key here is that Mrs. Davidson's classroom was not really divided up into "subjects" so much as it was into problems, puzzles, brain teasers, challenges, games, word problems, and intellectual obstacle courses, many of which require kids working together toward a solution. What counts is the little kids showing, with pride, that they are every bit as smart as the big kids and the big kids showing, with pride, that they are king, and all of them understanding (although they'd fight you rather than admit it) that they need one another to learn and to dream.

Like the iPod experiment, what this classroom story shows is that kids want to learn and can propel themselves to all kinds of learning as long as there is a payoff, not in what is won or achieved in statistical terms, but what is won and achieved inside, in the sense of self-confidence and competence. Learning, in this sense, is skill and will, an earned conviction that, faced with a challenge ahead, this past achievement will get one through. You can count on your ability to learn, and nowhere is that more important than when what you've learned in the past no longer suffices for the future. That is the glistening paradox of great education: It is not about answering test questions. It is about knowing that, when tested by the most grueling challenges ahead, you have the capacity to learn what is required to succeed.

It is in this sense that unlearning is a skill as vital as learning. It is a skill you have to acquire, too. Unlearning requires that you take an inventory of your changed situation, that you take an inventory of your current repertoire of skills, and that you have the confidence to see your shortcomings and repair them. Without confidence in your ability to learn something new, it is almost impossible to see what you have to change in order to succeed against a new challenge. Lacking confidence in your ability to change, it's much easier to blame the changed situation—typically, new technologies—and then dig in your heels, raising a bulwark against the new. Confidence in your ability to learn *is* confidence in your ability to unlearn, to switch assumptions or methods or partnerships in order to do better. This is true not only for you, as an individual, but for whole institutions.

That's what those kids in a tiny town in rural Alberta learned as fourth and fifth graders pitted against the six graders (and vice versa). Mrs. Davidson had to fight the school superintendent every year. Her kids all knew that, too. They knew that she'd put herself on the line for their learning; she'd stand up to anybody, including the superintendent, on their behalf. She dared him, year after year, to fire her. In a different era, he might have, but every year, there'd be the ritual dressing down in his office. I was never in the room with them, but I have a feeling that it wasn't Mrs. Davidson who left feeling chastised.

Put those kids in a lecture hall, give them a standardized curriculum with standardized forms of measuring achievement and ability, and they learn a different lesson. They might well master what they are supposed to learn, but that's not education. When you think of learning as something external to yourself, learning becomes a levy—*an assessment, not an asset.* The assessment no longer matters after the schooling stops. The asset is a resource one draws on for a lifetime.

How can the lessons of this woman's extraordinary classroom be put to use? The model of learning in Mrs. Davidson's classes is probably as old as human history. It is the game. As in a game, there are parameters, what you need to know in order to succeed within the game's rules and requirements, subject matter, or methods. There is a specific target or goal, and you compete with others and against others to win. The real winner, always, is you, not because you earned the trophy but because you learned the inestimable skill of responding to a challenge.

||||||||||||||||||||||||||||||||

The Dark Knights of Godzilla sit, furrow-browed, poring over an ingenious series of cogs and ramps dwindling down to a small slingshot. They are devising a mechanism to power their machine. It's hard enough to do onscreen, when they are playing the video game LittleBigPlanet (LBP) on their PlayStation 3's, but now they face a boss-level challenge: They have to engineer an actual starter mechanism in real life, in the school lab. They have to take it from a blueprint all the way to completion in plastic, wood, and rubber, and then demo it in front of everyone.

That's what a boss-level challenge is in the gamer world. It is the last big test of your group. It is your way to show your stuff, to put together everything you've learned about working together, so you can move on to the next level.

Across the room, the I Don't Knows are transfixed by C-clamps, hacksaws, and wood saws. They made their Rube Goldberg contraption from cardboard and tape, but it fell apart, so they are calculating tensile strength, experimenting this time with wood and wood glue.

Vamelepiremenest ("the *t* is silent") takes a break from the *awesome* building project they are contemplating in order to feed Ameer's turtle some lettuce and tomatoes. Ameer's own Home Base has dozens of parts arrayed on the lab table. Maybe a robot isn't the right way to go for an automated device to power a faucet?[19]

These are eleven-year-olds, seventy-two of them in sixth grade at Manhattan's Quest 2 Learn (Q2L), a public middle school where all classes are taught on gaming principles. To succeed at the boss-level challenge, students must make actual, physical constructions based on two months of playing LBP, the popular video game that bursts with quests and gizmos and characters that you customize and guide through an elaborate maze of obstacles. The kids have been learning Newtonian physics as they work their avatars—Sackboy and Sackgirl—higher and higher through the game play, and now, as a final quest, they have to actually design and build ramps and faucets and page turners in the real world. They are taking the action off the screen and now designing and engineering the mechanisms in real life, all working together in teams. This is the last assignment before winter break at Q2L.

Q2L is the brainchild of Katie Salen, a game designer and a professor at

Parsons The New School for Design, a university-level art school in New York. A tall, athletic-looking redhead who has written books on the importance of games for learning, Salen likes to talk about the "ecology" of gaming, from "code to rhetoric to social practices."[20]

I am mesmerized when Katie speaks. I first met her around 2005 when I became part of the Digital Media and Learning Initiative at the John D. and Catherine T. MacArthur Foundation. They're the philanthropic foundation that gives the "Genius Awards" every year. The Digital Media and Learning Initiative is dedicated to promoting innovative ways to energize kids to learn. Katie presented the foundation with an outlandish proposal. Why not make a school based on the principles of games and gaming, on strategy and problem solving? Why not keep an eye out for bright kids who have been failed by conventional schooling? And the hardest challenge of all: Why not make this a public school? To make a point about the validity of games as an exceptionally challenging and rigorous learning method, she wanted her school to be college prep in its emphasis but to be open to accepting bright students who seemed to have learning disabilities, trouble in school, or other issues that might normally preclude them from a college-prep track. I must admit, I was skeptical. We all knew she had her work cut out for her. But Katie is a gamer and a game designer, and with all the doggedness of someone working to achieve a new level, she set to work on school boards and bureaucracies, rules and regulations, parents and principals. She did not stop, and a few years later she got her school. I don't know a better object lesson in how a gamer learns to get things done.

It has been gratifying to watch her success, and thrilling, too, to be able to sit back and see the world shower its appreciation on this great accomplishment. In fall of 2009, Quest 2 Learn opened its doors with an inaugural class of sixth graders. A partnership among Parsons, New Visions for Public Schools (an educational reform group), the Institute for Play (a nonprofit dedicated to game-based learning), and the Department of Education, Q2L garnered so much interest that it needed a lottery to select its first class. Serious games, serious fun, serious learning: That's Q2L's business.

The conjunction of strategy, fun, and learning makes games particularly rich for education. Kids at Q2L don't just learn math, for example. They learn to act like mathematicians, both in the video games they play and customize and create, and in the real world outside the game. They use math principles to create a gizmo that will open a book. Or they learn the mathematics of a

Möbius strip that they will then shape into fashion, such as the adjustable shoulder strap of a purse they design and make.

As Salen notes, the concept of gaming extends far beyond stereotypical video games in the same way that the concept of learning extends far beyond the standard configuration of a classroom or a multiple-choice test. "Gaming is play across media, time, social spaces, and networks. . . . It requires an attitude toward risk taking, meaning creation, nonlinear navigation, problem solving, an understanding of rule structures, and an acknowledgment of agency within that structure, to name but a few" of the basic elements of games.[21]

Just ask Lesli Baker, whose eleven-year-old son, Beauchamp, was diagnosed with attention deficit disorder. After difficulties at previous schools, the family applied for Q2L. He was one of the seventy-two kids to win the lottery and be awarded admission. It took a "leap of faith" to put him in a school based on games, but, his mom says, "It's a great match for him. He's really enthused about learning."[22]

And what exactly might Beauchamp be learning? Are you skeptical about how much kids can really learn playing games? Here's a sample of this week's blogs by his teacher, Alicia, at Q2L:

> *This week . . . we continued our Mini-Quest from Arithmus, this time showing the twins that we could multiply and divide with integers. . . . We also finished our training around "About or Exactly," which dealt with place value, rounding and beginning estimation. . . . We were then ready to crack the second piece of Prof Pie's Perfect Pie Recipe! This time Pie told us he had written his code in something called a key word cipher. We studied and practiced with key word ciphers and worked to complete the encoded message so we could figure out what page to scan next. . . . Next week we will begin a quest around parts of speech and how they connect to our language of code.*

Integers and grammar and computer code are all part of the same quest. Interestingly, these kids are not only tested by each boss-level challenge in their game worlds, but they also take all the tests required of sixth graders in public school by New York State. Those year-end tests are another boss-level challenge for which they are preparing.

What is obvious from Quest 2 Learn is that the school has taken the lessons of attention to heart. The game-based learning makes for a constant challenge, one calibrated to the individual child's learning abilities and progress, each level becoming more challenging after the previous one has been mastered, with constant disruptions and shifts, not from topic to topic but with all forms of knowledge integrated to the ultimate test: the boss-level challenge!

ALTHOUGH ONE IS HIGH-TECH AND one was in a one-room schoolhouse without electricity for much of its existence, the Quest 2 Learn classroom has much in common with Mrs. Davidson's schoolroom in Mountain View. Both, in different ways, are based on the same experimental, student-driven lessons in attentive learning that were at the root of the iPod experiment, and both have almost nothing to do with Ichabod's rote form of education. Intrinsic to inquiry-based learning, there's a "gamer disposition," which is to say a real commitment to learning that goes far beyond school to the cultivation of "risk-taking, critical reflection, collaboration, meaning creation, non-linear navigation, problem solving and problem definition, and innovation."[23]

Seeing new options is precisely what this world of gaming is about, and so is collaboration. These kids are learning their subject matter. They aim to go to college one day—but they are also learning invaluable skills that will serve them in the twenty-first-century world of constant change. They are interested, motivated, and challenged, and are learning the lesson that it is great to work hard at what you love most. *Liebe und arbeit,* as Freud said, "love and work," the keys to a successful life. In any century.

|||||||||||||||||||||||||||||||||

As I think back to my own childhood misadventures in learning, the insufferable torture of those interminable days in school, as I think of my son, Charles, now a lawyer with his own practice, whose life in school was no more endurable than my own, I have to believe Q2L has found an infinitely better way for kids to learn. Check out their curriculum and you see that they are hitting all the bases. They offer assessment, standards, and an explicitly college-prep program: Everything is undertaken with the goal of going to college. Yet many of their students have learning disabilities and have failed in other public schools. They seem to be thriving here. And, better, they are *learning*.

E. O. Wilson, the distinguished professor emeritus of biology at Harvard, thinks so too: "Games are the future in education. I envision visits to different ecosystems that the student could actually enter . . . with an instructor. They could be a rain forest, a tundra, or a Jurassic forest."[24]

That's right. They could explore together a terrain where all the forms of knowledge and play and social life and team-building and even "wellness," as it is called at Q2L in sessions at the start and end of each day, go together. And in a way kids love. Not all schools can be designed like Q2L, but that doesn't mean that the principles embodied here can't be embraced by more typical schools. So that raises a question. To appreciate how far Q2L has come, we need to look at the methods and challenges of a typical public school in America today.

||||||||||||||||||||||||||||||||

It doesn't snow all that often in North Carolina, but when it even looks like snow, the local schools close down. One recent winter it snowed a lot. There weren't other forms of child care. One of my coworkers had to figure out where her nine-year-old daughter could spend the day. Fortunately, we had the incomparable Sackboy and Sackgirl, the heroes of LittleBigPlanet, at the ready to perform as virtual babysitters.

Not a lot of university profs have a Sony PlayStation 3 set up in the office, but I am also the cofounder of a virtual organization called HASTAC, a network of educators and concerned others dedicated to new forms of learning for a digital age. HASTAC is a long acronym that stands for Humanities, Arts, Sciences, and Technology Advanced Collaboratory. Everyone just says "Haystack." We were gearing up for a new HASTAC/MacArthur Foundation Digital Media and Learning Competition. It's fun to give away $2 million a year to educators, entrepreneurs, and software developers for coming up with creative new ways of learning. That year, for one part of the competition, we were partnering with Sony and Electronic Arts to challenge competitors to come up with ways of turning commercial games that kids love into inspiring platforms for learning. If it seems to you that this is a bit like the old iPod experiment, you're right.

On this snow day, three nine-year-olds, a boy and two girls, spent the day—the entire day—playing LBP, with the occasional Cheetos break thrown in for good measure. By the end of the day, they were sweaty faced, excited, with the glazed and glassy eyes that you see in nineteenth-century illustrations of opium eaters. When they packed up their things to go home for the night,

I intercepted them, eager to hear all that they had learned in a day of the kind of interactive, challenging game play that we were pushing as the next frontier of learning.

"What did you learn today?" I asked.

"Learn?" They dissolved into helpless, hysterical laughter.

"You didn't learn anything?" I asked again, a little pathetically.

"No! No!" They laughed some more. "No way!"

They were adamant. They had just had the greatest day of their lives. How could they possibly have *learned* anything?

I said good-bye and went back to my office, wondering how we had failed. Three smart kids, and they hadn't realized that they'd spent a day making Sackgirl and Sackboy succeed against all kinds of odds, requiring ingenuity, basic science knowledge, collaborative team-building skills, and other marks of what, in our competition, we consider to be imaginative (and reimagined) learning. Apparently, their idea of learning was rather different from ours.

After letting myself feel discouraged for a little while, I decided it would be useful to visit their school to see why what they had done that day couldn't possibly be conceived of as learning. Thanks to my friend David Stein, the senior education partnership coordinator at Duke whose Duke-Durham program works in conjunction with public schools near the campus, I was able to visit several schools in town, including the school a little further away that these three kids attended. This school seemed to have nothing to do with the kind of utopia of learning games envisioned by Katie Salen and others. This school was not close enough to be part of the Duke-Durham partnership, although the principal I met clearly was eager to make that connection or to partner with another institution of higher education or a supportive local business or community organization. He said over and over how much difference even a modest boost in attention, resources, special programs, and volunteer teacher's aides could make to a school like his. Duke is a wealthy university in a poor and valiant town with its own unique history. Durham was once called the "Black Wall Street," when it was the center for black banking, insurance, and investment businesses under segregation, and the city has a history of a strong black upper-middle class and a proud African American working class too. Connections with the local schools and the local community are fostered at Duke and considered as important to the learning of Duke students as they are for Durham's aspirations.

The school the three kids attend is a magnet school featuring the arts and humanities. It is also officially deemed to be a "failing school" by the national standards-based educational policy known as No Child Left Behind. For the sake of privacy, we'll call this school Middleton Elementary. Everyone—the principal, the teachers, the parents, the children—is ashamed of its failing status.

Middleton's student body is a mixture of middle-class kids, a few children of professors and graduate students, and kids whose families live on exceptionally modest means or worse. Nearly 60 percent are on what in North Carolina is called the "free and reduced lunch plan," meaning the family income is so low that the federal government covers their lunch costs. Racially, the school mirrors Durham itself, in being roughly divided among Caucasians, African Americans, and Hispanics. As a magnet school, kids have to enter a lottery to get in, and parents tend to be more involved than at some public schools.

The physical building looks almost identical to the middle school I attended as a child—cinder-block construction, one story, long central corridors, smaller hallways to the sides. Like my own school, this one was built in the early 1950s, as part of the post–World War II baby boom. Everyone here is excited that a new building will be opening soon, but even in the meantime, I'm impressed that there is art everywhere on the walls—and it is good. The level of sophistication, both the talent of the kids and their knowledge of art history, is astonishing. Third and fourth graders do paintings in the style of Picasso, Kandinsky, Romare Bearden, and others. It is Black History Month, and the walls are adorned with photographs, quotations, and drawings from the civil rights movement, as well as lots featuring President Obama. This place has an undeniable spirit. It is trying hard not to fail.

However, there is also a palpable and off-putting rigidity and discipline, which makes the school very different from our day playing LittleBigPlanet. Kids walk single file in the halls, their hands clasped behind their backs. When David remarks on this "old way" of doing things, the principal is defensive, answering that he'd like to do it another way but he has to make sure the kids stay focused, that they do not waste a minute of precious class time. Implicit in the comment is the ever-present challenge the school faces. The kids have to pass their end-of-grade exams. The school can't afford to fail again. In the classrooms, the desks are crowded together in tight rows. The class size is in the high twenties and even thirties. The teachers wear determined expressions, intense as they look out over rows and rows of children.

Because they are a "failing school," they have to get 100 percent of their test scores up to the required standards by 2014, or under the provision of the national policy, the school will lose its public funding and will no longer be able to operate as a public school. It will either be shut down or privatized (put into the hands of a private for-profit oversight firm). Middleton has a special challenge. Durham was once a city that was only black or white, but it now has a sizable population of immigrants from Latin America. About a third or more of the kids at the school do not speak English at home. For many, an indigenous language like Quechua is spoken there, with Spanish as a second language. Yet they must take their end-of-grade exams in English. By state policy, there are no provisions for non-English testing. By what logic would failing a test in a language other than the one spoken in your home constitute a failure for you as well as for your teachers, your classmates, and your entire school? That is the burden I feel, despite the exuberant walls, despite the efforts of teachers who care and parents who do too.

I swear you can hear the word *failing* being whispered in the halls. There is a determination here that moves to the other side of *rigor,* close to *rigid:* There are EOG tests that *must* be passed this year. Or else. As kids walk between classes, in single file, their hands behind their backs, I can almost see the weight of this national policy bearing on them. This feels nothing at all like the playful day at our office, getting Sackboy through various levers and gears, over this and that obstacle, calculating how much effort it will require for him to swing over this or pass under that and deciding whether it would be better to just make a running leap to the next platform rather than trying to use a tool to make that happen. There's not much here like Mrs. Davidson's school in Mountain View.

These kids sit mostly in rows. I see one of the girls from our snow day sitting in her classroom. Her back is to me. She is looking intently forward. She's very smart, a very good student. I hate to say it, but she looks bored to death, nothing like the ball of energy she was on snow day.

On the door of the kindergarten class, there's a sign. It doesn't say "Welcome!" It doesn't say "Learning is fun here." The sign sets out the rules for "how to get along" in the classroom. "Do not talk. Do not hit. Do not eat. Do not chew gum." Most of the kids are too young to actually read, so I assume this sign is as much a symbol as it is a message. It is conveying an attitude of constraint and discipline. In this school, we pay attention by making rules, by fear, by rote, by paying attention to the teacher. This is what learning is. No wonder the kids on snow day felt as if they'd been granted a reprieve.

I hasten to add that once one moves past the unwelcoming sign, there is a lot of love in the kindergarten classroom, and there's a teacher trying very hard to do the right thing. She doesn't want this school to fail, her kids to fail.

She probably earns about $30,000 a year (the average starting salary for an elementary school teacher), and I'm guessing she uses some of that to buy school supplies for her students. Almost half of all new teachers leave the field within five years. The drop-out rate of teachers is a bigger crisis in this country than the drop-out rate of our high school students. It is *tough* being a teacher. Personally, I'd rather dig ditches than spend my days with eighteen little kids and my evenings on the phone with irate parents sure that I'm shortchanging their little Johnny and Janey. The current punitive attitude toward teachers is just plain offensive. *You* try being an elementary or high school teacher for a day and tell me how it goes!

As the principal was proudly showing us his school—and there was much to be proud of—I gently noted the negatives on that sign. He gave no indication of hearing, but a few weeks later, when a friend happened to be back at the school, she noticed the poster I'd mentioned at the front of the kindergarten, the one that said "Let's Get Along." The rules on the poster were different: "Use kind words. Be quick to forgive. Listen. Share. Encourage others. Take turns. Think before acting. Talk it over." Had my simple words made a difference? Or did the teacher have two different signs and I'd only seen one? I have no idea, but I was pleased to learn about the more positive message on the second sign, amid the same border of happy smiling rainbow children who, in fact, looked pretty much like the rainbow kids at this school.

When it snows again in North Carolina, the kids are ecstatic at the idea of another day joining Mom at our office, another day laboring away at LBP. This time, we're determined to ruin their day just a little by having them think about what they're learning. (I know, I know.) We rephrase the question. We tell them that we are running a competition where people will be able to create learning challenges for LittleBigPlanet. We need their help. Can they tell us about learning, what they are learning when they play?

Question and answer. We're a culture good at questions and at answers. It's two girls today, and they squirm with delight and rise to the challenge, as surely as they've maneuvered adorable Sackperson through numerous "platforming scenarios" (parched Mexican deserts, austere Japanese gardens, bustling New York City streets) using what the manufacturer calls "its robust physics engine." For the kids, it means making Sackboy or Sackgirl run, jump, push, and grab.

"So the way it's learning," one of the girls begins, her voice dropping with the seriousness of the task she's been set, "is because you learn *structure*." She lets the impressive word hang there a moment. She has our attention. "You learn how to build things and to calculate."

We follow up. "Is there anything you could do to make it an even better learning game?" we ask. We are planning to run a second competition wherein kids themselves, kids just a few years older than her, can design special learning levels for LBP. We'll be giving awards for the very best ones. Experienced users being able to customize the game by creating new levels is a reason it's won the Interactive Achievement Award for Overall Game of the Year, equivalent to an Oscar for Best Motion Picture, from the Academy of Interactive Arts and Sciences (AIAS).

"You could give Sackboy different *personalities*," the other girl says, another head-turning word. "One Sackboy could be a nerd, one could be smart, one could be silly, and then you'd have to play the game different depending on his personality . . ." Her voice trails off. She can't quite complete the thought, but I'm blown away.

They are learning a lot in school, even with all the rules; I can't forget that. Those teachers are struggling against odds of poverty, a mediocre facility, low pay, and, unbearably, the "lockdown" of being declared a "failing school" because its test scores don't measure up to the average set by No Child Left Behind. Not even Sackperson could rise to that challenge without some serious discipline and direction.

No wonder there were warning signs on the kindergarten door! "Be Prepared All Ye Who Enter Through This Portal!" They will be drilled. And drills—unlike *challenges*—never feel like fun. The kids have absorbed that lesson. If school drills are what constitutes learning, LBP isn't learning. Except that when you test the kids, as we did, asking them pointedly about their two days of big fun, you find that, lo and behold, they not only learned and had a great time, but they can answer with concepts like "structure" and "personality." If we were running our office on standards-based public school principles, we could have started with questions like that—teaching to the test, it's called—and, if we had worked hard enough at it, probably could have ruined a perfectly great snow day.

I ask my friend David if we can go to a school that he thinks exemplifies the best in public education, and he takes me to Forest View Elementary. It has the same demographic as the other school, the same racial and economic mix of kids. It's not a magnet school but a neighborhood school, with 60 percent free or reduced lunches. The school building itself is lovely, a much newer school than Middleton. And there are special touches everywhere. Many, I find out, are exactly what the principal at Middleton mentioned, those small "extras" that public schools don't have the budget for and that make a big difference. These happen to come from the Duke-Durham partnership program—recycled computers, old textbooks and children's books, supplies in the abundant art room, fossils and other cool things in the science rooms, even bright used tennis balls on the feet of the chairs, to ensure silence and unscuffed floors. David often comes over with sacks of them. I happen to run into one of my own students in the hall. He's talked to me before about his cousin who is autistic, and he says he's at Forest View as a volunteer, helping a teacher in a special-needs class.

We're there today to talk to the teachers about kids who might be eligible for a special science program called Boost, wherein kids in fifth grade will be doing actual research with graduate and medical school students and faculty at Duke, beginning in summer and continuing in the school year. Teachers have selected candidates, who have spent a pizza lunch with David, working in teams to solve problems David has set them, and have filled out an application with their parents. All this is being weighed again, in conversation with two science teachers eager to have their best and most deserving kids in this very special program. There are about five times more candidates than there are places.

Forest View is not a failing school, and I can feel the spirit here before I've even set foot inside. Not only is there art, and gardens, and garden art, and playgrounds, and color and brightness, but the warmth just flies at you from the second you step inside. Kids are joking in the hall, not walking in rows with their hands behind their backs. They seem to work in a lot of groups, not in rows, and so much is happening everywhere that I barely know where to look.

So we start at the beginning, in another kindergarten class.

When we enter, teacher Sarah Tichnor is finishing reading a story. She's sitting on a stool and the kids are on the floor, watching as she reads. They are in a little reading alcove carved out of the cavernous room, books on all sides, like a cozy little cave. The larger room is alive with life and spaces and animals and computers and interesting things, great stuff to look at and do things with. In this reading corner, there's a long row just of books, all kinds of books, many of them used and donated to the school.

"That's it for today," Ms. Tichnor says and the kids get up and go immediately for their mats.

"Oh, it's nap time!" David says cheerfully,

Suddenly eighteen pairs of eyes swing in our direction, bottom lips jutted out, and one little girl even puts her hands on her hips. Before we can react, the teacher has stood up, tall, and said to us, in that corrective and pedagogical tone usually reserved for preschoolers, "On no! This is a *kindergarten*. In kindergarten, we don't take naps. We're getting ready for silent reading."

The kids turn their backs on us and go look, seriously, at the bookshelf. Each child selects a book. Then, mat in hand, they each begin to spread out far from one another, all over the large room. The teacher has not said a word to them about no hitting, no talking, no punching, no joking.

"They spread out," she says to us in that same authoritative voice, "so that no one can distract them from their hour of silent reading."

The eighteen heads nod up and down definitively.

These are the luckiest children on earth, to be big enough to have a whole hour to do what big kids—kindergartners—do: read silently to themselves.

The independence, the maturity, the pride in learning are conveyed without a single negative. The individual kids place themselves far enough apart so *they* won't be distracted.

I'm very impressed. This teacher is setting up these kids for new challenges ahead, for independence, for new levels of learning, new pride not only in what they were learning but in how learning changed them as people. Silent reading isn't the punishment; it's the reward. Learning is for big kids. Who would want to be distracted from learning?

That's pretty much what I aspire to as an educator: not in teaching facts but in conveying to my students the passion of learning, far beyond my classroom, far beyond any graduation ceremony.

Those kids were coming from very modest, even impoverished, backgrounds and were at a neighborhood school. No worries. They were on fire with learning. *"This is kindergarten"*—and all the world of future wonders that entails.

||||||||||||||||||||||||||||||

These lessons in setting up a classroom came back to me when it came time for me to return to teaching after several years in my R & D job as a university administrator. Now suddenly I found myself with a chance to put into practice all of the ideas that had been percolating in my mind about what it takes to design a class for the way we live now. I was determined to remember the lessons of intellectual respect and independence we'd learned from the iPod experiment. In a world of helicopter parents, hovering overhead, and boomerang kids, returning to the family basement because they can't find a job after college, those lessons are hard-won. The age and context were different, but I wanted my students, too, to swell with pride in their own accomplishment: *"This is a university."*

I decided to offer a brand-new course called This Is Your Brain on the Internet, a title that pays homage to Daniel Levitin's delightful and inspiring book *This Is Your Brain on Music,* a kind of music-lover's guide to the brain.[25] Levitin argues that music makes complex circuits throughout the brain, requires all the different kinds of brain function for listening, processing, and producing in various forms, and makes us think differently. Substitute the word *Internet* for *music* and you've got the gist of my course.

Because no one knew me in the classroom anymore and there was no word of mouth about the kind of teacher I was, I advertised the class widely. In what department, exactly, does This Is Your Brain on the Internet belong? Its main home was ISIS, our new program that brought together the computational sciences, the social and human analysis of the role of technology in society, and multimedia arts. I advertised everywhere else too, and I was delighted to look over the class roster of the eighteen students in the seminar and find more than eighteen majors, minors, and certificates represented. Score!

Next, I took on the syllabus. I created a bare-bones suggested reading list that included everything from articles in specialized journals such as *Cognition* or *Developmental Neuropsychology* to pieces in popular magazines like *Wired* or *Science* to novels and memoirs. There were lots of Web sites too, of course, but I left the rest loose. This class was structured to be peer-led, with student interest and student research driving the design. "Participatory learning" is one term used to describe how we can learn together from one another's skills, contributing to a collective project together. "Cognitive surplus" is another term used in the digital world for that "more than the sum of the parts" form of collaborative, customized thinking that happens when groups think together online.[26]

We used a method that I call "collaboration by difference" that I had pioneered while vice provost and that has become the primary principle of HASTAC, the network dedicated to new forms of learning for a digital age that I cofounded with Professor David Theo Goldberg and other educators in 2002. Collaboration by difference is an antidote to attention blindness. It signifies that the complex and interconnected problems of our time cannot be solved by anyone alone and that those who *think* they can act in an entirely focused, solitary fashion are undoubtedly missing the main point that is right there in front of them, thumping its chest and staring them in the face. Collaboration by difference respects and rewards different forms and levels of expertise, perspective, culture, age, ability, and insight, treating difference not as a deficit but as a point of distinction. It always seems more cumbersome in the short-run to seek out divergent and even quirky opinions, but it turns out to be efficient in the end and necessary for success if one seeks an outcome that is unexpected and sustainable. That's what I was aiming for in This Is Your Brain on the Internet.

In addition to a normal research paper, I had the students each contribute a new entry or amend an existing entry on Wikipedia, or find another public forum where they could contribute to public discourse. There was still a lot of criticism about the lack of peer review in Wikipedia entries, and some profs were "banning" Wikipedia use in the classroom. I don't understand this. Wikipedia is an educator's fantasy, all the world's knowledge shared voluntarily and for free in a format theoretically available to all for free, and that anyone can edit. No rational-choice economic theory of human nature or human motivation explains the existence of Wikipedia! Instead of banning it, I challenged my students to use their knowledge to make Wikipedia better. All conceded it had turned out to be much harder to get their work to "stick" on Wikipedia than it was to write a traditional term paper.

SPEAKING OF TERM PAPERS, LET'S stop for a moment to talk about student writing, an essential component of any academic program. It isn't as easy to make a judgment about a student from her papers as it might first appear. Given that I was teaching a class based on learning and the Internet, having my students blog was a no-brainer. I supplemented this with more traditionally structured academic writing, and when I had both samples in front of me, I discovered something curious. Their writing online, at least in their blogs, was incomparably better than in the traditional term papers they wrote for the class. In fact, given all the tripe one hears from pundits about how the Internet dumbs our kids down, I was shocked that elegant bloggers often turned out to be the clunkiest and most pretentious of research paper writers. Term papers rolled in that were shot through with jargon, stilted diction, poor word choice, rambling thoughts, and even pretentious grammatical errors (such as the ungrammatical but proper-sounding use of *I* instead of *me* as an object of a preposition).

But it got me thinking: What if bad writing is a product of the form of writing required in school—the term paper—and not necessarily intrinsic to a student's natural writing style or thought process? I hadn't thought of that until I read their lengthy, weekly blogs and saw the difference in quality. If students are trying to figure out what kind of writing we want in order to get a good grade, communication is a secondary point of the writing. What if "research paper" is a category that invites, even requires, linguistic and syntactic gobbledegook?

Research indicates that, at every age level, people take their writing more seriously when it will be evaluated by peers than when it is to be judged by teachers. Online blogs directed at peers exhibit fewer typographical and factual errors, less plagiarism, and generally better, more elegant, and persuasive prose than classroom assignments by the same writers. A longitudinal study of student writers conducted at Stanford by Andrea Lunsford, a distinguished professor of rhetoric, used the same metric to evaluate the quality of writing of entering Stanford students year after year. Lunsford surprised everyone with her findings that students were becoming more literate, rhetorically dexterous, and fluent—not less, as many feared.[27] Using the same evaluative criteria year after year, she could not prove any deleterious effects on the writing of the exceptionally gifted students at Stanford from exposure to the Internet. Pundits insist even elite students are being dumbed down by their writing and reading online, by their texting and Tweeting and their indulgence in other corruptions of standard English. Looking at the data, Lunsford found no hard evidence of decline among Stanford students. She found no empirical basis for hand-wringing or jeremiads about the digital generation going to the dogs.

My students were surprised to read these studies. As lifelong A students, they believed what their teachers had told them about the canine-like intellectual nadir their generation had achieved. Yet when I pushed a little more, they also readily admitted that they had many hang-ups about research papers. Many came to university with such a trove of past negative experiences around term papers that, they admitted, they often procrastinated until the last moment, then forced something to come together. We also talked about the matter of practice. A student typically writes research papers only in school, one per term, at an average (this is another study) of forty-two pages of papers for each semester's classes. During the same period, the average college student writes five hundred pages of e-mail.[28] That's a lot of practice at prose, even if they are not writing the finished product as in a formal essay. If this were golf or tennis, and they were warming up on the putting green or by hitting balls against a wall before a tournament, we'd expect improved results. Give the amount of time youth spend writing online, it's not all that surprising that their blogs were better, when judged by the most traditional English teacher standards, than their research papers.

But students weren't expecting that result. They all came into class assuming that I and other educational professionals would believe the Internet was "bad for you," whatever that means. In fact, they all downplayed their online proficiency at first, considering it (and especially gaming) as a sign that they weren't "smart." One student confessed only belatedly that he had attained the highest level in WoW (World of Warcraft). He came clean only after another student, a professionally trained modern dancer, stood up in class to illustrate to us the difference between a verbal and a kinetic description of movement. Another demonstrated how he won quite a lot of money by playing online poker, using a stunning background in both statistics and psychology to succeed. Only then did our Level 80 Warlock admit his own accomplishments.

The semester flew by, and we went wherever it took us. The objective of the course was to get rid of a lot of the truisms about "the dumbest generation" and actually look at how new theories of the brain and of attention might help us understand how forms of thinking and collaborating online maximize brain activity.

We spent a good deal of time thinking about how accident, disruption, distraction, and difference increase the motivation to learn and to solve problems, both individually and collectively. To find examples, we spent time with a dance ensemble rehearsing a new piece, a jazz band improvising together, and teams of surgeons and computer programmers performing robotic surgery—a choreography as careful and as improvised as that of the artists. We walked inside a monkey's brain in a virtual-reality cave. In another virtual-reality environment at the University of North Carolina, we found ourselves trembling, unable to step off what we knew was a two-inch drop, because it looked as if we were on a ledge over a deep canyon. A few students were able to step over the edge, but I could not. We were all fascinated to learn the body-gripping illusion doesn't work, though, unless you lead up to it by walking up a very slight incline that tells your brain you're climbing, yet another testimony to the fact that while we think we know the world, we really only know the world our body thinks.

One of our readings was *On Intelligence,* a unified theory of the brain advanced by Jeff Hawkins, the neuroscientist who invented the PalmPilot. I agree with many of Hawkins's ideas about the brain's "memory-prediction framework." My own interest is in how memories—reinforced behaviors from the past—predict future learning, and in how we can intentionally disrupt that pattern to spark innovation and creativity. Hawkins is interested in how we can use the pattern to create next-generation artificial intelligence that will enhance the performance, and profitability, of computerized gadgets like the PalmPilot. He believes we need to redesign artificial attention by pattern prediction, not by specific tasks.[29] The students and I were having a heated debate about his theories in class when a student discovered that Hawkins happened to be in our area to give a lecture. Mike, who was leading the sessions on Hawkins, attended the lecture, told Hawkins about the debate, and invited him to our class. I was in Chicago at a MacArthur Foundation meeting on the day of Hawkins's lecture, when suddenly my BlackBerry was vibrating with e-mails and IMs from my students, who had convened the class without me to host a special guest on a special topic: Jeff Hawkins himself debating the ideas of Jeff Hawkins. It felt a bit like the gag in the classic Woody Allen movie *Annie Hall* when someone in the line to purchase movie tickets is expounding pompously on the ideas of Marshall McLuhan and then McLuhan himself steps into the conversation to settle the matter.

It was that kind of class. This Is Your Brain on the Internet didn't look or feel like any course I'd ever taught before. It had no standard field, discipline, topic, syllabus, method, or conclusion, and, for that matter, no standard teacher either. The semester ended not with a body of knowledge mastered but with a certain skepticism about easy answers to questions like, Is multitasking good for us? or even, What is multitasking? My students became adept at taking *any* statement of fact and asking, What does that really mean? What are the intellectual blind spots of this study? What is not being taken into account because, in this framework, our attention blindness won't let us see it? When is it a two-inch ledge and not a canyon?

Nothing in my own traditional training had prepared me for this way of knowing what we don't know. There were times when I was nervous that I might not be leading them in the right direction. But each time I wavered, their confidence reassured me that we were on to a new way of paying attention, a new way of learning in a digital age. As with Mrs. Davidson, as with Katie Salen, this class connected to the real world; it was about testing assumptions, not accepting them; it was about collaboration and debate; and perhaps most of all, it was about gaining the confidence to feel that, however you are tested in the future, you can count on your capacity to learn and succeed.

Many students said it was the best class they'd had in four years of college. But it wasn't just a class. It was a different way of seeing. We'd seen a transformation not just in the classroom but in all of us.

"Jeff Hawkins thought it was odd that we decided to hold class when you weren't there," one student texted me. "Why wouldn't we? That's how it works in This Is Your Brain on the Internet."

Project Classroom Makeover. I heard the pride. Step aside, Prof Davidson: This is a *university*!

4

||||||||||||||||||||||||||||||||||

How We Measure

"Nonsense!"

"Absurd!"

"A lie!"

"Cynical and meaningless!"

"A wacko holding forth on a soapbox. If Prof Davidson just wants to yammer and lead discussions, she should resign her position and head for a park or subway platform, and pass a hat for donations."

Some days, it's not easy being Prof Davidson.

What caused the ruckus in the blogosphere this time was a blog I posted on the HASTAC site called "How to Crowdsource Grading," in which I proposed a new form of assessment that I planned to use the next time I taught This Is Your Brain on the Internet.

"The McDonaldization of American education."

"Let's crowdsource Prof Davidson's salary too."

It was my students' fault, really. By the end of This Is Your Brain on the Internet, I felt confident I'd taught a pretty impressive course. I settled in with my students' course evaluations, waiting for the accolades to flow over me, a pedagogical shower of student appreciation. And mostly that's what I read, thankfully. But there was one group of students who had some candid feedback to offer me for the next time I taught This Is Your Brain on the Internet, and it took me by surprise. They said everything about the course had been bold, new, and exciting.

Everything, that is, except grading.

They pointed out that I had used entirely conventional methods for testing and evaluating their work. We had talked as a class about the new modes of assessment on the Internet—everything from public commenting on products and services to leader boards—where the consumer of content could also evaluate that content. These students said they loved the class but were perplexed that my assessment method had been so twentieth-century. Midterm. Final. Research paper. Graded A, B, C, D. The students were right. You couldn't get more twentieth-century than that.

It's hard for students to critique a teacher, especially one they like, but they not only did so, they signed their names to the course evaluations. It turned out these were A+ students, not B students. That stopped me in my tracks. If you're a teacher worth your salt, you really pay attention when the A+ students say something is wrong.

I was embarrassed that I had overlooked such a crucial part of our brain on the Internet. Assessment is a bit like the famous Heisenberg principle in quantum mechanics: the more precisely you measure for one property, the less precisely you can measure for another. If you're looking for conventional achievement using conventional measures, then by definition you cannot at the same time be measuring by other criteria or measuring other qualities. In grading them strictly on content based on papers and exams, I had failed to evaluate them on all the really amazing ways that they had contributed to a semester of collaborative research, thinking, and interdisciplinary teamwork—exactly what the course was supposed to be about.

I contacted my students and said they'd made me rethink some very old habits. *Unlearning*. I promised I would rectify this the next time I taught the course. I thought about my promise for a while, came up with what seemed like a good system, then wrote about it in that "How to Crowdsource Grading" blog.[1]

"Idiocracy, here we come."

"Groupthink!"

"David Lodge of our generation—Wherefore Art Thou?" someone offered, apparently thinking that Prof Davidson was ripe for a witty British parody.

* * *

REALLY? FOR A COLUMN ON grading? I was taken by surprise again. I had underestimated just how invested nonacademics were in the topic of evaluation. My usual blog posts on the HASTAC site receive several hundred readers in the first week or so, with an audience largely of educators interested in cross-disciplinary thinking for a digital age. Ten thousand readers later, "How to Crowdsource Grading" had crossed over to a different, wider audience. It also garnered write-ups in academic news outlets like *The Chronicle of Higher Education* and *Inside Higher Ed,* where, by the end of the month, it ranked as the "most read" and "most commented on" article. The Associated Press produced an article about my blog, and that piece turned up in local newspapers from Tulsa to Tallahassee. For a moment or two, "crowdsourcing grading" even reached the pinnacle of celebrity. It became a trending topic on Twitter.

Not all the comments were negative. Many were exceptionally thoughtful and helpful. But a lot of people were unhappy with me. It threatened to be the iPod experiment all over again.

"Slacker!"
"Lazybones!"
"Prof Davidson is an embarrassment."
"I give the professor F for failure to account for real-world appropriateness and efficacy."

This is what happens when you listen to the A+ students in This Is Your Brain on the Internet.

LEARNING TO GIVE AND TAKE feedback responsibly should be a key component of our training as networked citizens of the world. When I built the class that aimed to leave behind a dusty set of old pedagogical tools, I'd given up many of my props as a prof, but not the most basic one: my gradebook. Like Ichabod Crane's switch or Obi-Wan Kenobi's lightsaber, the gradebook is the symbol of pedagogical power. Testing and measuring is always the most conservative and traditional feature of any educational enterprise—and the most contested.

It is artificial, too. In the world of work beyond graduation, there's no such thing as a "gentleman's C"—or even a hard-earned B+. At work, you succeed or fail, and sometimes you're just fired or laid off, no matter how good you are. In life, there is no grading on the curve or otherwise. We know that success in business does *not* correlate with one's GPA in college. Many other factors are just as important as a perfect academic record in my own field of academe, where one would think grades would be the prime predictors of success. They are not. We just tend to treat them as though they are, as if tests and measurements and grades were an end in themselves and not a useful metric that helps students and teachers all stay on track.

My new grading method that set off such waves of vitriol strove to be fair and useful, and in method it combined old-fashioned contract grading with peer review. Contract grading goes back at least to the 1960s. In it, the requirements of a course are laid out in advance and students contract to do all of the assignments or only some of them. A student with a heavy course or work load who doesn't need an A, for example, might contract to do everything but the final project and then, according to the contract, she might earn a B. It's all very adult. No squirming, no smarmy grade-grinding to coax the C you earned up to a B.

But I also wanted some quality control. So I added the crowdsourcing component based on the way I had already structured the course. I thought that since pairs of students were leading each class session and also responding to their peers' required weekly reading blogs, why not have those student leaders determine whether the blogs were good enough to count as fulfilling the terms of the contract? If a blog didn't pass muster, it would be the task of the student leaders that week to tell the blogger that and offer feedback on what would be required for it to count. Student leaders for a class period would have to do this carefully, for, of course, next week they would be back in the role of the student, blogging too, and a classmate would be evaluating *their* work.

Many of those in the blogosphere who were critical of my crowdsourcing grading idea were sure that my new grading philosophy would be an invitation for every slacker who wanted an easy A. These critics obviously hadn't seen my long syllabus, with its extensive readings in a range of fields, from the computational sciences to literature and the multimedia arts, and with its equally long writing and research requirements. I have not yet had a single student read over my syllabus and think he was in for a cakewalk.

I also liked the idea of students each having a turn at being the one giving the grades. That's not a role most students experience, even though every study of learning shows that you learn best by teaching someone else. Besides, if constant public self-presentation and constant public feedback are characteristics of a digital age, why aren't we rethinking how we evaluate, measure, test, assess, and create standards? Isn't that another aspect of our brain on the Internet? Collective responsibility—crowdsourcing—seemed like a fine way to approach assessment.

There are many ways of crowdsourcing, and mine was simply to extend the concept of peer leadership to grading. Crowdsourcing is the single most common principle behind the creation of the open-source Web. It works. This is how the brilliant Linux code—free, open, and used to power everything from servers to supercomputers—was originally written. As we'll see in a later chapter, businesses, even traditional ones like IBM, use crowdsourcing more and more as a way to get a read on their employees' opinions, to frame a challenge, or to solve a problem. Sometimes they create a competition and award the winner a prize while the company receives the rights to the winning idea. Crowdsourced grading was an extension of the peer leadership we'd pioneered throughout the course.

From the fury of the remarks, it was clear I had struck a nerve. My critics were acting as if testing and then assigning a grade to the test scores had real-world value. It doesn't. Grading measures some things and fails to measure other things, but in the end, all assessment is circular: It measures what you want it to measure by a standard of excellence that you determine in advance. We too often substitute test results for some basic truth about a whole person, partly because we live in a test-obsessed society, giving more tests (and at a younger age) than any other country. We love polls and metrics and charts—numbers for everything. Studies have shown that if you include numbers in an article, Americans are more inclined to believe the results. If you include a chart or table, confidence goes even higher. And if you include numbers, a chart, and an illustration of a brain—even if the article isn't really about the parts of the brain—the credibility is through the roof.[2] That's the association we make: Numbers equal brain.

Given how often we were tested as kids, it is no surprise that we put a lot of faith in test results of one kind or another. But in fact, grading doesn't have "real-world appropriateness." You don't get an A or an F on life's existential report card. Yet the blogosphere was convinced that either I or my students would be pulling a fast one if the grading were crowdsourced and students had a role in it. That says to me that we don't believe people can learn unless they are forced to, unless they know it will "count on the test." As an educator, I find that very depressing. As a student of the Internet, I also find it implausible. If you give people the means to self-publish—whether it's a photo from their iPhone or a blog—they do. They seem to love learning and sharing what they know with others. But much of our emphasis on grading is based on the assumption that learning is like cod-liver oil: It is good for you, even though it tastes horrible going down. And much of our educational emphasis is on getting one answer right on one test—as if that says something about the quality of what you have learned or the likelihood that you will remember it after the test is over. At that failing school I visited, everyone knew the kids who didn't speak English well were ruining the curve. Ironically, Spanish was a required subject but not one on which kids were tested—and so it was done as an afterthought, and very few of the Anglo kids could speak or read Spanish. Blinded by the need to succeed at a multiple-choice test, no one had the energy to create peer-learning teams in which those immigrant kids could teach their classmates how to speak some Spanish and *everyone* would learn more about respect and regard for one another's abilities in the process, including how to communicate across difference and in ways not measured by a bubble test.

Grading, in a curious way, exemplifies our deepest convictions about excellence and authority, and specifically about the right of those with authority to define what constitutes excellence. If we crowdsource grading, we are suggesting that young people without credentials are fit to judge quality and value. Welcome to the Internet, where everyone's a critic and anyone can express a view about the new iPhone, restaurant, or quarterback. That democratizing of who can pass judgment is digital thinking. As I found out, it is quite unsettling to people stuck in top-down models of formal education and authority.

Explicitly or implicitly, any classroom is about both content and method, information that you pass on and ways of knowing the world that you put into action. If a teacher wants a hierarchical class, in which she quizzes students on the right answers, she reinforces a process of learning in which answers are to be delivered and unquestioned. What a teacher—like a parent—has to decide is not only what content is worth knowing, but which ways of knowing should be modeled and encouraged.

Learn. Unlearn. Relearn. How do you grade successful unlearning? It is a contradiction to try to model serious intellectual risk-taking and then use an evaluation system based on conventional grades. My A+ students were right. How you measure changes how you teach and how you learn.

In addition to the content of our course—which ranged across cognitive psychology, neuroscience, management theory, literature and the arts, and the various fields that comprise science and technology studies—This Is Your Brain on the Internet was intended to model a different way of knowing the world, one that encompassed new and different forms of collaboration and attention. And more than anything, it courted failure. Unlearning.

"I smell a reality TV show," one contributor, the self-styled BoredDigger, sniffed on the Internet crowdsourcing site Digg.

That's not such a bad idea, actually. Maybe I'll try that *next* time I teach This Is Your Brain on the Internet. They can air it right after Project Classroom Makeover.

||||||||||||||||||||||||||||

If we want our schools to work differently and to focus on different priorities, we still have to come up with some kind of metric for assessing students. And it should come as no surprise that the tests we use now are just as outmoded as most of the ways we structure our classrooms. They were designed for an era that valued different things, and in seeking to test for those things, they limited the types of aptitude they measured. While twentieth-century tests and the accompanying A-B-C-D grades we use to mark them are quite good at producing the kind of standardized, hierarchical results we might expect for a society that valued those qualities, they often fail to tell us anything meaningful about our kids' true potential.

If grading as we know it doesn't thrill us now, it might be comforting to know that it was equally unpopular when it first appeared. Historians aren't exactly sure who invented testing, but it seems as if the concept of assigning quantitative grades to students might have begun at Cambridge University, with numerical or letter grades supplementing written comments on compositions in a few cases by a few dons sometime at the end of the eighteenth century. It was not really approved of, though. In fact, quantifying grades was considered fine for evaluating lower-order thinking, but implicitly and explicitly was considered a degraded (so to speak) form of evaluation. It was not recommended for complex, difficult, higher-order thinking. You needed to hear oratory, watch demonstrations, and read essays to really understand if a student had mastered the complexity of a subject matter, whether classics or calculus.[3] At Oxford and

Cambridge, to this day, there is still heavy reliance on the essay for demonstrating intelligence and achievement.

In the United States, Yale seems to have been the first university to adopt grading as a way of differentiating the top twenty or thirty "Optimi" students, from the middling "Inferiores" and the lowly "Pejores."⁴ The practice, with variations, spread to other universities, including Harvard, William and Mary, and the University of Michigan. By the middle of the nineteenth century, public school teachers were affixing grades to students' essays (this was before the advent of multiple-choice "objective" exams), largely as a convenience and expedient for the teachers, not because such a grade was considered better for the students.

Assigning a number or letter grade to the attainment of knowledge in a complex subject (let's say, history) as measured by a test is not a logical or self-evident method of determining how much someone understands about history. There are a lot of assumptions about learning that get skipped over in the passage from a year's lectures on the controversial, contradictory, value-laden causes and politics of, for example, World War I, to the B+ that appears on a student's transcript for History 120, World History 1900–1920. That letter grade reduces the age-old practice of evaluating students' thoughts in essays—what was once a qualitative, evaluative, and *narrative* practice—to a grade.

The first school to adopt a system of assigning letter grades was Mount Holyoke in 1897, and from there the practice was adopted in other colleges and universities as well as in secondary schools.⁵ A few years later, the American Meat Packers Association thought it was so convenient that they adopted the system for the quality or grades, as they called it, of meats.⁶*

Virtually from the beginning of grading in the United States, there was concern about the variability of the humans assigning grades. What we now know as grade inflation was a worry early on, with sterner commentators wagging censorious fingers at the softies who were far too lenient, they thought, in giving high grades to the essays they were marking. Others were less concerned

---

* It's interesting that if you dig at all into the U.S. Department of Agriculture literature, you find that even grading slabs of sirloin or chuck is a less clear-cut matter than meets the eye. There are uniform federal standards of meat quality that are debated regularly and changed occasionally. A truth-in-labeling law had to be instituted to guard against misrepresentation of what the grade even means. If it's that tough grading beef, imagine what's behind testing our kids.

about the "objectivity" of the tester and more concerned with efficiency. If you were testing not just a few privileged students at Yale or Harvard, but thousands and thousands of students, could you really be reading and carefully marking their long, handwritten, Oxbridge-style ruminations on life or geography or mathematical principles?

Because uniformity, regularity, standardization, and therefore objectivity (in the sense of objective measurement of grades) were the buzzwords of the first decade of the twentieth century, the search was on for a form of testing that fit the needs of the day. How could we come up with a uniform way of marking brain power that would help sort out the abilities of the industrial labor force, from the assembly line worker to the CEO?

Thus was born the multiple-choice test, what one commentator has called the symbol of American education, "as American as the assembly line."[7] It is estimated that Americans today take over 600 million standardized tests annually— the equivalent of about two tests per year for every man, woman, and child in America. There are apparently more standardized tests given in schools in the United States than in any other country. Americans are test-happy. We use standardized tests for every level of education, including, in some cases, preschool. We use them for driver's licenses and for entry-level positions in business and industry, for government and military jobs.[8] Americans test earlier and more often than anyone else. We even have standardized tests for measuring the validity of our standardized tests.

So where did standardized testing come from anyway? That's not just a rhetorical question. There is a "father" of the multiple-choice test, someone who actually sat down and wrote the first one. His name was Frederick J. Kelly, and he devised it in 1914. It's pretty shocking that if someone gave it to you today, the first multiple-choice test would seem quite familiar, at least in form. It has changed so little in the last eight or nine decades that you might not even notice the test was an antique until you realized that, in content, it addressed virtually nothing about the world since the invention of the radio.

Born in 1880 in the small farming town of Wymore, Nebraska, Kelly lived until 1959. A lifelong educator, he had seen, by the time of his death, the multiple-choice test adapted to every imaginable use, although it was not yet elevated into a national educational policy, the *sole* metric for assessing what kids were learning in school, how well teachers were teaching them, and whether schools were or were not failing.

Kelly began his career at Emporia State University (formerly Kansas State Teachers' College). In 1914, he finished his doctoral dissertation at Teachers' College, entitled *Teachers' Marks, Their Variability and Standardization*. His thesis argued two main points. First, he was concerned about the significant degree of subjective judgment in how teachers mark papers. Second, he thought marking takes too much of a teacher's time. He advocated solving the first problem—"variability"—with the solution of standardization, which would also solve the second problem by allowing for a fast, efficient method of marking.

Inspired by the "mental testing movement," or early IQ testing, Kelly developed what he called the Kansas Silent Reading Test. By that time, he had progressed to become director of the Training School at the State Normal School at Emporia, Kansas, and from there, he went on to become the dean of education at the University of Kansas. "There has always been a demand on the part of teachers to know how effectively they are developing in their children the ability to get meaning from the printed page," Kelly wrote. "Nothing is more fundamentally important in our school work than the development of this ability."[9] For Kelly, "effective teaching" meant uniform results. In this, he was a creature of his age, prizing a dependable, uniform, easily replicated product—the assembly-line model of dependability and standardization—over ingenuity, creativity, individuality, idiosyncrasy, judgment, and variability.

Thus was born the *timed* reading test. The modern world of 1914 needed people who could come up with the exact right answer in the exact right amount of time, in a test that could be graded quickly and accurately by anyone. The Kansas Silent Reading Test was as close to the Model T form of automobile production as an educator could get in this world. It was the perfect test for the machine age, the Fordist ideal of "any color you want so long as it's black."

To make the tests both objective as measures and efficient administratively, Kelly insisted that questions had to be devised that admitted no ambiguity whatsoever. There had to be wholly right or wholly wrong answers, with no variable interpretations. The format will be familiar to any reader: "Below are given the names of four animals. Draw a line around the name of each animal that is useful on the farm: cow tiger rat wolf."

The instructions continue: "The exercise tells us to draw a line around the word cow. No other answer is right. Even if a line is drawn under the word cow, the exercise is wrong, and nothing counts. . . . Stop at once when time is called. Do not open the papers until told, so that all may begin at the same time."[10]

Here are the roots of today's standards-based education reform, solidly preparing youth for the machine age. No one could deny the test's efficiency, and efficiency was important in the first decades of the twentieth century, when public schools exploded demographically, increasing from about five hundred in 1880 to ten thousand by 1910, and when the number of students in secondary education increased more than tenfold.[11] Yet even still, many educators objected that Kelly's test was so focused on lower-order thinking that it missed all other forms of complex, rational, logical thinking entirely. They protested that essays, by then a long-established form of examination, were an exalted form of knowl edge, while multiple-choice tests were a debased one. While essay tests focused on relationships, connections, structures, organization, and logic, multiple-choice exams rewarded memorization rather than logic, facts without context, and details disconnected from analysis. While essays allowed for creativity, rhetorical flourishes, and other examples of individual style, the Silent Reading Test insisted on timed uniformity, giving the most correct answers within a specific time. While essays stressed coherent thinking, the Silent Reading Test demanded right answers and divided knowledge into discrete bits of information. While essays prized individuality and even idiosyncrasy, the bywords of the Silent Reading exam were uniformity and impersonality.[12]

What the multiple-choice test *did* avoid, though, was judgment. It was called objective, not because it was an accurate measure of what a child knew but because there was no subjective element in the grading. There was a grade key that told each teacher what was right and what was wrong. The teacher or teacher's aide merely recorded the scores. Her judgment was no longer a factor in determining how much a child did or did not know. And it *was* "her" judgment. By the 1920s, teaching was predominantly a woman's profession. However, school administration was increasingly a man's world. It was almost exclusively men who went off to earn advanced degrees in schools of education, not so they could teach better but so they could run schools (our word again) *efficiently.*[13]

The values that counted and prevailed for the Silent Reading Test were efficiency, quantification, objectivity, factuality, and, most of all, the belief that the test was "scientific." By 1926, a form of Kelly's test was adopted by the College Entrance Examination Board as the basis for the Scholastic Aptitude Test (SAT).[14] Because masses of students could all take the same test, all be graded in the same way, and all turned into numbers crunched to yield comparative results, they were ripe to become yet another product of the machine age: statistics, with different statisticians coming up with different psychometric theories about the best number of items, the best number of questions, and so forth.[15] The Kansas Silent Reading Test asked different kinds of questions, depending on whether it was aimed at third graders or tenth graders. The reason for this was that Kelly designed the test to be analyzable with regard not just to individual achievement, as an assessment tool that would help a teacher and a parent determine how the child was doing, but also as a tool that would allow results to be compared from one grade to another within a school, across a range of grades within a school, and then outward, across schools, across districts, across cities, and divisible in any way one wanted within those geographical categories. If this sounds familiar, it is because it's almost identical to our current educational policy.

This story has a bittersweet coda: It is clear from Kelly's own later writings that he changed his mind about the wisdom of these tests. He didn't write much, but what he did write doesn't mention the Kansas Silent Reading Test. Far from enshrining this accomplishment for the historical record, his later writing passes over it in silence. It seems as if his educational philosophy had taken a decidedly different turn. By 1928, when he ascended to the presidency of the University of Idaho, he had already changed direction in his own thinking about the course of American education. In his inaugural presidential address, "The University in Prospect," Kelly argued against what he saw as the predominant tendency of post–World War I education in America, toward a more specialized, standardized ideal. His most important reform at the University of Idaho during his presidency was to go stridently against the current of the modern educational movement and to create a unified liberal arts curriculum for the first and second years of study. His method emphasized general, critical thinking. "College practices have shifted the responsibility from the student to the teacher, the emphasis from learning to teaching, with the result that the development of fundamental strengths of purpose or of lasting habits of study is rare," President Kelly said, announcing his own blueprint for educational reform. He railed against specialization at too early a stage in a student's educational development and advocated "more fundamental phases of college work." He insisted, "College is a place to learn how to educate oneself rather than a place in which to be educated."[16]

Unfortunately, his message ran counter to the modernization, specialization, and standardization of education he himself had helped start. Faculty in the professional schools at the University of Idaho protested President Kelly's reforms, and in 1930, he was asked to step down from his position.[17] Yet his test soldiered on and, as we have seen, persists to the present day in the end-of-grade exams that measure the success or failure of every child in public school in America, of every teacher in the public school system, and of every public school in America.

Once more, the roots of our twenty-first-century educational philosophy go back to the machine age and its model of linear, specialized, assembly line efficiency, everyone on the same page, everyone striving for the same answer to a question that both offers uniformity and suffers from it. If the multiple-choice test is the Model T of knowledge assessment, we need to ask, *What is the purpose of a Model T in an Internet age?*

IF THE REACH OF STANDARDIZED testing has been long, so has been the social impact of this form of testing. Not far from the uniform and efficient heart of the standardized achievement test was the eugenic soul of the IQ test. The dubious origins of IQ testing have been entwined with the standardized, multiple-choice achievement tests for so long that it is easy to confuse the aims and ambitions of one with the other. In fact, the very *origins* are mixed up and merged.[18] Although more has been written about IQ tests, it is not well known that the first ones were intended to be not objective measures of innate mental abilities (IQ) but tests for one specific kind of aptitude: an ability to do well in academic subjects. While Kelly was trying to figure out a uniform way of testing school achievement, educators in France were working on testing that could help predict school success or failure, as a way of helping kids who might be having difficulty in the French public school system.

In 1904, two French psychologists, Alfred Binet and Théodore Simon, were commissioned by the Ministry of Public Education to develop tests to identify and diagnose children who were having difficulties mastering the French academic curriculum.[19] They were using the word *intelligence* in the older sense of "understanding" and were interested in charting a child's progress over time, rather than positing biological, inherited, or natural mental characteristics.

Like Kelly in Kansas, who began his testing research around the same year, the French psychologists were of their historical moment in seeking efficient and standardized forms of assessment: "How will it be possible to keep a record of the intelligence in the pupils who are treated and instructed in a school . . . if the terms applied to them—feeble minded, retarded, imbecile, idiot—vary in meaning according to the doctor who examined them?" Their further rationale for standardized testing was that neurologists alone had proved to be incapable of telling which "sixteen out of twenty" students were the top or the bottom students without receiving feedback from the children's teachers.

Note the odd assumption here that you need a test to confirm what the teacher already knows. The tests were important because the neurologists kept getting it wrong, merely by plying their trade and using their scientific methods. Historian Mark Garrison asks, "If one knows who the top and bottom students are, who cares if the neurologists can tell?" He wonders if the importance of the tests was to confirm the intelligence of the students or to assert the validity of institutional judgment, educational assessment, and scientific practice of the day.[20]

Binet himself worried about the potential misuse of the tests he designed. He insisted they were not a measurement, properly speaking. He argued that intelligence comes in many different forms, only some of them testable by his or by any test. His understanding of different skills, aptitudes, or forms of intelligence was probably closer to that of educator Howard Gardner's concept of "multiple intelligences" than to anything like a rigid, measurable standard reducible to a single numerical score.[21]

His words of caution fell on deaf ears. Less than a year after Binet's death in 1911, the German psychologist William Stern argued that one could take the scores on Binet's standardized tests, calculate them against the age of the child tested, and come up with one number that defined a person's "intelligence quotient" (IQ).[22] Adapted in 1916 by Lewis Terman of Stanford University and renamed the Stanford-Binet Intelligence Scale, this method, along with Binet's test, became the gold standard for measuring not aptitude or progress but innate mental capacity, IQ. This was what Binet had feared. Yet his test and that metric continue to be used today, not descriptively as a relative gauge of academic potential, but as a purportedly scientific grading of innate intelligence.

Before his death, Binet protested against the idea that his tests measured *hereditary* brain power or innate intelligence. In midlife, he had become more introspective and cautious about science than he had been as a young man, having done some soul-searching when intellectual areas he had championed (spiritualism, hypnotism, and phrenology—the study of head bumps) had proved suspect. Largely self-taught, Binet had become cautious about overstating the importance of his diagnostic testing.

He would have been appalled and disgusted by the misuse of his test that began in 1917. In a story that has been told many times, the president of the American Psychological Association, Robert Yerkes, convinced the military to give the new Stanford-Binet IQ tests to more than a million recruits to determine who had sufficient brain power to serve as officers, who was fit to serve overseas, and who simply was not intelligent enough to fight in World War I at all.[23] This massive sampling was then later used to "prove" the mental inferiority of Jews, Italians, eastern Europeans, the Irish, and just about any newly arrived immigrant group, as well as African Americans. Native-born American, English-speaking Anglo-Saxons turned out to have the highest IQ scores. They were *innately* the most intelligent people. Familiarity with English or with the content was irrelevant. There were Alpha tests for those who could read and Army Beta tests for those who couldn't, and they were confusing enough that no one did well on them, raising the eugenic alarm that the "decline" in American intelligence was due to racial and immigrant mixing. Almost immediately, in other words, the tests were used as the scientific basis for determining not just individual but group intelligence. The tests were also adopted by U.S. immigration officials, with profound impact on the Immigration Restriction Act of 1924 and other exclusionary policies.[24] They were also used against those of African descent, including native born African Americans, to justify decades of legal segregation, and against Asians and Native Americans as the basis for inequitable citizenship laws.[25] Of course, the argument had to run in the opposite direction when applied to women who, from the beginning, showed no statistical difference from men in IQ testing. Magically, the same tests that "proved" that different races had unequal innate intellectual capacities nevertheless were not held to "prove" that different genders *were* equal.[26]

ONE FINAL FACTOR HAS TO be added into this discussion of how we measure: the very statistical methods by which we understand what test scores mean. There is nothing intrinsic about measurement. Yet after a century of comparative testing, finding the implicit "mean" and "median" of everything we measure, such forms of testing are so woven into the fabric of our culture that it is almost impossible to imagine assessment apart from distribution curves. It is as if someone cannot do well if someone else does not do badly. We hold to the truth of grades and numbers so firmly, as if they tell the objective truth about how we are doing, that you would think that, from time immemorial, all human mental activity came with a grade as well as a ranking on humanity's great bell curve.

Yet the statistical assumptions about how we measure were also invented quite recently, and also as part of the industrial age. Many of the most familiar features of modern statistics were devised by Sir Francis Galton (1822–1911), a cousin of Darwin's and proponent of social Darwinism, a misapplication of evolutionary theory that argued that those at the top of the social heap—aristocratic members of the upper classes—had gotten there through their excellent inherited qualities. Galton believed that to ensure "survival of the fittest," rich aristocrats should receive government subsidies to encourage them to have more children while the British poor should be sterilized. If that seems a reversal of Darwin's ideas of how the strongest in a species survive, well, it is.

Galton was a fervent believer in what he called eugenics, selective breeding as applied to humans. Sir Francis also coined the phrase "nature *versus* nurture" and believed that inherited, natural characteristics were what mattered. He believed that the human inheritance pool was being "weakened," and he was positive he had the statistical measures to prove his eugenic contentions. He was the first person to develop the form of the questionnaire in order to gather what he saw as empirical evidence, and he developed the influential statistical ideas of standard deviation, statistical correlation, and regression toward the mean.[27]

These ideas about statistics aren't inherently problematic, but given what we know about how attention works, it shouldn't be surprising to see how they can be used to support interpretations that reinforce dominant values. Because we see not the world but the particular view of it conditioned by our past experiences and past learning, it is predictable that we view deviation as bad, a problem rather than simply a change. If we establish a mean, deviation from the mean is almost inevitably a decline. When we measure the decline against that mean, it looks like scientific proof of failure, weakness, a downward trajectory: crisis. But what is interesting is how in the very first statistical methods and the very first tests—multiple-choice or IQ—there was already a problem of decline (in standards or intelligence) that had to be addressed. From the beginning, the scientific assessments were designed to solve a problem presumed (not proved) to exist.

Not everyone who uses statistics or objective forms of testing does so to prove that nature favors the ruling class. For example, Horace Mann championed both common schools and testing as more fair and unbiased than the system of privilege operating in the private schools and elite universities of the day. But it is depressing to see how often methods purported to be quantitative, objective assessments have been used for ideological purposes justified by those metrics.

THREE FEATURES OF HOW WE measure came together during the machine age and embody the machine-age ideals of uniformity and efficiency: statistical methods of analysis, standardized or multiple-choice testing, and IQ testing of "natural" capabilities. Together those three have defined virtually all conversation about learning and education for a century.

This is not to say that educational philosophers and reformers haven't protested these methods, metrics, and conclusions. Hundreds have. In *The Mismeasure of Man*, the late Harvard paleontologist Stephen Jay Gould showed how quantitative data can be misused not to measure what they purport to test but to support the preconceived ideas upon which the tests themselves are constructed.[28] He suggested there is an inherited component to IQ—but it is inherited cultural privilege, supported by affluence, passed on from parents to children, that is the greatest educational legacy, not genetics.[29] Recently a group of educators trained in new computational methods have been confirming Gould's assertion by microprocessing data from scores earned on end-of-grade exams in tandem with GIS (geographic information systems) data. They are finding clear correlations between test scores and the income of school districts, schools, neighborhoods, and even individual households. As we are collecting increasing amounts of data on individuals, we are also accumulating empirical evidence that the most statistically meaningful "standard" measured by end-of-grade tests is standard of living, as enjoyed by the family of the child taking the exam.[30]

University of Virginia psychologist Timothy Salthouse has been conducting research in a different vein that also should sweep aside any lingering perceptions that the current, formal mode of testing captures the whole truth of our learning potential. Most psychologists study intelligence or aptitude by administering one test to a sampling of subjects. In 2007, Salthouse used a different method. He gave a battery of sixteen different intelligence and aptitude tests to subjects ranging from eighteen to ninety-seven years of age. He included the most important cognitive and neuropsychological tests designed to measure everything from IQ to attention span. By using many tests, Salthouse put into his own complex experiment all the variables—the categories, biases, assumptions, and hierarchies—implicit in each test but inconsistent across them.

Typical studies focus on one activity or characteristic and use that measure as if it is an absolute against which other qualities are gauged. By using multiple tests, Salthouse gave us the tools to see in a different way.

He did something else ingenious too. Most studies of intelligence or intellectual aptitude give a subject a test one time. They operate from a presumption that one person tests the same on all occasions. We know that isn't true. We know our scores vary. In the case of a major, life-changing test such as the SAT, you might do badly one year, and so take it again later. Or you might even shell out some money to take the Kaplan tests that prep you on how to take that test better, and then you might take it again. Salthouse tested the tests by having each of his subjects take the tests more than once. He repeated the same tests on different days, with no new prepping. He also had the same subject retake various kinds of tests under different conditions. He then compiled all the statistical results from all of these multiple tests by multiple people, aggregating the different scores and then crunching the numbers in various ways, seeking out different variables, including such "extrinsic" factors as time of day a test was taken, whether it was the first or a subsequent time that the same kind of test was taken, and so forth. With new computational tools, it is certainly possible to take data with various different structures and crunch it every which way to see if anything interesting emerges.

After analyzing all the data from all the tests in many different ways, Salthouse found wide variability *within the same individual,* even when the individual was being tested for the same characteristics. The variety depended on such factors as what day and what time of day a test happened to be taken, as well as what type of test the person was taking. An individual scored differently on the same test taken on two different days and scored differently, again, depending on the kind of test he or she took. Salthouse found that "the existence of within-person variability complicates the assessment of cognitive and neuropsychological functioning and raises the possibility that single measurements may not be sufficient for precise evaluations of individuals, or for sensitive detection of change."[31] Not sufficient? With all the data crunched, he was able to demonstrate that the within-person deviation in test scores averaged about 50 percent of the between-person deviation for a variety of cognitive tasks.[32] Or as one popular science reporter put it, "Everyone has a range of typical performances, a one-person bell curve."[33] That is a decisive statistic, one that should have everyone rethinking how we measure.

Note to self: Remember not to take important tests on Mondays or ever in the late afternoons. Note to those who have substituted standardization for high standards: Even what we presume is standardized isn't even particularly consistent for measuring one individual. Yet we put so much weight on test scores that whole lives can be changed by having a bad day and doing badly on one. Your child's excellent magnet preschool is suddenly off limits because of a mediocre test score, and a whole domino effect of subsequent achievement based on past achievement can be set in motion.

For the most test-happy nation on earth, Salthouse's experiment poses a conundrum. Critics of standardized tests often say that they reduce learning to the practice of "teaching to the test"—that is, teaching not so students understand but so they do well on the test. The Salthouse study, and the critique of testing that goes all the way back to its origins with Kelly and Binet, suggests it may be worse than that. We may be teaching to contradictory, inconsistent, and inconclusive tests of lower-order thinking.

||||||||||||||||||||||||||||||||

It's only recently that our policy makers seem to be getting the message, but for now we're still stuck, trapped in a system that isn't working for kids or their teachers. Everyone knows this. We don't know yet how to get out, though more and more people have finally started thinking of new ways to solve the problem. Even Diane Ravitch, the influential educator and policy maker who helped shape national educational policy under presidents George H. W. Bush and Bill Clinton, has recently renounced her earlier faith in standardized testing.[34] In the end, she says, the only people to have profited by our national standards-based reform are commercial providers of tests and testing services, all of whom are now concerted in their lobbying efforts to keep end-of-grade tests mandatory across the nation.

Assessment is a particularly volatile issue in America now because, since January 8, 2002, we have had a national policy that is, for all intents and purposes, wholly shaped by standardized testing. The No Child Left Behind Act (Bill 107-110), typically abbreviated as NCLB, was proposed by President George W. Bush soon after he came into office and was led through Congress by Senator Ted Kennedy. It received overwhelming bipartisan support in Congress and justified a substantial influx of funding into public education. It is based on the theory of "standards-based education reform," which relies on standardized testing to measure individual and school performance results. It equates those test results with better outcomes in education. No Child Left Behind adjusts federal funding to the test scores achieved by students on state-prescribed multiple-choice exams. School districts in some states can be penalized with lower funds if their kids don't achieve a certain standard, as set by individual states, on the tests.

What are the values that No Child Left Behind reinforces? Memorization, mastering a definitive body of knowledge that an institution insists is worth knowing, understanding that some subjects "count on the test" and other areas aren't worth testing at all. Everything about No Child Left Behind reinforces the idea of knowledge as a noun, not knowing as a verb. Exactly at the moment that kids have the tremendous opportunity to explore online on their own, when they need to be taught higher-order critical thinking and evaluation skills to understand what is credible, how you think, how you know what is or isn't good information, and how you use information to drive sound conclusions, we subject them to machine-readable, multiple-choice, end-of-grade bubble tests. It's not that there is anything terrible about them, but they are redundant for the most academically gifted students and are an irrelevant distraction from real and necessary learning for the kids who don't do well in school. Except for preparing one for trivia-based games such as *Who Wants to Be a Millionaire,* it's hard to imagine what might be the real-world utility of testing based on responding to any individual question by selecting from a sampling of possible answers.

Online, kids have to make choices among seemingly infinite possibilities. There's a mismatch between our national standards of testing and the way students are tested every time they sit by themselves in front of a computer screen. If the school bell was the symbol of American public education in the nineteenth century, summoning America's farm kids into the highly scheduled and efficient new world of industrial labor, should the multiple-choice test really be the symbol of the digital era? If so, our symbol underscores the fact that American public education, as currently constituted, is an anachronism, with very little relevance to our children's lives now and even less to their future.

In 2006, two distinguished historians, Daniel H. Cohen and the late Roy Rosenzweig, worked with a bright high school student to create H-Bot, an online robot installed with search algorithms capable of reading test questions and then browsing the Internet for answers. When H-Bot took a National Assessment of Educational Progress (NAEP) test designed for fourth graders, it scored 82 percent, well above the national average. Given advances in this technology, a 2010 equivalent of H-Bot would most likely receive a perfect score, and not just on the fourth-grade NAEP; it would also ace the SATs and maybe even the GREs, LSATs, and MCATs too.

Cohen and Rosenzweig created H-Bot to make a point. If calculators have made many old arithmetic skills superfluous, search functions on Google have rendered the multiple-choice form of testing obsolete. We need to be teaching students higher-order cognitive and creative skills that computers cannot replicate. Cohen and Rosenzweig want far higher standards for creative, individual thinking. "We will not be among the mourners at the funeral of the multiple-choice test," they insist. "Such exams have fostered a school-based culture of rote memorization that has little to do with true learning."[35]

‖‖‖‖‖‖‖‖‖‖‖‖‖‖‖‖‖‖‖‖‖‖‖‖‖‖‖‖

"So, Prof Davidson, if you're so smart, how would you change assessment today?" the blogosphere well might ask. "Well," I might answer, "having spent time with so many extraordinary teachers, I have a few ideas."

First, I would stop, immediately, the compulsory end-of-grade exams for every child in an American public school. The tests are not relevant enough to the actual learning kids need, they offer little in the way of helpful feedback to either students or their teachers, and there are too many dire institutional consequences from failure resting on our children's shoulders. The end-of-grade exam has become a ritual obligation, like paying taxes, and no one wants our kids to grow up thinking life's inevitabilities are death, taxes, and school exams. That's a disincentive to learning if ever there was one. Wanting to improve learning on a national scale, across the board, in every school district, north and south, east and west, urban and rural, rich and poor, is a fine ideal, but the current end-of-grade testing has not had that result, nor does it even do a good job of measuring the kinds of skills and kinds of thinking kids need today. This national policy has been around less than a decade. Let's admit the experiment has failed, end it, and move on.

In a digital age, we need to be testing for more complex, connected, and interactive skills, what are sometimes called twenty-first-century skills or twenty-first-century literacies (see the appendix). Of course this includes the basics, the three *R*s, but for those, any kind of testing a teacher devises to help her measure how well her class is doing is fine. Parents and students all know when a teacher isn't doing her job. We don't need multiple-choice tests to prove she's not, with consequences for the whole school. Use the funds from all that grading to get a teacher who isn't succeeding in the classroom some additional mentoring, maybe even a teacher's aide. If she still isn't working out, maybe then, as a last resort, suggest this may not be the job for her. Our crisis in teachers leaving the profession is greater than that of students leaving, and no wonder, given our current punitive attitudes toward them.

I'm not against testing. Not at all. If anything, research suggests there should be more challenges offered to students, with more variety, and they should be more casual, with less weight, and should offer more feedback to kids, helping them to see for themselves how well they are learning the material as they go along. This is called adaptive or progressive testing. Fortunately, we are close to having machine-generated, -readable, and -gradable forms of such tests, so if we want to do large-scale testing across school districts or states or even on a national level, we should soon have the means to do so, in human-assisted, machine-readable testing-learning programs with real-time assessment mechanisms that can adjust to the individual learning styles of the individual student. Just as it has taken almost a hundred years to move from Frederick Kelly's Kansas Silent Reading Tests to the present tests created by the Educational Testing Service (ETS) and other companies, so too are we at the beginning of tests for a digital age that actually measure the skills our age demands. It is possible that within a decade, each learner—whether a schoolchild or a lifelong learner—could be building up his own private ePortfolio, with badges and credentialing built in, of all learning, all challenges, all there to measure accomplishments at the end of a school grade and still there several years later to help refresh a memory or to be shown to a future employer, all indexed and sortable at will.[36]

But if we are on the verge of new forms of testing for a digital age, we should think about what we want to test, as well as how we have to change our institutions of education, testing, and government policy that are currently reinforcing methods of assessment and even the kind of "itemized" learning that is firmly rooted in the last century.[37] Because we are on the brink of new computerized, individualized, adaptive, self-assembling testing that will be useful to learners and easier on teachers, what exactly could we be testing for that eludes us in the item-response method?

We've made headway on that score too, with more and more educators, policy makers, parents, and students themselves arguing that we urgently need to be thinking of interconnected, not discrete, twenty-first-century skills. Instead of testing for the best answer to discrete questions, we need to measure the ability to make connections, to synthesize, collaborate, network, manage projects, solve problems, and respond to constantly changing technologies, interfaces, and eventually, in the workplace, new arrangements of labor and new economies. For schools this means that in addition to the three *R*s of reading, writing, and arithmetic, kids should be learning critical thinking, innovation, creativity, and problem solving, all of the skills one can build upon and mesh with the skills of others. We need to test students on how critically they think about the issues of the digital age—privacy, security, or credibility. We could write algorithms to test how well kids sort the information that comes at them, how wisely they decide what is or is not reliable information. It would also be easy to assess the effectiveness of their use of new technologies and all the multimedia tools not only at their disposal but more and more necessary for their future employment. If you can't get on Twitter, you haven't passed that test. Just as you can build copyright restrictions into online source materials, you could as easily test the ways kids remix or download, testing for creativity and also for their sensitivity to the ethical or legal use of the intellectual property of others.

If change is the byword of a great era of technological innovation such as our own, we need to have methods of assessment that show not how well kids learn how to psyche out which answer among five has the highest probability of being right but how well they can apply their knowledge to novel situations. How flexible and adaptable are they; how capable of absorbing and responding to feedback? I am sure that one reason we've gone from 4 to 320 reality TV shows in a decade is that *American Idol, So You Think You Can Dance, Project Runway,* or *Top Chef* do a better job of teaching sound judgment and how to react well (or badly) to feedback than most of our schools.

We need to measure practical, real-world skills, such as how to focus attention through project and time management. There is no punch clock in do-it-yourself culture, so where do kids learn how to manage themselves? Similarly, where do they learn, online or face-to-face, at school and in the future workplace, how to work together? Every employer says working well with others is a key to success, but in school every child's achievement is always weighed against that of everyone else's. Are you in the top 1 percent or the bottom? How does that teach collaboration? Once we figure out how to teach collaboration, how do we measure it? Kids shouldn't have to end up at their first job with a perfect report card and stellar test scores but no experience in working with others. When you fail at that in your first job, you don't get a C. You get a pink slip and a fast walk to the exit door.

We also need to be able to measure how well young people communicate, including with people who may not share their assumptions and background. In social groups, online and face-to-face, we often live in a relatively homogeneous world.[38] But in the workplace, we increasingly face a globalized workplace and a new world of digital communication, which has the potential for interaction with others who may not even share the same language. Through computer translating software, we can exchange *words* but not necessarily deep and differing cultural values, which is why business schools today insist that "culture" and "context" are the two most important features of global executive education. Executives need to know how to reason effectively, how to use systems and network thinking to understand the relationship between one problem and its solutions and other problems that may arise as a result. Bubble tests cannot begin to teach students how to analyze the way parts of a system interact with other parts of a complex system. We need to teach them how to make sound judgments and determinations about what is or is not credible, especially in a digital age when information comes unsorted.[39]

If all that sounds impossible, it is mostly because we have spent the last one hundred years believing that a multiple-choice test can tell us more about how kids are learning than can Mrs. Davidson, Katie Salen, or thousands of other inspiring teachers, past and present, who know a great student or one who is having difficulty when they see one, without benefit of any kind of test, machine-readable or otherwise.

We have to unlearn a lot of what we've come to believe about how we should measure. We need to think about how unnatural those tests are for assessing "unschooled" ways of learning, whether in everyday life or in the workplace. We don't give Baby Andrew an item-response test at the end of his first year of life to see if he can crawl or walk. We observe his progress toward walking. We don't subject a new employee to a standardized test at the end of her first year to see if she has the skills the job requires. Why in the world have we come to believe that is the right way to test our children?

If we want national standards, let's take the current funds we put into end-of-grade testing and develop those badges and ePortfolios and adaptable challenge tests that will assist teachers in grading and assist students, too, in how they learn. Lots of online for-profit schools are developing these systems, and some of them work well. Surely they are more relevant to our children's future than the bubble tests.

THERE IS ONE FEATURE OF the present end-of-grade exams that I would like to keep: the pacing of the school year. As I learned over and over, most schools teach real subjects and real learning from September to March, then stop what they are doing in order to help students prepare for the lower-order thinking required for the end-of-grades. Let's build on that structure and reverse it. Instead of dumbing down to psyche out the test, insert a boss-level challenge at the end of the year. Let every student know that when March comes, everything they will learn in their classes will be put into an important, practical application in the world. Let them think about that every day. How could what I am learning today about the American Revolution be turned into something that could help someone else? How could my geography lessons be of interest in the world? Spend some part of a day, maybe once or twice a week, talking to students about how what they are learning matters in the world, and let them be thinking of how that would translate into a collaborative end-of-grade challenge. When March comes, instead of putting away the exciting learning to study for the tests, instead of nightmares over whether they will fail or pass this year, whether their schools will be deprived of funding because of their poor performance, let them lie awake at night dreaming of projects.

Here's one. Let's say a fifth-grade social science class spent a lot of the year focusing on the heritage of the local community. What if they decided they wanted to make a webinar on what they learned for the fourth graders who would be coming through the next year? That girl with the green hair would be enlisted to do the artwork. Rodney would be doing any math calculations. Someone might write and sing a theme song. Someone else would write the script. Budding performers would be tapped to narrate or even act out the parts. The computer kids would identify the software they needed and start putting this together. The teacher would help, would bring in businesspeople from the community, teachers from community colleges or local universities, or maybe college students who could aid here. And the kids would then go out and interview their parents and grandparents, local leaders, the police, the community organizers, maybe even the mayor, and all of this would be part of their real, exciting end-of-grade "test." It would put into practice everything they have learned, would use all of their skills, and unlike the waste of time of the current cramming for lower-order thinking on the end-of-grades, it would mean each year ended with a synthesis of everything, showing each child what he or she could do in the world. Parents would have to be involved. Instead of the usual bake sale or ritual recital, an end-of-year "idea sale" would be held. Parents and anyone from town, including all those interviewed (get out the mayor!) would tour the school auditorium viewing what the students created. The kids could write press releases for the local newspapers and news programs. On the school's interactive Web site, they would collect feedback from local Web developers or public historians or just a neighbor or two, so next year they could make an even better project.

This may sound implausible and impractical, but it isn't. It's happening in informal after-school programs all over America and could certainly happen in a school near you. We know students learn best when their schooling relates to their lives in their homes and communities. Make that connection work for schools as well as for students by connecting schools to community colleges and universities in the town, to retirement communities, to local businesses and non-profits, to libraries and civic centers. The possibilities are endless.

I know of a great project like this. It's called Hypercities (http://hypercities .com) and it works out of UCLA and the local public schools in Filipinotown in Los Angeles. It was one of the winners of the 2007 HASTAC/MacArthur Foundation Digital Media and Learning Competition. It is "a collaborative research and educational platform for traveling back in time to explore the historical layers of city spaces in an interactive, hypermedia environment." Hypercities began working with students and community members in L.A.'s Historic Filipinotown (HiFi), with students interviewing their relatives, doing research in local archives, and restoring a map of an area of the city that no longer exists as it did fifty years ago, at the heart of America's Filipino immigrant community. The students found photographs, documents, and even forgotten film footage; they also mastered the methodologies of history and software code writing, as well as multimedia editing, sound and video production, and other skills.[40]

Would a project like Hypercities reinforce all the standardized knowledge that students would cram for on a current end-of-grade exam? No. Thank goodness. First, we know that students tend not to retain that testable knowledge. Second, we know they don't really know how to apply it beyond the test. And third, Kelly was right: It *is* lower-order thinking. What I am suggesting is reserving the last two months of school not for dumbing down but for inspiring the most unstructured, collaborative learning possible. It teaches students project management, achievement, and how to apply learning. It means that all the "cramming" that comes before it is not in order to pass a test but in order to teach kids how to think. What questions do I need to know? Taken over from spy talk, "need to know" is gamer language for that which is necessary in order to succeed in the next challenge. Games work so well and are so infinitely appealing because they reinforce the idea that the more we know, the better the game is. And that's a life lesson every teacher wants to instill.

There is nothing better that students can take home over summer vacation than a sense that what they have learned the previous year has meant they were able, with the help of lots of other people, including that alienated girl with the green hair and that kid who counts on his fingers, and lots and lots of people beyond the walls of the school, to make something important happen, to meet a challenge. All of the great teachers I have met have been implicitly or explicitly preparing their students for the challenges ahead. That, not the end-of-grade exam, is where we want to focus their attention. Eyes on the prize, and that prize is an ability, for the rest of their lives, to know they can count on their knowledge, their skills, their intelligence, and their peers to help them meet the challenges—whatever those may be—that lie ahead.

5

||||||||||||||||||||||||||

The Epic Win

Duncan Germain, a first-year teacher at Voyager Academy, a publicly funded charter school in Durham, North Carolina, teaches Creative Productions. It's not your typical school subject (which means I like it already), so it's sometimes hard to tell what exactly is going on in this twenty-four-year-old's classroom. Computers form an L along the back and side of an alcove. By a row of windows, long wooden poles, thick as I-beams, lean into the corner, partially covered by a tarp. A box of Popsicle sticks sits on a desk with another computer, along with index cards, Post-it notes, rulers, and scissors, and plenty of glue guns, too. The room is full of books. Beneath a Smart Board is an unrecognizable object, clearly handmade, about three feet high and three feet wide, constructed from wooden planking cut into various lengths and then glued together, with something like pilings on each side and an irregular V-shaped gap down the middle. Sturdy screw-eye hooks top each of the pilings, as if waiting to be attached to something, although I'm not sure what. Mr. Germain starts to write. Between the pointer finger and thumb of his left hand, there's either a tattoo or something he's written there this morning in dark blue ink. It reads: "What would Ender do?"

Now there's a tough question. Ender is the hero of the science fiction saga beginning with *Ender's Game* (1985), by novelist Orson Scott Card. Ender is a tactician and a genius, the ultimate leader and protector of Earth. In this series set far in the future, the inhabitants of Earth must decide how to fend off invasions by hostile beings inhabiting solar systems light-years away. Ender's childhood games prepare him to be that ultimate leader, holding the destiny of entire civilizations in his powerful command. Mr. Germain believes we all need to take our education seriously and that every child, including each sixth grader at Voyager Academy, needs to view learning as the key to a future whose destiny she holds in her own hand. Like Ender. That's how Mr. Germain teaches, with the intensity of someone who believes that, individually and collectively, we are responsible for our civilization's survival. When I ask him what his standard for success is, his answer is simple: "Perfection." I'm taken aback, but as I watch the class unfold, I understand that he challenges each child to find a perfection of his or her own. No aiming for the A, B, C, or D here. The bar is set at "How high can we go?"

As his students come into the room, they all pause and read what Mr. Germain has written on the board: "What's your specific role in your group today? What's your job?"

They take turns grabbing markers and they scrawl answers on the board, initialing what they write. They are chatting quietly, a school rule for entering and leaving the room. They make their way to their seats, arranged in groups of four chairs around tables in two rows down the room.

While the kids are answering the questions on the board, Mr. Germain tells me he often begins a day like this, focusing their attention with questions that succinctly restate problems or dilemmas they expressed the day before. He condenses their concerns, but always as questions that require them to answer, focusing on solving what bothered them the day before. Yesterday, some of the kids were frustrated that others weren't collaborating as well as they might have. So today's class begins by addressing that leftover anxiety before moving on to the set of challenges to be addressed today.

Voyager Academy is a very special school. It accepts only a hundred kids a year, as determined by a lottery system, and the kids know they are lucky to be here. It's been open only two years, so everyone here has been somewhere else, and they know the difference. They are all here by choice. At the same time, there are other forms of diversity. The proportion of kids with learning disabilities is higher at Voyager than the norm, as it is with many charter schools. Because of the smaller class size and the attention to hands-on learning, many parents who feel as if their kids have been falling behind in traditional schools try Voyager. In that it is kin to Q2L, though the two schools take different approaches to reach similar ends.

As with any charter school, this one has its own forms of accountability and must abide by those, in addition to the state-approved goals it must meet in order to continue receiving public funding. It charges no tuition, and there's a considerable waiting list of parents hoping their kids might get in. Kids at Voyager also take standard North Carolina end-of-grade (EOG) tests, but they are dealt with as an obligation, an add-on; it's as though they are not part of the educational philosophy of the school but an encumbrance imposed by regulation.

Every teacher I meet at Voyager has a story about the EOGs—what the exams fail to test; how they distract everyone from real learning; how in April everyone, from the kids to the principal, starts losing sleep over them; and how much disruption they cause to the real education happening throughout the school year. But they must be taken seriously if the school is to continue. "Failure" according to those standards isn't an option, even though the learning method designed to produce success in standardized testing is frowned upon, in practice and principle, by everyone I meet.

So in what, then, does a class called Creative Productions offer instruction, and how does it assess? Evaluation in Mr. Germain's class is entirely test-based, but his method has almost nothing in common with the assessments of the EOG. One could almost say that the *subject* of Creative Productions is assessment. This class doesn't have a set topic. Rather, the purpose is to take what the kids learn in their other classes and find ways to apply it to some kind of real-world experience. It's a class about usefulness, taking curriculum concepts out of the abstract. The kids come up with their own ways to work toward solutions and to test the validity of their solutions in real-world and experimental situations. They analyze their own results and write essays and formal abstracts that document those results.

The project they are working on is building a bridge from Popsicle sticks. That's what that wood construction under the Smart Board is. It's the "valley" that their bridge needs to span. Mr. Germain made it himself, and the students are excited because this is the first time they've seen it, the first chance they've had to test their own creations on it.

For this project, Mr. Germain has let the kids create their own groups, and they can work in a group of any size from one to five. Some kids chose to work alone, and so long as the student is able to provide a justification for working individually, Mr. Germain approves it. Part of this lesson is to analyze advantages and disadvantages of different ways of working: independent versus group thinking, delegation of tasks, how some people step up to challenges and others let you down, and how you make the group work well together. Anxieties over collaboration were at the heart of the questions Mr. Germain wrote on the board, and when he raises the subject, ten or twelve hands shoot up, and the conversation starts in earnest over the best ways of working together.

The students offer a variety of concerns. For some, working collaboratively produces information overload. They have a hard enough time managing themselves, knowing when they are over their own capacities. One little girl is jumping up and down saying loudly, "I feel *jumpy* today, I'm too *jumpy* today." Another girl at her table looks at her sternly and says, in her best maternal voice, "Well, stop being jumpy or we won't finish our experiment today either." The speaker is the girl who wrote, in answer to the "What's your job?" question, "To get us finished." The jumpy girl wrote "Operating the glue gun" as her answer to the question, and she turns to the maternal gal at the table and says, "I'll stop jumping and get the glue gun." And amazingly, with no intervention from Mr. Germain, she stops jumping and gets the glue gun.

"Who else wants to talk about their roles today?" Mr. Germain asks.

"I want to revise our blueprint," a boy says, and Mr. Germain points to a stack of blueprints. "OK, you can go get it now and work on it for a while before we start a group project," Mr. Germain says.

Another girl has written "Leading the group" as her answer.

"That's the hardest job of all. You know that, right?" Mr. Germain asks the tall, blond girl who has taken on that leadership role for her group. She nods seriously as her eyes slide in the direction of a boy in her group who is stretched out as far as possible while still sitting on a chair, his arms raised, his legs straight, his toes pointing. He's yawning. Mr. Germain's eyes follow her gaze toward the boy and then he looks back at her. "You can do it. I'll help," he says, his voice quiet and calm.

Throughout the class, Mr. Germain keeps one eye on the boy who was stretching. He's doing well today, but I learn he's smart and energetic enough to turn the class upside down with his antics. He's been learning, lately, how to tell for himself when he's in a disruptive mood, and he has a deal going with Mr. Germain. If he feels he cannot control himself, he's allowed to just walk away and go work by himself at the computer. He doesn't have to ask permission. All he needs to do is take himself out of the situation where he'll be disruptive. It's a public pact: Everyone knows it, including the tall girl. Mr. Germain has given this boy the responsibility of controlling himself by knowing how he can be least disruptive to the group. Today it's working.

As Mr. Germain goes around the room asking each group to talk about the management challenges ahead, he gives me the project plan to look at. Each student fills out such a plan at the beginning of a new class project. On the project plans, students fill in name, class period, and date, and then answer the very first question: How will they complete their project—alone, with a partner, or with a toon? A *toon* is short for "platoon," a term for the group one depends on for strength and ingenuity, those partners who offer the vision that you cannot, on your own, even see is missing. Though it's a military term, Germain took the shortened version of it from *Ender's Game*: It's the way Ender organizes his troops.

The project plan is divided into sections: who, what, where, when, why, and how. The plan covers two full sides of a sheet of paper and asks questions about the materials needed and their source; the backup plan in case the students' project plan doesn't work; those people to whom the student will turn for feedback and help; the research sources and design of the project; and what principles require research. And then, in case I'm not impressed enough already, the sheet presents a list of three areas where the student has to provide justification for the project. I'm looking over a project plan filled out by a quiet girl with olive skin and curly brown hair. Her sheet reads:

I can *justify* this project because it connects to

1. Standard course of study: measurement, scale, structure, scientific method.
2. Habits of mind: thinking flexibly, asking questions, problem solving, applying past knowledge, rejecting poor solutions, repairing, revising.
3. The outside world: understanding the principles whenever we go on a bridge, whenever we need to improvise and make a bridge.

This checklist looks like something one might find in a Harvard Business School book, the answers as good as an executive might offer up. When I ask him how he came up with his project-plan form, Mr. Germain notes that his father is a management consultant who often works with businesses adjusting to the shock waves of the new, digital global economy. Of course! I should have known. A program of learning for CEOs grappling with the twenty-first century is a great lesson plan for twelve-year-olds about to face the challenge of building a bridge from wooden sticks—and vice versa. Skills these kids learn in Creative Productions will serve them for a lifetime.

In this class, the students are testing themselves and one another all the time. For the Popsicle-stick bridge project, they began by finding the best articles they could locate on the engineering principles of bridge construction for the four main kinds of bridges. They each read some articles and then wrote abstracts for their toon and their classmates to read. Based on their reading, each toon chose a bridge type and then did further research on that one kind of bridge. They then had to ask themselves what they *didn't* learn in their research and had to come up with an empirical experiment, based on scientific principles, to find out an answer to their question. Is it better to make the bridge wider or narrower? How wide or narrow? Is it better to build with two layers or one, laying the Popsicle sticks on a diagonal or horizontally? In the classroom, there are heavy iron weights to test the tensile, yield, and breaking strengths of their structures. Each toon then writes up its experiment and their test results on index cards, with diagrams and conclusions.

Mr. Germain has distributed stacks of cards with their experiments to each table. The students read one another's experiments and findings carefully, then write out a summary of the research and an analysis of whether it seems convincing or inconclusive. They make value judgments on whether the experiment was well conceived or not. And then, after doing this for three or four experiments, they confer on whether any of the empirical evidence makes them want to modify their blueprint in any way. It's astonishing hearing them discuss the relative merits of research methods of the various studies conducted by their classmates. They act as frenetic as any sixth graders in their manner and attitudes, but the questions they raise about scientific method would make a university lab director pleased.

All of this activity leads up to the building of bridges that will be suspended over the model canyon built by Mr. Germain. They will then subject their bridges to all kinds of tests to see if they break. Germain, who has won awards for the Lego battleships he has created, is building his own bridge, too. It is the standard of perfection to which the kids in the class must aspire. When all the bridges are done, the kids will decide which bridge is the best. And there's a prize for the best. Mr. Germain has put up a $100 prize, from his own money. If his bridge is the best, he will keep $100. If one of the groups win, they will receive this money.

But that will lead to yet another lesson. How will they divide the money? Mr. Germain tells me they've been studying monetary systems in other classes, everything from percentages, interest rates, collateral, and so forth to whole economic systems that govern different countries around the world—capitalism, socialism, communism, exchange and barter economies versus cash and credit economies. If one of the toons wins, they will have to put these economic principles to work. One for all and all for one? Or does the team leader take the biggest reward and pay wages to the others? Do you compensate your classmates in any way? You may need their help for the next project. The tests never end in Mr. Germain's class. Each test presents its own set of complex issues, requiring its own set of skills and knowledge. It is all part of the responsibility of winning, and the class has had long discussions about borrowing, funding bridge building, financial impacts bridges have on communities, environmental consequences, and so forth.

It's at this point that the penny drops. In Creative Productions, each challenge leads not to an end point but to another challenge. It's like the game mechanics of LittleBigPlanet on the snow day in my office or like the tests Mrs. Davidson set her kids to that always led to new tests being devised by those who lost that week's challenge. Game mechanics are based on the need to know: What you need to know in order to succeed is the most important kind of knowledge, and that's not something someone else can tell you. You have to decide what you need to know and how to obtain that knowledge. The world isn't multiple-choice. Sometimes defining the question is even harder than answering the question you come up with. This is as true in life as it is in Mr. Germain's classroom.

I ask my question out loud, "You're really operating on game mechanics here, aren't you?" Duncan Germain smiles. He confides that he's one of the longest-continuing American participants in something called parkour, or freerunning, a noncompetitive physical discipline that originated in France, in which participants run along a route in man-made, typically urban, environments, negotiating every obstacle using only their well-trained bodies. Duncan doesn't look like some muscle-bound athlete, but when I watch his parkour videos on YouTube, I see him scramble up walls, jump across the gap from one rooftop to another, jump off a building and land in a roll, and pop up again to run, jump, climb, dodge, and all the rest. It's easy to see why parkour prides itself on needing brains as much as bodies to negotiate spaces that most of us would insist could not be traversed without ladders or ropes.

The philosophy of parkour is that all the world is a challenge to be respected and met. You train in groups so you can learn from one another not only techniques but your own weaknesses. Left to your own devices, you can miss your own weaknesses until it's too late. You rely on your training partners to help you improve your skills so you don't take foolhardy risks. Duncan says he's fallen seriously only once in six years, and that he didn't hurt himself badly in the fall. Parkour requires you to pay attention to all the things you normally take for granted. It tests you in situations where nothing can be taken for granted.

As in good teaching or good parenting, the challenge of Mr. Germain's hobby-sport is calibrated to the skills and understanding of the player, and the challenge gets harder only when you have succeeded at the easier challenge. The standard? Perfection.

At the end of a long day at Voyager Academy, the kids sit quietly in their homerooms, waiting for someone to pick them up and take them home. The school rules require silence during this hour of the day. Kids read or look out the window or fiddle, silently.

A boy, smaller than most of the others, with dusty blond hair and a habit of not looking at you when he speaks, asks if he can spend this homeroom time working on his bridge. It is elegant, intricate, and impeccable in its construction. Mr. Germain is going to have to work hard to beat this one. The boy takes a handful of Popsicle sticks over to the fabrication station and uses a mallet and a chisel to cut and bevel them in precise ways so he can use them to form a lattice of supports and struts.

I ask about the boy after he goes off to a far corner of the room to work on incorporating these new pieces into his beautiful bridge. Duncan says he was a bit of a lost child, someone who seemed not to have any talents, who was failing everything, every exam, every class, falling so far behind that there seemed to be no way to rescue him. From the way he holds himself, including how he avoids eye contact with others, I suspect that he might have something like Asperger's or another attention-spectrum disorder. But this bridge assignment has his complete attention, and the result is marvelous to behold.

Is this child a failure because he cannot begin to answer A, B, C, or D in multiple-choice tests in his other classes?

Or is he a master bridge maker?

The answer may well be both, but right now we have no system in place to recognize, foster, or build upon what this child is accomplishing.

Over the PA system, a voice says a boy's name, and this one looks up. He seems sad, reluctant to leave, but a mother is waiting for him outside. He asks Mr. Germain if he can bring his bridge home so he can work on it there. This is a child who manages to lose homework assignment after homework assignment. Mr. Germain looks him squarely in the eyes, insisting on the eye contact. "Of course you can bring it home," he says, calmly, "as long as you are sure you can count on yourself to bring it back again tomorrow."

The boy hesitates. He stands up and turns in a little circle. Mr. Germain stays with him but doesn't speak; he's letting the boy work through this. I see the boy's face as he looks up at his teacher. He could not be more serious, more intent. He nods.

"Yes," he says thoughtfully and decisively. "I can count on myself."

Right now, I can see only the back of Mr. Germain's head, but I know that he must be smiling with pride. His student has passed a very important test. His score? Perfect.

Duncan Germain at Voyager Academy may or may not have studied attention blindness, but it is clear in everything he does in his class that he has mastered the idea that you need to pay attention in different ways at different times and that no one can ever see the whole picture on his own. Extended to learning, that means instead of failure being the default, adaptation is. The issue isn't whether you have learning disabilities because, in a larger sense, everyone does. No one does everything perfectly all the time. Given that inherent lack, it's just not interesting or useful to label some disabilities and not others. Far better to diagnose what the issues are and then find the right tools, methods, and partners to compensate for those so that you are able to contribute in the unique ways that you can. Or as in the iPod experiment, better to set the challenge and leave the solution open-ended rather than prescriptive. That's the parkour credo too: The way you overcome the obstacles in your environment is by adapting your movements to the environment. In freerunning, there are no abstractly right or wrong answers. The freerunner finds solutions to problems as they present themselves.

This is a very different way of thinking about human ability than the one we have perfected and measured and scaled for over the last one hundred years, in which right was right and wrong was wrong and never the twain shall meet. We have spent decades building the institutions and the metric for evaluating what constitutes "normal." We have been so exacting in that assessment that we've created a system in which fewer and fewer people measure up. Recently I spent an afternoon with a teacher at an elite private high school who noted that 26 percent of the students at her school have been officially labeled as having learning disabilities, which means they all have special classes, tutors, and adaptations made by the school to help them compensate for their flawed learning styles. *Twenty-six percent.* The teacher chuckled ruefully and said, "Of course, the real number isn't twenty-six percent but one hundred percent." By that she meant that all of us have different ways of learning the world, only some of which manifest themselves in the conventional classroom setting.

I'm convinced we're in a transitional moment and that in a decade, maybe two, we will all be seeing the "long tail" of learning styles, just as we are recognizing the long tail of consumer goods and information made available to us in every imaginable form courtesy of the World Wide Web. What is impressive about all of the extraordinary teachers I have been able to spend time with, exemplified by Mr. Germain, is how they have already begun to maximize all manner of different human talents. They see ways that we have the potential, collectively, to organize our unique abilities to some end, to some accomplishment, which will elude us if we continue to hold to a rigid, individualistic, and normative model of achievement.

We have spent a lot of time and money building elaborate systems for rewarding individual excellence and punishing individual failure, as if worth had an on/off switch. What we are beginning to see are the ways that we can take our individual energies and work toward goals that accomplish something greater than we imagined on our own. As we shall see, some visionaries are also beginning to use digital interaction to come up with new ways of working together to envision solutions to seemingly intractable problems. That next step, of imagining a possible new kind of future together, is, after all, what Ender would do.

||||||||||||||||||||||||||||||||||

If Jane McGonigal had been there to witness the scene with the boy and his bridge at Voyager Academy, she would have described it as an epic win. That's gamer-speak. In *Urban Dictionary,* an online user-created dictionary of contemporary lingo, an *epic win* is defined as "an overwhelming victory over an opponent, a victory that's been snatched from the jaws of defeat, an unexpected victory from an underdog." Or in the words of another contributor, an epic win is "an incredible success so fantastic that long poems would be written about it in the future."[1]

McGonigal is a game designer and a social evangelist who believes that "reality is broken" and that the best way to fix it is by playing games. With a PhD from the University of California–Berkeley, she has had epic wins in designing games for business, the arts, entertainment, social causes, environmental purposes, and political movements. For McGonigal, games are the solution to many of our problems; if we were all gamers, we could cure a lot of problems in society. She believes that if from infancy on we all played games as if our life depended on them—especially multiplayer collaborative games—we would learn that challenges never stop, and that it is worth risking utter failure in order to have an epic win. She believes that, once you've experienced an epic win, you have the confidence to be fearless and unstoppable, not just in the next game challenge, but in the world.

In real life, we don't have enough chances to conquer our fears or to practice leadership in situations where the results aren't potentially catastrophic, she believes. Games, though, provide an opportunity with a built-in safety net, a place where we're not limited in the options available for exploration. Anyone who has had a leadership role in an epic win by a complex group organized online, in a challenge such as World of Warcraft, has learned much about human nature, about how people act in groups for a certain purpose, and about their own individual capacity to persuade, organize, and lead. McGonigal thinks we should try harder to organize complex, diverse, and unknown groups of people to come together for specific, challenging purposes defined by common interests and goals—something we don't often do because we don't believe we are capable of it. We defeat ourselves, and this self-defeat is brought on by a lifetime of bad or weak habits, reinforced by a formal education that tells us, first, that you succeed by pleasing others and, second, that making change—especially organized group change—is insolent, risky, subversive, dangerous, and guaranteed to fail. She would say one reason that we cheer for the underdog in spectator sports as in life is that, in our hearts, we have been schooled (literally) to feel like underdogs. Through game play—in which a need to know focuses our attention, motivates our search for skills and knowledge, and tells us how to deploy them—we develop the confidence to feel not like underdogs but like talented, skilled, competent winners.

Like Mr. Germain, McGonigal believes that gamers are experts at attention. They know that concentrated effort, combined with strategic and collaborative thinking that pays attention not just to the main task but to everything important happening around the peripheries as well, is what leads to great results. With the right strategy, the right planning, the right leadership, the right practice, the right experience, and the right blend of courage and concentration, you can experience epic wins, not once in a lifetime but over and over again.

McGonigal quotes Albert Einstein's dictum, "Games are the most elevated form of investigation." She advocates games for all forms of teaching and all forms of learning, insisting that game mechanics apply to all actions online and off, from the simplest to the most complex global social actions. She advocates using gaming to teach collective decision making and responsible action in the world, and she hopes to introduce students to the joy of activist learning.[2]

This might sound like a wild-eyed idea, or at the very least something of a stretch, but McGonigal has already found remarkable success in bringing her ideas to life. While in her twenties, she changed the world of marketing strategy by designing an alternate-reality game and viral marketing campaign for the 2004 release of *Halo 2,* at the time a cutting-edge game developed for Microsoft's Xbox. As part of the project, called I Love Bees, McGonigal first sent jars of actual honey through the mail to fans of previous alternate-reality games (ARGs). With the jars were letters that directed the recipient to the I Love Bees Web site, where a clock was counting down for an unspecified, mysterious purpose. Simultaneously—but with no connection drawn whatsoever—movie theaters were showing trailers for *Halo 2* that included the Xbox logo and the URL Xbox.com. If you went to that site, you saw a crude message on the screen, one that looked as if some hacker had dismantled the Xbox site and was forcing a redirect to a hacked site about beekeeping.

It was a puzzling and intriguing setup, and after a month—when buzz had built up to deafening levels in the ARG community—McGonigal finally leaked rumors and then confirmations about a connection between ilovebees.com and *Halo 2.* As more and more clues emerged over the course of several months, a growing community of ARG aficionados began working together to solve ever more elaborate mysteries that unfolded in both virtual and physical worlds, involving phone booths and cell phones, GPS systems and Web sites, and more. In the tech world, this spontaneous forming of an online community to solve a problem together is called *hive mind.* It's one of the most inspiring forms of collaboration indigenous to the Internet, where people who may not even speak the same language can come to a site and work together on anything from code to design. By the time the I Love Bees Web site was finally launched, it was already drawing over a million visitors a month, a community of visitors who shared certain gaming values and interests. When *Halo 2* launched, it had a huge appreciative audience waiting to buy it. The launch was an unprecedented success, and at a fraction of normal marketing costs.

McGonigal next turned her attention to the art world, helping to create audience-participation multimedia games for the Whitney Museum of American Art and the Museum of Contemporary Art, Los Angeles. She showed conclusively how the lines between art and technology, games and aesthetics, and the real world and the virtual world were blurring. She also demonstrated how artists, technology designers, curators, museums, and communities could come together to co-create and even co-curate new-media art forms in imaginative ways.

This all sounds amazing, but it pales in comparison to McGonigal's recent work, which turned gaming strategies toward pressing real-world problems. Few issues vex our present moment quite as much as energy, and with her role in designing the simulation World Without Oil, McGonigal has helped to put the strengths of massive multiplayer online gaming (MMPOG) to challenges of the world's declining oil reserves.

World Without Oil launched on April 30, 2007, and concluded on June 1, 2007. It imagined the first thirty-two days of a cessation in the production of oil worldwide. Over a hundred thousand players eventually contributed to the game, completing 1,500 complete scenarios, all with different global and local consequences. Those scenarios are all archived and can be downloaded. They are so realistic and so well informed by research, experience, and the science and sociology of environmentalism that they are now used by educators in courses at all levels and by policy makers who use them to train for informed decision making. The World Without Oil Web site argues that, because of its success, we now have proof that "a game can tap into the awesome problem-solving capabilities of an Internet 'collective intelligence' and apply them to real world problems."[3]

In March 2010, McGonigal launched a new game, Evoke, which the World Bank Institute described as a "ten-week crash course in changing the world." Evoke is designed to engage hundreds of thousands of people in a select set of problems. It is designed as a social networking game that empowers young

people all over, but particularly in Africa, to solve problems such as food short-
ages or drought survival. You participate by mobile phone because even the
poor in Africa have inexpensive mobile phones. Within its first three weeks,
fourteen thousand people were playing Evoke. Each challenge is divided into
three parts: learn, act, and imagine. Participants research a solution, act on that
solution, and then imagine what the result would be if their action could be
scaled up to a whole community, tribe, or nation. Participants choose winners
for each challenge. What the winners receive is online mentorship from world-
famous authorities in the fields of social policy and business, who give them
specific advice, suggestions, and connections that inspire them to scale their
idea. Game organizers hope that from among the Evoke graduates (as winners
are called), a new generation of leaders will be identified, trained, encouraged,
and supported as they attempt to find realistic and creative solutions to seemingly intractable problems.[4]

McGonigal insists that many problems are intractable only because we
keep trying the same failed solutions. A massive game that virtually anyone can
play offers the promise that you do not have to be rich or famous to contribute
and do good in the world. It also offers the hope that small actions, when aggre-
gated, can make a significant difference.

If reality really is broken, as McGonigal believes, can games actually fix
it? Certainly social-activist gaming is an inspired form of attention—one in
which the need to know is coupled with a collaborative seeing that makes visible
all the possibilities around the peripheries. It exposes the unique abilities that
members of the human "toon" can contribute. If we understand games in this
way, then, yes, games may well be able to help us solve real-world problems in
the future.

|||||||||||||||||||||||||||||||||

Why games? First and foremost, because right now a lot of people play them.
According to a September 2008 study conducted by the Pew Internet and
American Life Project, 97 percent of American kids aged twelve to seventeen
play digital games.[5] That's just about all of them.

Maybe there's a reason why games have captured so much of our kids'
attention in this digital age. For the last fifteen years the default position has
been to dismiss games—typically video games—as a waste of time at best and

dangerous at worst. But maybe instead of dismissing them, instead of assuming we understand the reasons behind games' tremendous popularity among kids, we should be curious. Why games? Why now? Let's take games seriously for a moment and see what we can learn from them.

To be sure, throughout human history, games have been considered an important tool for teaching complex principles and honing sophisticated forms of procedural thinking—in other words, a form of cognition that sees all parts of a problem, all the possibilities for winning and for losing, and that responds with the best set of responses (procedures) to maximize success. Games have long been used to train concentration, to improve strategy, to learn human nature, and to understand how to think interactively and situationally. In many games, the objective is to move toward a goal while remaining aware of the action happening around the edges. Games teach the skills of winning and the skills of losing and what to do with the spoils. Procedural thinking is key in ancient strategy games such as Go or chess or in the most successful modern board game ever invented, Monopoly, released by Parker Brothers at the height of the Great Depression, in 1935, and played by around 750 million aspiring capitalists since. This type of thinking is taken to an entirely new level of complexity in massively multiplayer online role-playing games, the biggest and most important of which is World of Warcraft, released by Blizzard Entertainment in 2004 and boasting nearly 12 million monthly subscribers by fall of 2008. What we might find surprising is that World of Warcraft isn't just for kids. As of 2010, the average age of a player is thirty-two.

Games are unquestionably the single most important cultural form of the digital age. The average gamer in America now spends about ten thousand hours online by the age of twenty-one—about the same amount of time he might spend in school from fifth grade to high school graduation. That's also the "magic number" often cited as the amount of time one needs to invest in a difficult new enterprise to go from mediocrity to virtuosity, whether playing the violin or learning a new language. Most parents would be appalled that their child could fritter away so many hours on games, but game designers like McGonigal would counter that we should be harnessing that energy in the service of learning, in the classroom and outside it in the world.

Given how much heat video games continue to take, McGonigal has an uphill battle. After all, aren't video games what we *blame* for destroying our children, for dumbing them down and leading them astray? For over a decade, we've had report after report saying video games are bad for kids. Yet what's becoming increasingly clear is that much of what we know about the impact of video games is shaped less by the games themselves than by historical events that shaped our culture's views of online gaming.

Until the Pew report came out in 2008, one had to go all the way back to the 1980s and 1990s to find real studies of games. By "real studies," I mean ones that were designed to measure actual experience of game play and what those experiences and effects might be, rather than studies distorted from the outset by fears and value judgments. Like Francis Galton's end-driven calculations, most of the game studies from the early twenty-first century are framed not to find out *if* games are good or bad for our children, but *how* bad the games are for them. In other words, there is a definite negative bias in the design of most of the classic studies of the harmful effects of video games conducted in the last decade.

But strangely, that bias is *not* present in studies done during the previous decade. Why is that? In a word: Columbine. That horrific school shooting, with its images of young boys armed to the teeth and dressed like assassins in violent video games, on a quest to pop off their schoolmates, brought much of the research on the cognitive, learning, and attentional benefits of video games to an abrupt halt. In the years after Columbine, virtually all studies of games begin with the assumption that the effects of games are intrinsically negative; typically, the point of the study is to find out more about how pervasive this negative effect is, what repercussions it is having, and implicitly how knowing this can help us cure the problem. Games were in a category of known negatives, like dropping out of school, breaking and entering, or bullying.

I am not suggesting that the studies were false or wrong. It is simply that, as we know from Duncan Germain's sixth graders at Voyager, most scientific studies begin with the statement of a problem. That's what a hypothesis is. The study then tests that hypothesis. We know from studies of attention blindness that what we focus on is what we see. If our studies focus on the negative impact of games on individuals and groups, we will find out a lot about the relative harm caused by games on individuals and groups.

Studies of video games, in other words, tend to be structured with gaming assumed to be the known problem to be addressed, if not solved. The point of the study is not to find out everything one can possibly discover about how video games are influential. Since Columbine, we've spent a lot of research dollars and time looking for the negative moral, social, psychological, and even physiological effects of video games. We've been trying to find out how bad "the problem" is, in what ways, and what is necessary to cure it.

Yet if we go back and look at the early research on gaming from the 1980s, now mostly forgotten, we see that the whole structure of the research around video games took a different form. Back then, television was considered too passive a medium to be good for the brain. It was considered too silly to help one's concentration and too debased or violent to be good for one's moral development. In fact, as would later happen with games, many studies of television "prove" it is bad for brain development, attention, and morality. From this perspective, the active play of first-generation video games seemed promising. It was a potential solution to the couch-potato syndrome.

So in the 1980s we find literally dozens of reports that ask about the potential developmental, cognitive, and learning *benefits* of video games. And, lo and behold, these studies find answers to that question, and mostly positive answers.

These studies are based on early video games such as Space Invaders (1978) and Pac-Man (1980).[6] A study from 1986, for example, shows how playing video games improves not just visual perception but visual judgment—our ability to extrapolate from what we see to an appropriate decision or an action. Playing video games improves several different ways of processing visual data, improving peripheral visual recognition, spatial accuracy, and focus. Most striking, it does so not only *during* game play but also afterward, once the game has been turned off. It is as if repetitive playing of video games improves our ability to see across time and space, to both see and predict patterns while moving, and to be able to see things that are moving around us. It helps to train our eyes for better tracking, in other words, not just in the game itself, but in the real world too.[7]

Other research found improved hand-eye coordination, enhanced physical dexterity, and more precise fine motor movements as a result of game play, again not only during the game but after it. Still other studies pointed to improvement (sometimes quite dramatic) in reaction time, short-term and working memory, rapid evaluation of positive and negative inputs, and in games played with others, cooperation and team-building skills.[8] Other studies record improved depth perception and night perception, dynamic spatial skills, and the ability to translate from two- to three-dimensional judgment—a crucial skill in transforming any kind of blueprint into a physical object (whether it's an architectural plan or a dress pattern).

Significant increases were also measured in what is called the mental rotations test, a component of IQ tests that requires extrapolating what a three-dimensional object will look like if it is rotated on its axis. That was big, exciting news back in the distant twentieth century, when gaming was being championed by numerous scientists for its potential to make us smarter.

What about attention? In the studies of the late 1980s and early 1990s, video-game players were found to respond faster and more accurately than nonplayers to both the easiest and most difficult attention-stimulus tests.[9] As psychologists C. Shawn Green and Daphne Bavelier note, game playing greatly increases "the efficiency with which attention is divided." Simply put, that means that gamers showed improved ability in what we would now call multitasking. They showed improvement relative to their own multitasking skills before they played games, and they improved relative to nongamers, too.[10] These early studies also revealed that game playing makes gamers more able to respond to unexpected stimuli: a new antagonist appearing on your digital visual field, a child running out into the road in front of your car, or a sudden drop in your company's capitalization.

In the 1980s, when studies of the positive benefits of video games were most prevalent, the term *video games* meant arcade games. The Internet wasn't yet publicly available, and access to these early games was relatively limited. One often played in public and was more or less supervised. Even after the 1989 release of Nintendo's Game Boy in the United States, game playing was not a subject of public concern. Some educators warned that too much time on the Game Boy could potentially be "addictive," but scientists were mainly engaged in testing the device's potential benefits. Cognitive psychologists and physiologists emphasized the importance of reinforcing good behavior, and games offered immediate feedback for good or bad game play. Psychologists were eager to know how repetitive play with such positive reinforcement might even lead to improvements in various human capacities, such as visual perception and spatial understanding. Educators wondered if the games could be useful for teaching 3-D concepts, such as advanced mathematics. The scientific community buzzed with excitement about improvements on spatial and motor tasks from game play, and researchers were even finding beneficial physiological changes. Game play seemed to regulate glucose levels and dopamine production, and have other positive metabolic impacts. Like the arcade games before them, those Game Boys seemed all to the good.[11]

Although schools are notoriously slow to adopt new technologies for classroom use, educators applauded the benefits of these early games. A Game Boy was considered an acceptable purchase for parents eager to contribute to the learning abilities and well-being of their child.[12] This general approval of handheld gaming devices' benefits contributed to their sales and was aided by the fact that, in the late 1980s, Japan was synonymous with smart, next-generation, visionary electronics—and really smart kids. It was thought that America was slipping relative to Japan, and some were sure that the flashy electronics coming out of Japan were at least part of the key to that country's ascendancy and our decline.

When you go on YouTube and look at the Nintendo commercials from the time, you notice how exuberant they are in their insistence that the Game Boy can "go anywhere"—in the back jeans pocket, inside the medical lab coat, in the thick-gloved hands of astronauts in space. A source of intelligence and power as well as fun, the commercials all imply, these devices don't have to be just for kids. Adults can benefit from them too. Or in the tagline from a popular TV commercial of the early 1990s, "You don't stop playing because you get old—but you could get old because you stop playing."[13]

THE INTRODUCTION OF VIOLENT QUASI-REALISTIC narrative began to alter the public attitude toward video games. Cute shapes and colored boxes were suddenly replaced by first-person shooter games set in simulated real-life settings. The story lines in these games made visible and even interactive our society's values of power, dominance, control, and violence. Games often lack nuance, rendering a society's assumptions and prejudices in stark terms of good and evil. Who gets the points and who loses, who dies and who moves on to the next level? Some games do this with serious, educational, thoughtful intent. Others simply require shooting people who are weaker (at the game and in society) than you are.

Coupled with the hyperrealism of a new generation of strategy games, the school shooting in Columbine, Colorado, indelibly changed American attitudes toward youth, schools, and our digital age. On April 20, 1999, when Eric Harris and Dylan Klebold walked into Columbine High School and began murdering their classmates in calculated cold blood, adults raced forward to find an external cause for such terrifyingly inexplicable youth violence. Along with a preference for loud rock music as embodied by the cross-dressing Marilyn Manson, it turned out that the adolescent murderers were devotees of the first-person shooter games Wolfenstein 3D and Doom, both of which featured graphic violence that made the days of Pac-Man seem a distant past.

By the turn of the new millennium, the verdict was in. Kids coming of age in humanity's fourth great information age were being destroyed in every way and on every level by their obsession with computer games. Although the official reports on Columbine by the Secret Service and the Department of Education emphasize both the extreme rarity of targeted school shootings (one in a million), and the absence of any one indicator that a child will move from persecuted to persecutor, the mainstream media were quick to cite video games as a cause of school violence. While the Secret Service report paid far more attention to access to weapons than to any presumed influence from violence on the news, in the papers, or in other media (whether television, movies, or video games),[14] the finger-pointing at video games was intense. Was this because mass media perceived video games to be a threat to their own commercial interests? Or just an ages-old tendency to blame social problems on new technology? Probably a bit of both.

The parents of some of the victims at Columbine filed a class-action lawsuit against Sony America, Atari, AOL/Time Warner, Nintendo, ID Software, Sega of America, Virgin Interactive Media, Activision, Polygram Film Entertainment Distribution, New Line Cinema, and GT Interactive Software. The complaint alleged, "Absent the combination of extremely violent video games and these boys' incredibly deep involvement, use of and addiction to these games and the boys' basic personalities, these murders and this massacre would not have occurred."[15] The lawsuit was eventually dismissed, on the grounds that computer games are not "subject to product liability laws," but within the popular imagination, the causal connection has remained powerful.

By the year 2000, games were being scapegoated as the cause of many supposed youth problems. There was a general condemnation of violent video games and, more, a consensus that Resident Evil, Grand Theft Auto, and Final Fantasy had turned nice boys and girls into know-nothing misfits capable of pathological violence. References to serious disorders of attention and learning disabilities were part and parcel of the critique. It's not surprising that, along with the 2002 passage of No Child Left Behind, with its emphasis on multiple-choice testing and a standardized body of required knowledge, there was also a widespread disparagement of video games and talk of banning them in schools or for kids. The language was all about containment: restriction, enforcement, toughening up, regulating, and controlling seemed to be necessary for America's wayward younger generation.

As is often the case, these social anxieties were quickly transformed into studies by the scientific community. Research funding that had previously gone to studying the potentially positive benefits of games on attention, motor skills, and visual perception was replaced mostly by funding for research into the potential negative side effects of the games. Research funding from many different agencies, from the National Institutes of Health to the Department of Education to private nonprofit and philanthropic foundations, was funneled to studies of games' harmful effects. Following Columbine, study after study posited corrosive moral, social, and attentional effects of video games—and, by extension, the Internet—on our children's lives.

In a transformation that has happened often in the last 150 years of scientific and empirical study, the paradigm had switched. Typically, a major social or technological attitude change occurs and research follows it. Then there is another change, and a new line of research is developed in a different, sometimes competing direction. In the first decade of the twenty-first century, we see pundits claiming that technology destroys our brain, especially in children. We see the tendency in a spate of books with alarmist titles and subtitles, such as Maggie Jackson's *Distracted: The Erosion of Attention and the Coming Dark Age*, Mark Bauerlein's *The Dumbest Generation: How the Digital Age Stupefies Young Americans and Jeopardizes Our Future (or, Don't Trust Anyone Under 30)*, or Nicholas Carr's *The Shallows: What the Internet Is Doing to Our Brains* (the follow-up to Carr's earlier article "Is Google Making Us Stupid?").[16] The assumption here is that kids are being ruined by technology. They are disabled, disordered—and disorderly. They can't sit still anymore. They need special care and special treatment. They need special diagnosis, special *testing*.

Not surprisingly, the number of reported cases of attention disorders—and the variety of the diagnoses—soared in the United States during this decade. Some studies suggest that up to 25 percent of those in high school today have some actual form of or a "borderline" tendency toward attention deficits or learning disorders.[17] The language in these studies is similar to that of infectious disease, with video games, like mosquitoes, passing the malarial attention infection from child to child. No one is safe.

With these rising numbers as well as with the hyperbolic press, we have to wonder whether we really know that our kids are being overwhelmed by an attentional "epidemic" driven by video games. Based on what evidence?

Very little, actually.

First, there is little to suggest that "kids today" are more violent because of video games. There isn't even evidence that kids today are any more violent than kids yesterday. Demographically, the opposite is the case. This is the *least* violent generation of kids in sixty years, perpetrating less violence against others or themselves. Because violence of both types correlates strongly with poverty, the declining numbers are even more significant if one adjusts for changing poverty rates as well. Youth violence is still too high, of course, and young people between eighteen and twenty-five remain the group most likely to meet violent deaths. But the numbers of such deaths are lower now proportionally than they were in 1990.[18]

The numbers have also declined for depression, mental health disorders, and self-inflicted violence. The most comprehensive longitudinal surveys are offered by UCLA's Higher Education Research Institute, which has been surveying entering college students, hundreds of thousands of them, since 1966. The institute's surveys indicate that the percentage of college freshmen describing themselves as "frequently depressed" has fallen from a peak of 11 percent in 1988 to 7 percent in 2005 and 2006. Suicide rates among youth fell 50 percent over the same time period and "self-destructive deaths" from drug overdoses, accidental gun accidents, and violent deaths classified as having "undetermined intent" have declined more than 60 percent in the same population over the last thirty-five years. Other surveys indicate that this generation has the least drug use and the lowest rate of arrests over that same time period.

By no quantitative, comparative measure are the digital-era kids, those born after 1985, bad, endangered, vulnerable, or in other ways "at risk," even in numbers comparable to earlier generations. The kids are all right. *And* 97 percent of them play games.

If kids cannot pay attention in school, it may be less because they have ADHD and more because we have a mismatch between the needs and desires of students today and the national standards-based education based on the efficiencies of a classroom created before World War I. Are kids being dumbed down by digital culture or by our insistence on success and failure as measured by testing geared to lower-order and largely irrelevant item-response skills? Are video games the problem, or, as I suspect, is the problem our widespread abdication of the new three *R*s (rigor, relevance, and relationships) in favor of antiquated testing based on rote memorization of random facts that has little to do with the way kids *actually* read, write, and do arithmetic online—or will need to do it in the workplace?

||||||||||||||||||||||||||||||||

Of course some video games are appalling in their violence. And of course there are horrific school shooters who play video games. If 97 percent of teens are now playing games, there is likely to be, among them, someone seriously disturbed enough to perpetrate a tragedy. Or to win eight Gold Medals in swimming at the 2008 Olympics. Or to do just about anything else. If nearly everyone is playing video games, we're not talking about a social problem anymore. We're talking about a *changed environment.*

The Pew study that generated the 97 percent figure is the first in-depth ethnographic study in the twenty-first century of how kids *actually* play games, not of how we think games affect them. It shows that kids who spend time playing video games report more, not less, time for their friends; 65 percent of teens say that they play games with others in the same room, and 76 percent claim they help others while playing and get help too. And they seem to spend more, not less, time on civic engagement than kids past; 44 percent report learning about social problems from their games, 52 percent say they learn about moral and ethical issues from their games, and 43 percent say games help them make decisions about how their community, city, or nation should be run. Absorption in games doesn't contradict social life, civic engagement, focus, attention, connection with other kids, or collaboration. On the contrary, it seems to promote all of the above. As Professor Joseph Kahne, the leader of the Pew study notes, "We need to focus less on how much time kids spend playing video games and pay more attention to the kinds of experiences they have while playing them."[19]

By all statistical measures these digital natives happen to be the happiest, healthiest, most social, most civic-minded, best adjusted, and least violent and self-destructive teens since large demographic surveys began at the end of World

War II. They're having fun and learning too.[20] Yet as a society, we tend to act as if we do not believe this is true: We limit our kids, test and retest them, diagnose them, medicate them, subject them to constant surveillance, and micromanage them as if they are afflicted with a debilitating national disease.

DESPITE THE TURN THINGS TOOK after Columbine, much of the research on video games done in the 1980s and 1990s has had a tremendously positive and lasting impact—just not for kids. One of the great ironies, and perhaps one could even say tragedies, of the early twenty-first century is that the fear of video games, however well intentioned, has made them virtually off limits to educators even as adults have benefited profoundly from them.

Many of the findings on the cognitive benefits of game play have been translated with great success from the potential of games for teaching kids to the potential of games for teaching adults. For example, game mechanics have been incorporated with very positive results in learning simulators that have been used for over a decade to supplement professional training. Learning simulations are now commonplace in the training of architects, engineers, mechanics, airplane and ship pilots, machine operators, draftsmen, surgeons, dentists, coal miners, racecar drivers, nurses, lab scientists, astronauts, and dozens of other professionals.

The military has also been quick to employ sophisticated games for training, testing, and recruiting its members. America's Army is widely regarded as one of the best, most exciting, and most cognitively challenging of all games. It is entertaining but it is also a learning game. The military firmly believes (and has for upwards of two decades now) that superior gamers make superior soldiers—and vice versa. The military is one sector that has continued to do research on the cognitive benefits of games, but its attention has focused, of course, on adults rather than kids.

The impressive findings from the studies of games done in the 1980s and 1990s also prompted the development of a variety of tools for so-called lifelong (i.e., adult) learning. You can now buy programs and devices based on game mechanics that will help you learn everything from a second language to musical sight reading, bird-watching, knitting, and even introductory computer skills.

There has been another powerful use of this early research as well. Several of these studies point to the far-reaching, positive effects for the elderly of playing video games. This goes back to that early television commercial warning that you don't have to stop playing games when you get older but that you can grow old if you don't continue to play Nintendo. Game mechanics are the basis for hundreds, perhaps thousands of simulation and rehabilitation programs used for everything from stroke recovery to postoperative adjustment after eye or ear surgery to cognitive adjustment to computerized hearing aids and cochlear implants. They've also been used to help patients understand and use prosthetic limbs or electronically enhanced wheelchairs.

On the nonprofessional medical level, there has been a boom industry in games designed specifically to aid the middle aged and even the elderly in memory retention and enhancement. Brain-Age and other games for oldsters promise to make your brain young again. Research suggests that, if used correctly, to stretch rather than to habituate patterns, these interactive, level-based games can, indeed, show benefits.

Imagine what schools would be like today if the design work that has gone into fighter-pilot simulators and the splendid interfaces in geriatric games had gone into challenging and stimulating learning games for schoolkids? Quest 2 Learn gives us a precious glimpse, and we can take heart in the fact that the school has grown so popular as to necessitate a lottery system. I hope the day will come soon when kids can be guaranteed an education that good without having to hit the jackpot.

But take heart, games are on the march. Since 2007, when I began codirecting the HASTAC/MacArthur Foundation Digital Media and Learning Competition, I've seen not just dozens but hundreds of projects that use digital means to advance and enhance classroom learning. One of the winners of our first competition recreated America's Army as Virtual Peace. Working with the makers of America's Army, Timothy Lenoir created a simulation game that is fast, challenging, and fun—but which focuses not on violence but on conflict resolution.[21] Students, teachers, game developers, computer scientists, and policy experts worked together to develop the scenarios as well as to create templates for future games that will be available, along with their open source code, to educators and the public. To win at Virtual Peace, you have to understand all of the variables of the countries in conflict, the roles of all of those negotiating the conflicts, the culture of argument and negotiation among the parties, the languages of the parties involved, and many other factors. Lenoir is now working with the renowned global health physician-activist Paul Farmer and his organization, Partners in Health, to create a version of Virtual Peace set in Haiti. It will be used by relief workers, doctors, educators, both from Haiti and abroad, to help understand the challenges of renovating that devastated country, and it will be available for educators to use in high school civics classes.

THE TIME NOW IS RIGHT to go back and reconsider all those pre-Columbine studies. We need research on the benefits to people of *all* ages offered by this challenging, test-based gamer way of thinking. If gaming absorbs 97 percent of kids in informal learning, it can be an inspiring tool in schools as well—including, potentially, an assessment tool. I don't mean measuring how well the game scores of U.S. students stack up against those of other nations. I mean that games could provide teachers, students, and parents with an ongoing way of measuring learning, with the game mechanics themselves designed to give feedback to the students about how they are doing as they are doing it. If Sackboy and Sackgirl don't do well, they are stuck on the same level for the next round of game play. That's boring. Any kid will tell you that boredom—that is, avoiding it—is a great motivator in the life of a kid. So to escape boredom, you have to learn more, try harder, and then you not only "win," but more important, you progress to the next level of play, with ever more scenarios, ever more challenges to entertain you but also keep you learning.

We might all have benefited over the last two decades if we'd done a better job of thinking through the positive and creative ways we might actually use the technology to which we are tethered for so much of our waking lives. It's salutary, surely, to think that this computer interfacing with my fingertips is actually improving me in ways cognitive and even social.

Games provide something else, too: *flow.* Mihaly Csikszentmihalyi, the distinguished psychology professor heading the Quality of Life Research Center at the Claremont Graduate University, has written extensively about the sublime happiness that emerges from certain creative activities that engage you so completely that you lose track of time and place. Flow activities, as Csikszentmihalyi calls them, assume that the brain multitasks on its own. Such activities provide multiple forms of multisensory stimuli in situations that require total mind-body activity, testing your mental, emotional, and physical skills simultaneously. As you become more proficient in these activities, your brain is suffused with blood flow, pumped through with energy, clarity, and optimism.

Csikszentmihalyi has concluded that TV watching leads, in a neuro-anatomical way, to low-grade depression. By contrast, he defines as ideal flow activities playing chess, rock climbing, dancing to intensely rhythmic music (specifically, he says, rock music), and performing surgery.[22] To this list, many ardent gamers would add game playing as a fifth all-consuming flow activity. To those who do not game, the teenager playing World of Warcraft may well seem distracted, inattentive, and unproductive. But when that teenager is playing well, he is *in the flow* and far more efficient, attentive, engaged, creative, and *happy* than Dad in the armchair across the room watching Fox News at the end of the workday, with the newspaper in his lap.

In a recent issue of the *Harvard Business Review,* two major thinkers have even argued that the teenage gamer might signal the skills needed for potential leadership in politics and commerce for our age. John Seely Brown, one of the earliest and perennial visionaries of the information age, and Douglas Thomas, a communications professor at the University of Southern California and himself a champion gamer, have noted that multiplayer online games are "large, complex, constantly evolving social systems." These researchers have defined five key attributes of what they term the "gamer disposition":

1. *they are bottom-line oriented* (because games have embedded systems of measurement and assessment);
2. *they understand the power of diversity* (because success requires teamwork among those with a rich mix of talents and abilities);
3. *they thrive on change* (because nothing is constant in a game);
4. *they see learning as fun* (because the fun of the game lies in learning how to overcome obstacles); and
5. *they "marinate" on the edge* (because, to succeed, they must absorb radical alternatives and imbibe innovative strategies for completing tasks).[23]

It is hard to think of five better qualities for success in our always-on digital age.

Even if kids are natural users of the digital (and by *natural* I mean that they have learned how to interface with computers from a very early age), they are not perfect users. Yet game play is about working harder and harder, receiving constant feedback on your progress, and progressing to the next level when you've mastered the last one. It is about striving for perfection. There is no bell curve in game play. Given that motivation, it is possible that video games are an ideal preparation for the interactive, iterative, multitasking, and collaborative world into which our kids are coming of age, a world they will need to navigate, lead, and, by their leadership, transform.[24]

Given this generation's interest in everything from global finance to global warming, today's young digital thinkers could well end up saving our world. Their paradigm of success is less the old management model of business schools than the gamer ideal of daring and sometimes even gonzo strategizing—working together as a team toward innovation on the global scale. The gamer model of success is also remarkably collective, grassroots, communitarian, coordinated, and distributed. If it can be translated into the future world of work—in the form of digital and social entrepreneurship of the kind Jane McGonigal espouses—it is possible that youth will be able to strategize solutions that work and that, in the end, are also rewarding and possibly even fun.[25]

And their motivation is the "magic circle," another gamer term for the way the mechanics of a great game capture your attention so completely that you never want to stop. Games resemble the endless curiosity of a small child asking, "Why?" Everything is a puzzle to be solved. Everything is a source of curiosity. Every answer leads to another question—and another quest. Can these kids who have inherited our digital age fix the realities that are broken? Do their passions and experiences prepare them for the future they have inherited and that, very soon, they will help shape?

|||||||||||||||||||||||||||||||||

Now a septuagenarian, Alvin Toffler is still thinking about the future, just as he was back in the 1970s. When recently asked what he thought was the "most pressing need in public education" right now, he answered, "Shut down the public education system." He noted that Bill Gates thinks the same thing.[26]

I'm not sure if we need to shut our schools down, but we need a time-out and we need a time of systematic unlearning. We need to rethink what we want from education from the bottom up. We need to build a system of education that prepares our kids for the twenty-first century, not the twentieth.

Everything about education today is based on a hunt and peck mode of understanding—A, B, C, D, all of the above, none of the above. As we've seen, compared to the creative ways the best teachers teach, our classrooms are generally among the most intellectually straitjacketed environments our kids enter. What a missed opportunity! Graded, segregated by topics and subjects, and

with EOG assessments based on multiple-choice answers and rote memorization of facts, most education is still stuck, institutionally and instructionally, in
the ideology and the methodology of the industrial age. We have done little to
transform education in view of the world of information at our kids' fingertips.
We've done little to make them understand, like those kids playing LittleBig-
Planet on the snow day, that exuberant, engaged fun also counts as learning.
We're underestimating our children. We are shortchanging rather than challenging them.

If kids are enthralled with games, why not make exciting ones that also
aid and inspire learning? Good learning is about inspiring curiosity. Good gaming is the same. We fret a lot about how our kids are falling behind, especially
in science and math. Yet fostering standardization is exactly inimical to the
scientific, inquiry-based, deductive and inductive reasoning skills that good science demands. With ever more rigid standards, all we are doing is killing the
spark of curiosity that virtually every child brings into the kindergarten classroom. I would argue that we are especially strangling the exploratory, learn-by-
doing, testing, retesting, modifying, rejecting, revising, and trying again form
of thinking that *is* the scientific method.

Game play—procedural, strategic thinking—is far more conducive to
inspired science making than is cramming for end-of-grade tests. The more we
learn about children's cognitive development, the more we realize that kids have
far greater complexity of thought at earlier ages than we previously assumed.
Game play not only requires complex social and conceptual mapping, but also
the mathematical skills that we (in the West) previously thought could not
be learned until adolescence. New work coming out of a series of preschools
being pioneered in some of the poorest urban areas in cities, including Boston,
Washington, and Nashville, has been showing that conventional ideas about
children's readiness to learn geometry, reading, language, and even self-control
in the classroom are simply wrong.[27] Cognitive neuroscientists have shown
that kids can learn these concepts as long as the concepts are incorporated into
inquiry-based learning projects, like learning games.

This really should not be so shocking, for in countries like China or India,
high-level math-based methods are taught as part of rhyming games even at
the toddler stage. The math then becomes more and more complex as the child
plays the games more often and becomes more adept at them. There are verbal
and physical variations on Nim, the ancient mathematical strategy game that

began in Asia (some say China, others India) and that moved throughout the world the way cinnamon and indigo did, as part of ancient intercultural flows. In Nim, players take turns removing objects from a heap in a way that leaves the opponent with the last piece. Complex calculating skills are also taught in the play of three- and four-year-olds among many indigenous peoples in Latin America and elsewhere, often using systems of knots made in cords, with accompanying songs, dances, and other forms of play.

The key component in these examples, as with all the examples of excellent education we've seen, is learning as play and tests as challenges. What better gift could we give our children than to learn to enjoy as they grapple, to aspire higher when they stumble, to find satisfaction as the challenges become ever greater? As the adage goes, the person who loves to work never has to.

All over America there are exceptional teachers who, against odds, are trying to find ways to motivate and inspire kids even as they have to prepare them for the end-of-grade tests that embody the antithesis of motivated, inspired, challenge-based, interactive learning that students experience in their daily digital lives. The HASTAC method of collaboration by difference, which we used in the iPod experiment and which I used in This Is Your Brain on the Internet, elevates to a university-level pedagogical method the chief exploratory and interactive mode of the digital age. It's the same method Katie Salen uses at Quest 2 Learn gaming school in New York, the same method used by the late Inez Davidson in her three-room schoolhouse in Mountain View, Alberta, and by Duncan Germain in Creative Productions. We can learn from all these teachers.

Our kids *are* all right. Despite our national educational policy, many kids *are* getting the kind of dexterous, interactive learning they need. Online, in games and in their own creative information searching, they are learning skills that cannot be replaced one day by computers. They are mastering the lessons of learning, unlearning, and relearning that are perfectly suited to a world where change is the only constant. They are mastering the collaborative methods that not only allow them to succeed in their online games but that are demanded more and more in the workplace of their future.

Part Three

||||||||||||||||||||||||||

Work in the Future

6

||||||||||||||||||||||||||||

The Changing Workplace

If there's one place where the question of attention raises even more of a furor than when it comes to our kids, it's at work. From the spate of books and articles on distraction in the contemporary workplace, you would think that no one had accomplished an honest lick of work since that fateful day in April 1993 when Mosaic 1.0, the first popular Web browser, was unleashed upon an unwitting public. The Internet, the pundits warn, makes us stupid, distracted, and unfit to be good workers. Yet according to the international Organisation for Economic Co-operation and Development (OECD), which has been measuring these things since World War II, the opposite is the case. Worker productivity has increased, not decreased, since the advent of the Internet and continues to increase even despite the global economic downturn.[1] In addition, every study of American labor conducted in the last five years shows that we work more hours now than our hardworking parents did—and they worked more than their parents did too. Most of us *know* this has to be true, and yet few of us really *feel* like we're doing better than grandfather, that icon of the tireless American laborer. "Work flow" just doesn't feel productive the way "work" was purported to feel. Why? Why do most of us feel inefficient when we are working as hard as we can?

As we have seen, most of the institutions of formal education that we think of as "natural" or "right" (because they are seemingly everywhere) were actually crafted to support an idea of productive labor and effective management for the industrial age. We have *schooled* a particular form of attention based on a particular set of values for the last one hundred years, with an emphasis on specialization, hierarchy, individual achievement, the "two cultures," linear thinking, focused attention to task, and management from top down. Those values may well have matched the needs of the industrial workplace. Now that the arrangements of our working lives are changing in a contemporary, globalized workplace, those modes of training aren't doing us much good. Why are we still hanging on to institutions to support a workplace that no longer exists the way it did a hundred years ago? If the current world of work can be frustrating and perplexing, it is for the same reason that parents, teachers, and even policy makers feel frustrated with schools: There is a mismatch between the actual world we live, learn, and work in and the institutions that have been in place for the last hundred years that are supposed to help us do a better job in that world.

The best way to answer the question of why we feel inefficient despite working to our limits is to take our anxieties seriously. We need to analyze their source and propose some solutions. The era we are living in is way too complicated to reduce to the current "Internet is making us stupid" argument or to its utopian counterpart ("Internet is making us brilliant"). We all know that neither generalization tells the whole story of our working lives. Both hyperbolic arguments reenact the attention blindness experiments from one or another point of view.[2] If you insist the Internet is the source of all trouble in the workplace, you miss one of the most remarkable achievements in all of human history, not only the invention of the Internet and the World Wide Web, but also the astonishing transformations we have all undergone, as individuals and as societies, in less than two decades. Who even thought change of this magnitude was possible so extensively or so soon? On the other hand, if you think only in terms of the "wisdom of crowds" and the bountiful possibilities of "here comes everybody," you trivialize the very real challenges that nearly all of us face as we try to adapt life-learned habits of attention to the new modes of communication and information exchange that have turned our lives around and, in some cases, turned them upside down.

Instead of going to either extreme, in thinking about how we might succeed in the contemporary workplace and what might lie ahead for our work lives in the future, let's take seriously the causes of distress in the digital workplace and take the time to examine how the new workplace requires different forms of attention than the workplace we were trained for—or even than the workplace for which, ostensibly, we are training our children. We will also look at people who have developed methods, small and large, for improving their workplaces and whose insights might help us transform our own workplaces to take maximum advantage of new, interactive and collaborative possibilities.

It's about time. We need a Project Workplace Makeover as desperately as we need to make over our classrooms. If your office is anything like mine, it is long overdue for renovation. I am guessing that yours, like mine, retains the basic architecture, organizational structure, personnel rules, management chart, and conventional work hours of the twentieth-century workplace. Except for the computer in the office, you would think the Internet had never been invented.

No wonder we so often feel distracted and unproductive! How could it be otherwise when there is such a radical mismatch between workplace and work? We've created structures to help us focus on a workplace that no longer exists. We're preserving old standards for measuring our achievements and haven't yet made the changes we need to facilitate our success in a global economy that, everyone knows, has rearranged some of the fundamental principles of how the world does business. Is the real issue of the twenty-first century, as some would have it, that the Internet makes us so shallow we can no longer read long novels?[3] Is that the real crisis of the contemporary workplace? I don't think anyone who works for a living really believes that. But as a culture, we certainly are paying a lot of attention to this very literal-minded way of measuring what our brain does or does not do as well as it purportedly used to. Once again, our metric is too belated and our optic for looking backward is short-sighted. We are so fixated on counting basketballs that we're not seeing the gorilla in our midst.

The gorilla isn't the Internet, but rather a fundamental contradiction between how we actually are working today and the expectations we have for our workplace and our workday. About 80 percent of us who are employed still spend some or part of our day away from home in an office. But although the walls of that office may not have moved, what an office is, in functional terms, has altered dramatically. More and more of us spend our days at our "desktops" rather than our desks. We might close the office door, but that merely shuts out the distractions in the corridor. It's our desktop itself that connects us to everything on the World Wide Web. The center of our work life is a machine whose chief characteristics are interactivity and ease of access. The Internet connects us to everything that is necessary to our work at the same time that it connects us to everything that isn't about the task we are doing in a given moment. A workplace is supposed to support us in our work endeavors, to help us focus on our work in order to support our productivity. Yet very little about the workplace of the twenty-first century has been redesigned to support the new ways that we actually work together. That is some gorilla!

LET'S STEP BACK A MOMENT and think about what a computer actually is before we even begin to tackle the question of attention in the contemporary workplace. Because the information revolution has happened so quickly and so thoroughly, it is easy to take it for granted. Let's slow the hyperbolic dumber/smarter Internet conversation way down and really think about the nature of the machine that has come to dominate our lives at work and at home.

To start with, a computer is not like any other piece of office equipment. Most machines in the workplace were designed with a single primary function. A typewriter types, a copy machine makes copies. A computer computes, of course, but it is also the repository of all of our work all the time, plus a lot of our play. The same computer gives us access to spreadsheets, the five-year audit, the scary corporate planning document, our client files, invoices, work orders, patient histories, and reports, as well as forms for ordering new pencils or even the mechanism for summoning the tech guy to come fix the computer. Whether we work in an automobile repair shop, a newspaper office, a restaurant, a law firm, a hospital, or even a lonely fire lookout tower in the Grand Tetons, more and more of our work depends on bandwidth. I used to joke that the only businesses today that weren't part of the digital world were in Amish country. Then, a few weeks ago, I happened upon an Amish grocery store, part of a small Amish community in rural North Carolina. The whole-grain flour I bought was hand-milled using techniques unchanged in the last two hundred years, but the cash register instantly tracked my purchases, updated the store's inventory, and sent the data to suppliers shipping Amish goods worldwide. So much for that joke.

The computer connects us to every single work task we do, past and present, but it also links us to our colleagues and their questions and quibbles. At the same time, it keeps us attached to every joy and problem in our personal life, to family, to community, to entertainment, shopping, and hobbies, to all of our habits in the workplace but also outside it. It's a lot to ask that a person stay on task when the very same keyboard holds all the distractions of everything else we know we should be doing, plus all the temptations of everything we know we shouldn't.

We've inherited a sense of efficiency modeled on attention that is never pulled off track. All those little circles to fill in neatly on those multiple-choice exams are the perfect emblem of our desire to instill a sense of "here's the problem" and "here's the solution," as if the world of choices is strictly limited and there is always one and only one right answer to be selected among them. If ever that model of workplace attention and problem solving sufficed, it is hard to imagine anything less like the limitless digital possibilities we juggle now. Single-minded focus on a problem doesn't remotely correspond to the way we *must* work if we are going to succeed in the twenty-first-century workplace.

It is not that the Internet is distracting. It is a fact of twenty-first-century life that we now work in a jumpier, edgier, more shape-shifting way than we did a decade ago. That way is required by, facilitated by, and impeded by the Web. Broadcast media are being displaced by participatory media, with the number of daily newspapers declining and the number of blogs escalating. Encyclopedias are replaced by Wikipedia. We're now up to 2 billion Google searches a day—and 4.5 billion text messages.[4] To say we live in an interactive era is a gross understatement. It's just missing the point to say all that connection makes us unproductive when connection is, for most of our endeavors, what also makes us productive. The principal mechanism of our productive labor is also the engine of our distraction.

That would seem to be an impossible pairing. Yet we're doing it, and according to the OECD, we're doing it well enough to rack up higher scores for our productivity than ever before. How are we managing this? Sometimes it's by burning the midnight oil, returning to our work after dinner to wrap up the report that got waylaid somewhere between the worrisome financial projections from head office, the plans for Mandy's baby shower, and the crazy YouTube videos everyone was sending around the office listserv today.

So those are the two interconnected traps of our workplace, the Scylla and Charybdis of contemporary attention. On the one hand, having the mechanism of distraction so close at hand hurts our efficiency at work. On the other hand, having those very same properties bundled in one computing device ensures that work invades our homes, our weekends, our vacations. All the old markers of what was supposed to be leisure time now bleed into the workday, while work now bleeds into our leisure time. If the Internet distracts us at the office all day, it also sends work home with us each evening to encroach on our family and leisure time.

The complaint about too much distraction on the job is the flip side of worrying that there's no escape from work. The executive on the beach with his PDA or the lawyer recharging her laptop for the next leg of a transcontinental flight are as much symptoms of our transitional age as is the fear of too much distraction. If there is occasion for leisure as well as work on the laptop at the office and occasion for work as well as play on the laptop at home, where are the boundaries anymore? Where does work stop and leisure begin?

As we've seen with schools, the last hundred years were spent developing institutions that were suited to productivity in the twentieth-century workplace. If you have a job just about anywhere but Google, you are most likely working in a space designed for a mode of work that is disappearing. We're all Amish in the sense that we juggle contradictory ideas of the workplace these days. We've just begun to think about the best ways to restructure the industrial labor values we've inherited in order to help us maximize our productivity in the information age.

The science of attention is key to the workplace of the digital future. If we understand more about how we have been encouraged to pay attention for the last hundred years and how we need to pay attention now, we can begin to rethink work in the future. By looking at new ways that individuals and even corporations have begun to refocus their efforts for the digital age, we can start to rearrange the furnishings of the contemporary workplace in order to better manage the work-leisure challenges on our desktop—and in our lives.

||||||||||||||||||||||||||||||||

Gloria Mark knows more about how we pay attention in the information-age workplace than just about anyone. Professor of informatics at the Bren School of Information and Computer Sciences at the University of California–Irvine, she studies the many different ways we work effectively and the ways we don't. With an army of students, she's been clocking the work habits of employees, especially those who work in the information industry, measuring their distractions in hours, minutes, and even seconds. She's discovered that the contemporary worker switches tasks an average of once every three minutes.[5] Often that change isn't just a matter of moving from one job to another but requires a shift in context or collaborators, sometimes communicating with workers around the globe with different cultural values and who do not speak English as a first language. Perhaps not surprisingly, once interrupted, the workers she's studying take nearly twenty-five minutes to return to their original task.[6]

The high-speed information industry may be an extreme case, but that's her point. If we're going to understand attention in the contemporary workplace, why not dive into the belly of the beast and see how those who are designing the information age for us are managing? If they can't stay on task for more than three minutes at a time, what about the rest of us? Are we doomed? That is certainly a central concern of our era.

Mark's research doesn't stop with measuring multitasking but looks more deeply into *how* our attention is being diverted from one thing to another. She finds that, even in the most high-speed, high-tech, multitasking, app-laden, information-intensive workplaces on the planet, 44 percent of workplace interruptions come not from external sources but from internal ones—meaning that almost half the time what's distracting us is our own minds.

That is really interesting. Think about it. If I leave the office and don't feel that I've had a productive day, I might blame too much e-mail, annoying phone calls, or coworkers bursting into my office unannounced, but almost half the time I'm actually distracting myself, without any real external interruptions to blame. I need no goad for my mind to wander from one job to another or from work to something that isn't work—surfing the Web, gaming, checking out my Facebook page, or simply walking down the hall to the water cooler or staring blankly out the window.[7]

Mark's emphasis on different forms of mind wandering, including those I instigate all on my own, is far more consistent with the modern science of attention than the more instrumental arguments about what or how the Internet "makes" us. Biologically speaking, there is no off switch for attention in the brain. Rather, our attention is trained in one direction or another. If we're paying attention to one thing, we're not paying attention to something else—but we are always paying attention to *something*. That's the way the brain is structured. Even when we're sleeping, our brain pays attention to its own meanderings. While awake, in a given moment, we may be focusing on what we should be focusing on, or we might be engaged in a reverie about tonight's dinner or the memory of last night's winning basket in the Bulls game. Fifteen minutes of daydreaming can go by in a flash because we're absorbed. We're paying attention. Time flies. Then, when we are interrupted out of our daydream, we see we've lost fifteen minutes toward the looming deadline, and we curse our distractions.

Suddenly it's a good time to blame the Internet. But blaming technology is actually a shorthand for saying that our customary practices—what we have come to think of as normal—aren't serving us in the comforting, automatic way we've come to rely on. We blame technology when we notice our habits aren't quite as habitual as they once were because we haven't yet developed new habits that serve us without our having to pay attention to them. Change is constant and inevitable. We blame technology when we aren't coping effortlessly with change. Like attention, change is what we notice when it's a problem.

It's precisely because we're always paying attention to *something* that we need a workplace optimally designed to help us focus in a way that facilitates the kind of productivity the workplace values. There is no attention or productivity in the abstract. What do we want out of the workplace? How can we structure the workplace to facilitate that goal?

We know it can be easier or more difficult to pay attention in certain situations than in others, and there's a cottage industry of industrial psychologists studying what does or does not contribute to our distraction in the workplace.[8] Given that the mind wanders all on its own a lot of the time, it makes sense to try to minimize external causes, then maybe rethink some of the possible internal motivators, too. For that, it's useful to analyze how our old industrial work habits might have been upended by our Internet connection.

Most of the physical spaces in which we work were designed in response to ideas of work developed in the late nineteenth and early twentieth centuries. Just as our methods of measuring success at school are intended to emphasize the separation of subjects, projects, age-based grade levels, achievement, ability, skills, and tasks, so, too, is the workplace constructed for separation of tasks and workers, not for connection or collaboration. Most workspaces take large, open spaces and divide them up with walls or prefab partitions. Even in offices with open, free-flowing spaces, desks (including piles of papers on desks) are often lined up to create symbolic separate workspaces.

Time, too, was divided up artificially in the old workplace. You didn't take a break when you were tired or go to lunch when you were hungry. The day was scheduled. You "got ready for work" and went to work. The prescribed workday was designed to separate time spent in the workplace from home, nature, and fun (leisure activities, entertainment, amusements, and so forth).[9] Both at school and in the workplace, break times—recess, lunch, coffee breaks—were officially separated from the serious parts of our day, often with regulations (whether enforced by the employer or championed by unions) about how much nonwork time one was to be allocated during a given day. That binary structure of *on* and *off*, work and break times, continues, for most of us, in the twenty-first-century workplace. The wired computer, however, defies segregation. Who needs a break away from the office to chat with friends, shop, or play games when all those opportunities exist a mouse click away?

The bricks-and-mortar industrial workplace was designed to minimize distractions from the outside world. Very few contemporary workplaces have rethought their mechanisms, rules, or even informal protocols in light of the myriad possibilities offered by the digital world that structures much of our day.

Even when we are doing great, either collectively (those higher and higher OECD scores for worker productivity) or individually, we live in a world where productivity is almost impossible for employers to quantify, especially if we measure by old markers. So much of what constitutes work in our times doesn't feel like measurable, productive labor.

Take e-mail. I don't know about you, but my laboring heart never swells with the pride of accomplishment after eight hours dealing with nonstop e-mail. The constant change in focus required to answer thirty or a hundred e-mails in the course of an average workday creates what management experts call a psychological toll, because it requires constantly refocusing and switching contexts. Each e-mail, with its own style and implied but unstated norms, is a miniature cross-cultural interaction that requires, again in the management jargon, code-switching.[10] No one who wants to keep a job sends the same style e-mail to her boss's boss and to that wacky marketing guy who forwards all the knock-knock jokes, but we haven't formulated even the most mundane conventions or rules for e-mail correspondence to help us out. Remember how prescriptive business letters were? Date on line fifteen, and so forth. Back in those good old days, when a business letter thunked into your in-box once or twice an afternoon, you had the leisurely pace of opening the letter, reading it, thinking of how to craft a response, researching the information needed, writing the response, typing it out, putting the letter into the envelope, and mailing it, with a waiting period before you had to deal with its consequences. That process might be the job of three different people, or a half dozen if we count the mail system and mail carriers, too. Now most of us do all these jobs many times a day. Even my doctor now examines me, decides if I might need medicine, and then, instead of scrawling a prescription on his pad, e-mails my personal information and the prescription to my pharmacy. I am certain that data entry was not part of his medical training.

We have not yet come up with etiquette, protocols, or even shared expectations for e-mail. Not only is there a stunningly greater amount of communication with e-mail than with what is so appropriately called snail mail, but no one agrees about timing. "Did you get my e-mail?" translates into "Why haven't you answered yet?" If I see that you've just updated your Facebook page from your iPhone on Saturday morning and send you a business query, should I expect an immediate answer, or are you justified in waiting until your weekend is over to send a reply?

To respond to the avalanche of information in our lives, some entrepreneurs are trying to create digital tools to help us to pay attention by managing those external disturbances from the perpetual digital flow. A software utility called Freedom, designed by developer and educator Fred Stutzman, lets you block the Internet for a specified period. If you need to work on your computer but want to be left alone for a few hours to focus only on the work on your screen, you can set Freedom to control the information flow. Or there's Rocket's Concentrate, which, for $29, you can download to your Mac and then load with only those programs you rely on for a given job. So, for example, if I have a work deadline looming, I can set Concentrate to open my preferred Firefox browser, but it might block my e-mail and close my access to distracting social networking sites such as Twitter and Facebook.

These devices may work well for some people, but I can't help thinking about Gloria Mark's research. If I use Freedom and Concentrate, how much will I be *worrying* about what may be coming in that I'm not seeing? Software programs might be useful to control the information flow to my computer but they aren't doing anything to control the flows of curiosity and anxiety that addle my brain.

We need not only new software programs but also revisions of workplace structures and rules. We also need more good, solid, empirical research on how we actually pay attention in the contemporary workplace and on what works and doesn't work to help us control our attention flow.

This is why I turned to Aza Raskin. Formerly head of user experience for

Mozilla Labs, he's now Mozilla's creative lead for Firefox, the world's second most popular browser. The Firefox story is a bit like David and Goliath. No one expected an open source, crowdsourced, collaboratively developed browser to do so well against the giant, Microsoft's Internet Explorer.[11]

I first met Aza Raskin in 2009, when he served as one of our eminent finalist judges for the international HASTAC/MacArthur Foundation Digital Media and Learning Competition. He was twenty-four or maybe twenty-five at the time, a wunderkind. He's a thoughtful, energetic person who has his mind on how to make the digital workplace a better and happier place for all of us. Because his expertise is in human-computer interaction, he is little concerned with abstract rhetoric about how we should or shouldn't pay attention or about which way to pay attention is best. His job is to design better, more efficient interfaces between our human abilities and current computer design. Rather than hectoring us on how we should or shouldn't be paying attention to our computer screens, he is trying hard to help us in that endeavor.

Right now, Raskin is developing new ways to improve the cumbersome system of tabs on our computers. His data gleaned from how millions of us keep track of our information during the course of a day shows that we rely on tabs more than bookmarks, but the tab system is a mess. It works OK for eight or ten, but, beyond that, our tabs run off the page, and text-based organization doesn't help us remember what we have open and how to get to that information easily. Raskin thinks it's evident that the product design of a standard computer has not begun to keep pace with the ways we use it. We're having to fight the machine's limitations just to do our business.

Raskin's work is all about removing those limitations. His charge is to figure out ways that the Internet can help us to do our work productively and even, he might say, joyfully. In fact, I'm not sure he would think that is a useful distinction, because cognitive science, as we've seen, shows conclusively that we do better at the things that make us happier, at the tasks that reward us emotionally and give us pleasure. So his challenge is to make the Internet more fun so we can be more effective.

Raskin scoffs at the idea that multitasking is something new under the sun. Computers require no more multitasking than, say, lassoing an injured bull in a field, fixing a power line downed during a fierce electrical storm, or keeping track of an infant and a toddler while making dinner. It's hard to imagine how any of those or other complex human activities count as "monotasking."

Attention *is* divided when various important things are happening around us to divide it. That's how attention works. So he sees it as his job to find the best ways to make the most of our cognitive capacities and everyday habits, given both the existing tools we have at our disposal and the better ones he and his team can imagine for us.

He's currently collecting data and experimenting on the ways we use visual and other sensory cues to stay focused. What if tabs weren't just strings of words but came with sounds or colors or icons or badges to help us recall what we opened earlier? He's studying how we actually use our computers, because humans, as we know from the attention-blindness experiments, are notoriously poor at knowing how we actually do what we do. Unless accompanied by a psychologist or ethnographer, most of us can't stand outside ourselves to see what makes us most efficient. Left to our own devices, we tend to go to the lowest barrier of entry, which can mean we gravitate to that which least disrupts our current habits, even if those habits don't work very well in the long run. He's convinced interface designers need to be a lot smarter about how the Internet can help us be smarter.

This, of course, prompts the one-terabyte question: How does the designer charged with remaking the Internet to make our lives easier organize his own workflow? Never mind how he organizes the innumerable data points from the billions of bits of information gleaned from 360 million Firefox users. How does he organize his own, personal workspace right now?

Raskin's main organizational computer trick is one I've heard others in the industry espouse too. He divides his tasks on two entirely separate computers. He keeps these two computers, each with its own separate work functions, separated by a small but meaningful distance. He insists that having a single machine for all functions doesn't suit the human brain very well. With the old physical typewriter, there was only one piece of paper in the machine at one time. The typewriter controlled our workflow whether we wanted it to or not. We haven't really figured out how much of a hodgepodge of old and new forms of work the computer really is. Maybe computers of the future will all come in multiples. Until then, Raskin improvises by using two computers, each loaded with different kinds of information, presorted for his work needs.

Instead of sorting by function, he sorts by hierarchies of what he wants to pay attention to. He uses his main computer to hold the major project on which he wants to concentrate on a given day. That's a first principle. He decides what he wants on the prize computer and leaves it on that machine until he is finished with it. That might be computer code he's writing, data he's crunching or analyzing, the results of an experiment he's running, or a scientific paper he's authoring. *Nothing else goes on that computer.* Adjacent to his main computer, there's an entirely separate one that connects him to the outside world—his work e-mail, his Internet connection. If he needs to move away from his programming to do some research on the Web, he physically turns to this second computer.

One thing Raskin emphasizes is that, although this seems simple, simplicity is good. It's an ideal for him. Behind this simple division is a lot of research on the brain and also the body. He notes that in the very act of turning, forcing himself to actually move his chair a little, his body moves too. He forces himself to physically move, to change postures. That switch is good physically and mentally. He might take twenty-five minutes on the secondary computer but, throughout that diversion, he constantly has the primary computer and the primary task there in his peripheral vision. When it's time to turn back to the main task, it is right there, uninterrupted, exactly where he left it.

"I'm still multitasking," he says, "but there's a hierarchy in my brain of what is unchangeable and important on the main computer, and then everything on the small screen is done in the service of that unchanged work. It helps a lot."

But that's not all. He has yet another computer set up at a distance, across the room or, even better, down the hall. This one requires him to leave his desk chair. He reserves it for all the fun things that might tempt him in a given day. His blog might be there and his active Twitter stream, perhaps personal e-mail, multiplayer or single-player games he likes, social networking sites, all the things he enjoys online. They are the enticement to fun so, to enjoy them, he has to get up physically from his desk and walk. "Ergonomically, that's important. I've made my temptation be a healthy thing for my body. What refreshes the body helps me be mentally alert when I return to my main screen."

He also has written a software program for his fun screen to aid and abet him in returning to the main, unchanging work screen. A programmed to-do list updates automatically and pops up only when he is using one of the procrastinating sites, not when he is in the middle of productive work. Ten minutes into joyful Tweeting, a message reminds him he has to send off some new programming to his associate in Toronto or Doha by 6:00 P.M. Time does fly when you're having fun, so the to-do list is programmed to remind him of passing time and of deadlines awaiting him.

He's recently come up with an even more ingenious way of controlling his procrastination. He has programmed the fun sites so they get slower and more spotty the longer he's on them. He's found that when he slows the sites (mimicking the annoyance of being on a bad connection), he gets frustrated at the site instead of at the blocker. In doing so, he takes the joy out of procrastinating! It's a brilliantly simple solution to one of the workplace dilemmas of our era. Like all Raskin does, whether to train his own work habits or all of ours, his motto is "I simplify."

Raskin, in other words, is not just a software developer. He designs better ways that we can interact with our machines. As we've seen, the revolutionary difference between the Internet and old mainframe computing devices is interactivity, and he believes that fundamentally the Internet is what we all make it. We do not have to be automata, machines controlled by our machines. Raskin urges us to take charge, for that, of course, is the Open Web credo. The Web R Us.

Not all of us have as many computers at our disposal as Aza Raskin does, but we can each learn something important from his emphasis on compartmentalizing his work and his play according to his own needs and in his goal of programming interruption and physical movement into his screen-dominated workday. I suspect that if consumers begin to clamor for devices that facilitate this kind of separation, the computer industry in a few years will be offering us relatively inexpensive options that make it possible—apps for different attentional styles, for different kinds of workflow, and maybe even for programmed work stoppage and messages that urge you toward that water cooler for a fifteen-minute break.

In our transitional time, when we haven't yet developed new habits, and in new and more flexible work situations—more and more of us working at home and without a foreman or middle manager standing over us telling us what has to be done *now*—learning our own ways of paying attention and customizing workflow to those ways is key. For one of my friends, humor is key. About six months ago, he brought an embarrassingly analog Felix the Cat kitchen timer (it has an actual bell) into his minimalist office at a high-tech firm and sat it defiantly beside his computer mouse. When he has a project due, his first task of the morning is to set the timer for an hour or ninety minutes, pull up the project on his desktop screen, and go for it until the bell starts ringing so crazily that Felix is grinning and waving his hands and the whole thing rocks and rolls.

That's his signal to give himself time to think about the work he's accomplished so far. He does this by going for his morning coffee and sipping it while taking a stroll around his entire building. When he returns to his desk, he then decides whether to wind up Felix the Cat again or, if the project is done, he opens his e-mail for the first time, sends the document off to his collaborators, and, with the timer ticking (really ticking) again, he sets himself a specific amount of time on e-mail until lunch. And so on.

Raskin is working on somewhat more sophisticated interfaces than the hilarious one my friend has come up with, but the point is the same, to recognize that we are people with bodies, not just brains and eyes and typing fingers. Because of the computer's evolution from the typewriter, information is currently dominated by text, but Raskin never forgets the whole person, which is why he's designed his office solutions to force himself to move and turn and twist and otherwise engage his body in the act of changing attention. Someone else might find auditory cues more helpful, anything from theme music assigned to different tasks or a cash-register *ca-chink, ca-chink,* satisfyingly recording each task completed. Others might find badges, such as the ones gamers earn for winning challenges, a useful way to measure progress. What about color? Would we even be able to keep seventeen colors straight, or would that rainbow text confuse us even more?

The answers to the questions are still being unpacked, but the point is that Raskin is asking them on this level. He is not using some twentieth-century measure to dictate how we use the twenty-first-century workplace. It's not our brain that's the issue. It's the antiquated design of the workplace, and even of computers themselves. Like Christian cathedrals built upon Roman ruins, the desktop is a remnant of an older culture that did not champion the values on which the World Wide Web is based: interaction, collaboration, customizing, remixing, iteration, and reiteration.

"We often think we've solved a problem when we've merely come up with a good answer to the wrong question," Raskin insists. He's big on unlearning. He calls it inspiration. Remember, Mozilla was built using an open, crowdsourcing model. That means, with any problem, if you get it wrong the first time, instead of being dismayed, you are challenged. When you get it right, you go to the next level. Open-source building is a lot like game mechanics that entice everyone playing to compete together so they can move to the next challenge. If the end product is enticing and the path exciting, everyone will pitch in, not only to solve the problem but to expand it and then solve the more complex problem too. That's how Firefox was created.

"You don't get people interested in riding the railroad by building the railroad," Raskin says. "And you don't do it by telling them they have to ride the railroad. You do it by building the great resort town at the end of the railroad that they want to get to. Then you build the first hundred miles of track and invite them to participate in building it the rest of the way."

It's a little like luring yourself out of your desk chair to go to the fun screen, thereby forcing yourself to do an invigorating stretch along the way. Simple. And also attractive. These principles are Raskin's alpha and omega. He even calls himself a maker of "shiny things." He believes simple, shiny things help us turn the digital office into a place of possibility and creativity, not anxiety and limitation. It's an opposite way of thinking, really, from much of the anxious discourse about the overloaded, distracting, and confusing contemporary workplace.

AZA RASKIN WAS BORN IN 1984. His father, the late Jef Raskin, started the Macintosh project at Apple in 1979, and his essay "Computers by the Millions" is one of the visionary think pieces of personal computing.[12] Unquestionably, Aza grew up imbibing the idea that the workplace of the future could be a shiny, simple space. I'm glad the way we use our computers is in his hands.

But what about the rest of us? What about those of us who were born *before* 1984? We weren't brought up on the seamless, continuous interactivity of the World Wide Web. We weren't brought up to see personal computing as either shiny or simple. Many people who do not count as "millennials" see the digital workplace as a morass and view the desktop as some siren tempting us away from the real business of work. And frankly, few of us have the power to requisition two extra computers to help realize ideal working conditions.

I don't want to dismiss our anxieties. As any good student of Freud will tell us, unless we understand root causes, we won't be able to come up with a cure. And most of us need a cure. The double-bind of the labor speed-up coupled with obsessive public and corporate attention to the "problem" of multitasking and distraction is an entirely unsatisfactory vision of the future of work.

From a historical standpoint, it's a problem whose roots are easy to see. After all, we inherited our twenty-first-century workplace from the industrial age. It makes sense that Gloria Mark and others have been called upon to monitor, clock, time, measure, and assess our efficiency. We have inherited an ethos of work designed for bricklayers, not information workers, and a model of attention designed with the same tasks in mind. We are inheritors of a workplace carefully designed to make us efficient for a world that no longer exists.

||||||||||||||||||||||||||||||||

After the panic of 1873, at the height of one of the worst depressions in American history, Frederick Winslow Taylor, scion of one of Philadelphia's wealthiest and most prominent Quaker families, walked into a pig-iron plant seeking work as an apprentice. Factory work was an odd career choice for a severely myopic graduate of the elite Exeter Academy who had seemed bound for Harvard Law.[13] Soon, though, Taylor wasn't just pursuing factory work as a career path; he was reforming it by dedicating himself to industrial efficiency. The day Taylor pulled out his stopwatch to clock his peers at their tasks was a watershed moment in the history of work. In two famous books compiling his "time and motion studies," *Shop Management* (1905) and *The Principles of Scientific Management* (1911), Taylor alerted readers to "the great loss which the whole country is suffering through inefficiency in almost all of our daily acts."[14] He documented the problem meticulously: "One bricklayer and helper will average about 480 bricks in 10 H[ours]," Taylor recorded on January 29, 1880. On the next day, he wrote that a laborer with a wheelbarrow full of loose dirt could "wheel it approximately one hundred feet exactly 240 times in a ten-hour day."[15] To solve the inefficiency problem, he set arbitrary (he would say "scientific") production quotas, penalized "malingerers" and "loafers," and rewarded the "soldiers" who exceeded the production quotas he set. Along with Henry Ford, who later turned out profitable Model T's on his assembly lines, Taylor created the modern industrial workplace, transforming how America and the rest of the world made things, whether in pig-iron factories, auto factories, or fast-food operations.

Management guru Peter Drucker calls Taylor's work "the most powerful as well as the most lasting contribution America has made to Western thought since the Federalist Papers."[16] If our schools are designed to train kids on how to do single, specialized tasks on time and to a schedule, it is to prepare them for the Taylorist workplace. If the modern manager and executive also feel the most important part of their jobs is to complete discrete tasks, that, too, is Taylor's legacy.[17] Taylor recognized that humans are more easily diverted from their tasks than machines. His antidote to this problem was to make labor as machinelike as possible. He argued that uniform, compartmentalized, and undeviating human operation *equals* productivity.[18] He also emphasized the importance of segregating one kind of worker from another, and especially in separating management from labor. He assumed, sometimes explicitly and sometimes implicitly, that laborers were inferior, possessing an "animal-like" intelligence that suited them to manual tasks. They needed strict supervision, structures, schedules, rewards and punishments, and a predetermined place in the movements of the assembly line in order to perform in a way efficient to industrial production.

It took a disaster to help extend Taylor's ideas beyond the factory to the modern office building, though the disaster itself took place two years before Taylor set foot in a plant. Whether or not the great Chicago fire of 1871 was actually started by Mrs. O'Leary's cow kicking over a lantern, America's hub city was left in ruins. Because the city connected the offices back east to the loam of the Midwest—cattle and corn—through both the financial operations of the commodities exchange and the physical nexus of the railroads, Chicago had to be rebuilt quickly, yet large questions loomed about how to rebuild. The fire caught the city at a turning point in architectural and urban design as well as in industrial philosophy. Retain Victorian edifices or zoom on into the modernist future? The construction would decide the fate of not just the city's architecture, but its character, even its future. Chicago's city officials and planners voted for modernism. Resources were invested in rebuilding the city as the urban workplace of the future. Led by architects Henry Hobson Richardson, Frederick Baumann, Dankmar Adler, and Louis Sullivan, Chicago became the first city of skyscrapers. While the factory and assembly line sprawled horizontally, the "Chicago skeleton," a steel frame structure that supported the full load of the walls, allowed buildings to be taller than ever before. In 1884–85, the ten-story Home Insurance Building opened in Chicago to enormous fanfare. Thereafter the urban landscape was rapidly transformed, first in Chicago and then everywhere.

As the skyscraper rose, Taylor's principles of management rose with it, as exemplified by the growing importance of the MBA. At the brand-new Wharton School of Business at the University of Pennsylvania, faculty and students began to theorize the best way to subdivide white-collar office work in a way analogous to what Taylor did for the assembly line. The first MBAs worked to define corporate structure that could all be realized and reinforced by interior and exterior building design: hierarchy, stratification, specialization of management roles and tasks, the creation of middle-manager roles, differentiation and coordination of managers and clerical workers, and measurement of individual and unit productivity and quality were all objectives. Because there's no foreman for white-collar workers, industrial theorists came up with literal ways to make the office hierarchy obvious. Top floors were delegated to top executives. Senior managers were awarded corner offices from which they could gaze out over the city, as if keeping watch. And to minimize distraction, the modern office became as devoid of ornamentation, decoration, luxury, leisure, or pleasurable amenities as possible. Offices became subdivided into areas, including financial departments, technology areas, and almost from the beginning, "human resources" (HR) departments to manage this new mass of people all in one physical, contained space.

Everything about the modern office building was designed to signal that "this is not your home; this is not fun; this is not personal; this is not about *you*." If you work in a city, the odds are that you spend the better part of each workday in a building that still adheres pretty closely to those one hundred year-old principles.

Since there was no pig iron to weigh in the office setting, new ways of measuring how effectively office workers performed had to be developed to help sort workers, to measure what they produced at their desks, and to help determine the practices for advancement and promotion. The new business schools championed the latest statistical methods, including deviation from the mean and multiple-choice achievement, aptitude, and personality testing, as well as IQ tests. Assessment skills became key to the MBA, which led to standardizing how we measure success and regulate work, from productivity to human-resource practices to profit margins.

There was a gender element too. By far and away the majority of business school professors, in the 1920s as in the present, were men. And more often than not, these men were training other men to be managers of women. During World War I, many office jobs that formerly had been held by men—secretaries, typists, receptionists, and phone operators—became "pink-collar" jobs assumed by the women who were not out fighting a war. When the soldiers came home, it was common for young, unmarried women to stay in those pink-collar office jobs, paid at considerably lower salaries than their male counterparts had earned. Men who would have occupied those positions previously spread out into the increasing ranks of the "middle manager."[19]

Business schools were almost exclusively the province of men. Being able to wield statistics—with numbers obtained from standardized testing—was part of the artillery of the MBA. The science of evaluation and assessment became key to the curriculum of the new business colleges and MBA programs.[20]

The spreadsheet (literally a "spread" across two facing pages as in a magazine or newspaper) became the white-collar equivalent of the stopwatch. Managers devised new ways to measure office productivity, to expose any deviation from attention to the task at hand. Buildings became as task-specific as workers, with spaces divided into business areas, recreational areas (restrooms, break rooms, cafeterias), areas of heavy traffic flow (central hallways), and then the quiet work areas that would be undisturbed, solitary, and sterile.

Whether applied to life at the assembly line or inside the new skyscrapers, efficiency was a harsh taskmaster. It required that humans be as uniform as possible, despite their individual circumstances, talents, or predispositions. Working regular hours, each person was assigned a place and a function; doing what one was told and not questioning the efficacy of the process were both part of twentieth-century work. But a problem increasingly reported in the modern office was self-motivation. With so much control exerted by others, there wasn't much reason for the office worker to think for himself, to exceed expectation, or to innovate. Regularity and regulation do not inspire self-motivated workers. Assigning tasks to workers and ensuring that they carried them out on time and with care fell more and more to the middle manager.

By the late twentieth century, HR departments, perfecting rules to regulate who must do what, when, where, and how, had grown larger than ever, to become one of the defining components of the modern workplace. According to the U.S. Bureau of Labor Statistics, over nine hundred thousand Americans are currently working as HR professionals.[21] As with testing, for HR, uniformity and standardization are the ideal, with workers' compensation laws and litigation practices providing the teeth to keep bureaucracies and bureaucrats productive. These rules come with time clocks and time cards, all prescribing the twentieth-century way of work.

DOES ANY OFFICE TODAY ACTUALLY work on the same principles as the HR department's rules? At HASTAC, our programmers and our server administrators work together constantly. Neither group reports to the other, but the relationship is mutually interdependent while evolving quite independently, except at the point of connection. You can't have a Web site without a server; servers exist because they are needed to host Web sites.

One New Year's Eve, Barbara, the newest programmer on our team, needed code transferred to the development server before she could complete a task. Brin, the server administrator, pitched in. I was the supervisor of the project, but I only found out after the fact that they had both been working on New Year's Eve. (I admit I was not happy about this.) They communicated with one another almost entirely on Jabber, an instant messaging system. One chatted from Southern California and one from North Carolina. The two of them have never met. And to this day, I have not met either one of them. *That* is a different kind of workplace and a different kind of worker than we have seen before.

What are the best labor practices in this twenty-first century, when the very idea of a *workplace* is being challenged and *workday* is basically 24/7? Social scientists study our contemporary "interrupt-driven" work life, but when work is this decentralized and distributed, the very concept of an interruption is irrelevant.[22] In fact, what was "interrupted" was their leisure time, no doubt when the participants all noticed one another on Twitter or some social networking site and responded and got the job done. Almost nothing about the New Year's Eve interaction at HASTAC "fits" in twentieth-century management theory, architecture, or interior design, and I cannot even imagine what statistical measures one would use to quantify or assess its efficiency. But it certainly is effective and productive in every way.

Is indefiniteness a bad thing? Are our twenty-first-century work lives really all that bad? Or to turn the question around, were work lives ever as uniform, task segregated, and uninterrupted as retrospect may make them seem? If I try hard enough, I can remember back to the distant twentieth century when work was not always humming along, all the attention in the right place all the time.

When we talk about workplace disruptions, there is an implicit idea that the work style we used to have was continuous, uninterrupted, natural. It wasn't. The twentieth-century workplace was completely and quite carefully constructed, with bricks and mortar, MBA programs, management philosophies, labor and productivity statistics, and other forms of regulation, all supported by child-rearing practices and an educational system designed to make the workplace—factory or office—seem more natural, right, and expected than it was.

Taylorism, with its emphasis on individual attention to a task, could be mind-numbingly dull and uncreative in its blue-, white-, or pink-collar manifestations. Given a choice—in our hobbies, for example—few of us take pleasure in doing one thing repetitively over and over again on someone else's timetable and with each action managed by someone else. There is little about human nature that is as time- and task-bound as Taylorism wanted us to be.

Certainly in agrarian life and in preindustrial times more generally, there was far less sense of either task or specialization and far less division between work and home, occupation and leisure time. Leisure time, as a concept, had to be invented, not least by the advertising industry, which turned leisure and recreation into big business. Even the creation of the National Park Service was justified in part so there would be natural areas set aside, in perpetuity, for "enjoyment" (as President Woodrow Wilson said) away from the workplace.[23] It's interesting that it was created in 1916, right when statistical methods were being developed, multiple-choice and IQ tests were being invented, public high school was becoming required, Model T's were being turned out on assembly lines on Taylorist principles, business and professional schools were being created, the labor movement was flourishing, and office buildings were reaching up to the sky.

As educator and philosopher John Dewey insisted upon over and over throughout his championing of integrated, practical teaching methods, none of these modern divisions are natural for humans. We had to learn these separations as part of the industrial economy. We used our schools to train children at a very early age to the mechanisms and efficiencies of twentieth-century work. We spent over a hundred years developing a twentieth-century way of thinking about ourselves as productive workers, which is to say as productive contributors to our industrial world.

Once the U.S. economy shifted so that not just the working poor but middle-class families, too, required two incomes to maintain a decent standard of living, with more and more women entering the workforce, the strict division of work and home became even more complex to maintain. With the Internet and the merging of home and work, leisure and work, play and work, into the computing devices that spin both our workplace and our home, we have entered into an era as complex as the one for which Taylor felt compelled to pull out a stopwatch to measure wheelbarrow loads, one for which an Ivy League school like Penn decided it had to create a separate, professional school of business.

The industrial age wasn't built in a day, nor will ours be. Reforming how we learn and work and play together in the digital age isn't going to happen overnight either.

The contemporary work environment is far from standardized from place to place. In the same building with HASTAC's headquarters, where our team thrives in the new digital world of merged work and leisure, there are numerous workers who keep nine-to-five hours pretty much as they would have done in the Taylorist past. Those workers are statistically much less likely than I am to take work home at night, but they are far more likely to be holding down a second job at another workplace away from home.[24] That is a condition of modern work and modern society that underscores the complexities of unequal work, unequal labor, all distributed in multiple ways even among people who greet each other at the office each day. We may inhabit the same "workplace," but often the physical space itself is the only thing our work lives share.

What are the possibilities for satisfying work in the distributed, dispersed, and ever-diverting workplace of the twenty-first century? And what happens if we think of the blurred distinction between work and play, home and office, not as a problem at all but as an advantage? What if the ideal twenty-first-century workplace *is* the home? And what if, like the best forms of learning, the most productive new forms of labor are also challenging, inspiring, exciting, and therefore pleasurable?

|||||||||||||||||||||||||||||||||

Chuck Hamilton rarely drives to work. He turns on his laptop, checks for messages on his smartphone, and he's there. He works mostly from home, exerting control over when he starts and stops his workday or adjusting it to the waking hours of his colleagues around the globe. Like about 40 percent of his firm's employees, Hamilton works remotely—which is to say he does not have an official office at company headquarters. The other stats from his firm also reflect changing trends in the contemporary workforce, such as the fact that over half of Hamilton's colleagues have been with the company fewer than five years. We can think of this as the corporation of the ever-changing future. His work style is so different from the old divisions of labor, the old separation of home and work or work and leisure, that he's had to coin a new term to describe his hybrid, blended, merged form of work. He calls it: @work@home@play.

What is the name of this frisky, visionary, youthful firm of the future? What company supports all these young new hires working away on their laptops on the patio, in their shorts and flip-flops?

Three letters: IBM. Yes, Big Blue. The symbol of the traditional, hierarchical way of business throughout the twentieth century is now pioneering virtual new models of the workplace that intentionally confound the old divisions among work and home and play. Once synonymous with conservative corporate values, with the spanking white shirt, the navy pinstripe suit, and buttoned-down management and personnel practices to match, IBM is one of the few business equipment manufacturing corporations to have made the transition from the machine age to the information age. It was founded in 1896 as the Tabulating Machine Company upon the success of its electrical counting machine, which had been used to process data from the U.S. Census of 1890. In the early twentieth century, IBM became the world's largest maker of business machinery—typewriters, punch-card machines, time clocks, copiers, data-processing equipment, and sorting equipment, and later became the major producer of room-size mainframe computers. Signified by the phrase "IBM and the Bunch," IBM dominated late twentieth-century office-equipment manufacturing.[25]

IBM went through a slump as technology shifted to Seattle and Silicon Valley and had to regroup and reassess its primary product line—business machines—not only as *things* but also as *information*. The mechanics of the Internet reside in hardware, software, and then also in less tangible commodities like information networks that you might not be able to "make," in the old sense, but that are essential to success in the new global business economy.

IBM's remaking of itself is one of the legends in modern business history, figured as the "elephant that learned to dance again."[26] This is not to say IBM has stopped making machines. Hardly. IBM produces the microprocessors used in just about every gaming console (Xbox, PlayStation, Nintendo, Wii, and so forth). But in a conceptual leap that would have befuddled its nineteenth-century founders, IBM now simultaneously maintains an open-source developer's Web site, developeWorks, with discussion boards, podcasts, wikis, and blogs where anyone in the industry can learn or contribute—*for free*.

What kind of business model is that for a proud old equipment manufacturer like IBM? The short answer is that it's a highly successful one. One symbol of its transformation is the discarding of the three-piece business suit as its dress code. "Business casual" is the company uniform of the rebuilt IBM—but who's checking, when 40 percent of employees work from home? Those homes can also be just about anywhere on the planet. A majority of the firm's four hundred thousand employees reside outside the Americas, and that majority grew substantially after 2004 with the sale of the personal computer division to Lenovo, a state-owned Chinese enterprise.

As Chuck Hamilton notes, quoting a recent IBM corporate presentation, "We are 'virtual' by nature and action, something that was not easy to achieve and is even harder to manage. We can see that social learning and informal connection of all kinds is becoming a sort of virtual glue for globally integrated companies."[27]

The corporation once synonymous with tradition and convention thrives in the digital world because it transformed itself. The question raised by the example of IBM is: How do you manage efficiencies without the familiar institutional structures governing (as it once did) the hours you work, how you arranged your office, and what you wore? Add to that context: How do you manage across diverse systems of government, economics, ideology, and culture, such as those one must negotiate in a partnership with the Chinese government and thousands of Chinese colleagues and employees? Over the course of a century, IBM developed corporate practices and policies to focus energy and attention and to maximize productivity. Yet evolving for continued success in this century meant jettisoning or refreshing many of these practices. How do you do it? How do you break old workplace habits so that you can learn how to pay attention in a work "place" that is diverse, scattered, decentralized, multicultural, and mostly virtual?

That's where Chuck Hamilton comes in. His title is virtual learning strategy leader, and he reports to the IBM Center for Advanced Learning, a global organization, although he lives in Vancouver, Canada. His job is to help IBM's worldwide employees work as effectively and productively as possible in their changing work environments. In the old IBM, his title might have been something like human resources manager or director of labor relations and employee training. Corporations always have had managers dedicated to keeping the workforce up-to-date, but Hamilton's job description doesn't line up precisely with the old management categories of accounting, marketing, operations management, or even human resources. He does a little of all of that, nestled under the surprising rubric of "learning strategies."

A partial list of topics on which he spends his time looks like this:

The remote worker
Endeavor-based work
Global teaming
Crowdsourcing
Mass collaboration—we versus me
Connectedness—the new classroom
Producers/consumers
@work@home@play[28]

This is the syllabus at IBM? Frederick Winslow Taylor must be turning in his grave.

Hamilton loves his job. His hours are long, and he works hard, but he says people are always telling him he has the "best job in the world." He is infectiously enthusiastic about the opportunities presented by the workplace of the twenty-first century. When I mention the drawbacks to a globalized and distributed workforce—the sense of dislocation, the work overload, the constant adaptation to change, the harm to communities, the exploitation of those less affluent, the lack of regulation of working conditions—he acknowledges those dangers, then adds, quickly, that the predigital workforce was also "distributed."

He is right. I live in North Carolina. Most of the industrial-age industries here—especially textiles and furniture making—were started during the 1890s because labor unions in New England and the Midwest were championing safety, wage minimums, shorter workdays, child protection, and vacation time. North Carolina was impoverished and advertised itself as a "right to work" state—meaning not unionized or regulated, the equivalent of a Third World country offering cheap labor to companies headquartered in wealthier places. In the late twentieth century, the textile and furniture industries left North Carolina for Asia out of a similar motivation.[29] Hamilton is skeptical of those who immediately call "globalization" a recent phenomenon or who think of it only as evil, without considering all the alternatives, the histories, the precedents, and the future: "There's a dark side to every light side," he says. "People often accentuate the dark side out of fear."

Although IBM may be ahead of other large corporations in its new ways of work, it's a harbinger of the future for many of us. The digital, distributed, fully globalized office isn't going away. Work will be even more dispersed in the future. So as virtual learning strategy leader, Hamilton's job is to keep asking, "How can I take all this and make something good?"

Hamilton believes we learn best when we learn together, and learning in the workplace of the future is continual. Instead of the boss with the stopwatch or the physical workplace separated by levels of power and authority telling you what to do at every moment, instead of an office building with its executives in corner offices with penthouse views, IBM's workplace arrangements are now remote and constantly shifting. Even the most basic feature of the workplace— the time clock—has changed dramatically in the last decade. Ninety-eight percent of employees at the particular IBM center where he works in Vancouver used to "badge in," meaning they held up their company name badge in front of a sensor when they walked into the office every day at 8:30 A.M. and then held it up again to badge out at night. Now about 11 percent badge in. They don't work any less, but they have more control over their workday, more ability to adjust it to the project they are working on and the people they are working with. That also means a new kind of self-control and self-regulation.

Where does one learn how to do that? Certainly not by twelve years of learning how to ace the end-of-grade tests or even by four years in a conventional college classroom. If you aren't required to badge in, you do your job when you are the most productive (morning person? night owl?) and when it best suits the workings of your team, wherever they may be. Very little in traditional formal education—from preschool to the conventional MBA—prepares us for this new decentralized workplace, although these are exactly the skills required for mastery in multiplayer online games.

This lack of preparation is at the heart of many of our difficulties in adjusting to the new workflow. If the workplace has changed and our training for that workplace has not, we not only have an unproductive mismatch but also a profound sense of dislocation and distraction. Except at the most innovative margins, we have not even begun to think about how we can train or retrain our focus and attention for the distributed digital workplace, for the method we've been calling collaboration by difference.

As we've seen, collaboration by difference is the open-source and open-access principle upon which the Internet and the World Wide Web were originally created and by which they continue to be governed. It is based on the idea that productive collaboration requires not just a lot of participation by many different kinds of people but a form of collaboration that is as open, unstructured, and flexible as possible, in its design, principles, and motivation. It is based on the Internet's many projects in massive collaboration, from Linux computer code to Mozilla Web browsers to Wikipedia to Craigslist. Rather than aiming at uniformity and standardization as regulated or enforced by institutional hierarchy, this form of crowdsourced collaboration is based on the idea that if you allow people to contribute in as many different ways as they want, offering points of view as distinctive as possible, the whole outcome is more innovative, stronger, better, and more ambitious than if you start with a goal or a mission and then structure each contribution as a deliberate step toward fulfillment of that goal. From the point of view of twentieth-century management theory, collaboration by difference is chaotic. It shouldn't work. Except that, well, the proof is in the digital pudding. What has been accomplished by mass collaboration—the Internet, the World Wide Web, Wikipedia, and so forth—is so astonishing that it's clear we are only beginning to understand its full potential.

The challenge, then, is to figure out how to change our institutions' structures to support these forms of collaboration based on difference. We need the best ways to train ourselves for participation, for productive interactivity, and even for the self-regulation necessary for collaborating with others for the success of the whole. To truly succeed, we need to start preparing, right now, for the kind of work required by IBM or for the kind of work that will exist even more in the Microsoft/Apple/Lenovo/Mozilla/Facebook/Twitter/Google/What Comes Next hybrid workplace of the future.

Chuck Hamilton is part of the small but growing contingent of leaders who are working to develop that preparedness. In an ideal world where both worker and workplace have struck the right balance between decentralization and efficiency, the benefits will be mutually reinforcing. Hamilton notes that flexibility serves IBM's global business operation. If some people at IBM prefer to work at night, some by day, that eases the differences in time zones across IBM's workforce. Flexibility is a key asset in achieving a more productive twenty-four-hour global work cycle. Without the old workplace rules, everyone at IBM is learning and adapting to new behaviors and patterns all the time.

Take attention, for instance. How one pays attention in a distributed environment has to change. None of the material conditions (silence, sterility, absence of distraction) or the personal ones (eye contact) help you focus when you are on a conference call, as is typical at IBM, with fifteen people in fifteen different locations. How do you stay on track, both personally and in a group?

Hamilton says this was a problem at first, but over time, IBM was able to create an entirely new kind of multiperson conversational culture, with its own methods and tools. The key, as is often the case in this kind of innovation, involves deemphasizing the typical hierarchy of a meeting through a clever use of technology. Instead of one executive leading the conversation while everyone else just listens, an IBM conference call now flows not only verbally but by text, too. Let's say fifteen people are on a conference call across Vancouver, Toronto, New York, Rio, and Beijing. Everyone chats using Sametime, IBM's internal synchronous chat tool, and has a text window open during the conversation. Anyone can be typing in a comment or a question (backchatting) while any other two people are speaking. Participants are both listening to the main conversation between whichever two people happen to be talking while also reading the comments, questions, and answers that any of the other participants might be texting. The conversation continues in response to both the talk and the text.

Hamilton admits that, when they began this practice, it seemed distracting. Now when he and one of his colleagues in Ontario join me in a conference call without any chat feature, he jokes that he finds his attention wandering occasionally as he wonders why Leslie isn't saying more. A few times he stops our train of thought to ask for her input then laughs and says that, if we had a back channel, we wouldn't have to interrupt ourselves. Over time everyone at IBM has become so proficient at conference calling with back channels that its absence seems like an impoverished, sluggish, frustrating, and unfair way of conversing. It almost feels *rude*. When fifteen people are on a conference call without a back channel, someone has ideas that aren't getting expressed. With same-time backchatting, while two people are talking, everyone else can be participating, responding to the topic, offering ideas and responses, coming up with new twists that can easily turn the conversation in a new direction without any rudeness or even interruption of the flow. Plus, you can save these useful text messages and refer to them later.

I joke that I can imagine some social scientist with a stopwatch clocking one of these fifteen-person backchatting sessions and proving that ideas aren't actually finished to completion, that people interrupt one another *x* number of times, that the multitasking is an illusion, that people are really paying attention only part of the time, and on and on. We've all read these studies as reported in the popular press. Hamilton protests that such a study would entirely miss how this process works once participants have become accustomed to it. There's no such thing as "an idea being interrupted," because each person's ideas are constantly being reshaped by the flow of feedback; it's a new way of thoughts melding together "synchronously," at the same time, in the manner of synchronized swimmers or dancers who could never do what they do alone, because they depend on one another for the final result. One cognitive consequence of this method is learning to pay attention to multiple inputs from others, including by reviewing the back channel later. These calls provide information, but they are also mental workouts training IBM employees for a new form of global collaboration wherein expertise flows in streams from those familiar with the needs and assets in a different part of the world.

Once again, we have a new form of communication being measured by old metrics that calculate only its least interesting (and most irrelevant) features. We don't have mechanisms yet for measuring the success of this back-and-forth form of collaboration. Because our statistical methods were designed for one form of productivity, we have very few principles or analytical tools for understanding collective, cumulative, synchronous thinking and problem solving. Who gets the credit for the idea? Who gets the merit raise or the bonus? In such multilayered conversation, who's the boss?

At IBM, they've all become so well schooled in backchatting that Hamilton says he now talks, texts, and reads text so seamlessly that, at the end of a conversation, he cannot recall who proposed what and doesn't really care. Everyone contributes to the process, and credit goes not to the person with the best idea but to the team that functions best. That's the "we versus me" item on his list. Conversations progress almost like "twin talk," with people finishing one another's sentences, filling in blank spaces, nudging the discussion one way or another, problem solving together. Rather than seeming like multitasking, the activity itself *flows*. Compressed, efficient, energizing, and deeply interactive, this synchronous flow, Hamilton insists, makes IBM employees know they can rely on one another, "one green dot away from each other, connected all the time."

Hamilton relates a story someone told him about driving kids to school. As they approached the school building, the kids madly IM'd away to their friends, all the while chatting energetically with one another. Then they arrived at school, and they lined up, went silent, and switched off—not just their electronic devices but their own personalities: They became less animated versions of themselves. They went into school lockdown mode—until recess. Out on the playground, the chatting and IM-ing began again. They switched off again for the rest of the school day and lit up once more when the school day ended. Hamilton jokes that when he hears about companies that still operate in the traditional twentieth-century way, he thinks of that school yard story. Something is off-kilter, we both agree, when school or the workplace, where we should be productively in the moment, prohibit behaviors that have already transformed our everyday lives.

If schools and workplaces create rules against practices that already shape everyday lives and everyday habits, they not only miss a tremendous opportunity but also *cause* disruption. Aren't they contributing to anxieties and fears about productivity and attention if they continue to rely on practices that are no longer part of our new habits? The "alienated worker" was a figure haunting the twentieth-century landscape. Are we trying to preserve that condition of alienation? If institutions of school and work fight changes that people happily have adopted in their lives, then perhaps the source of distraction in the workplace isn't technology—perhaps it is the outmoded practices required by our schools and workplaces.

Chuck Hamilton could not be more certain that these kinds of anachronisms and discrepancies are the real source of much twenty-first-century workplace stress. Holdovers from antiquated ways of working in a world that no longer exists leave an unproductive residue of alienation over everything else. He's not someone beleaguered and overwhelmed by the everyday world of contemporary work. On the contrary, he uses the word *play* a lot and, like Aza Raskin, believes that the best work—like the best education—has to be inspiring, challenging, exciting, *playful*.

Hard work is not necessarily the opposite of play. We know from studying video gamers and runners that the harder they work, the more adrenaline they produce and the more endorphins course happily through their bodies. Their brains light up like Christmas trees too, with neurotransmitters activating everywhere at once, including in the limbic and prefrontal areas associated with emotion and judgment. With every part of the brain engaged, concentration is also intense.[30] Hamilton notes that, as a musician, he is happiest playing with musicians who are better than he is. His own music improves and he's inspired by the challenge. He champions that model of work as well, thriving on the energy of others in the interactive workplace.

He's excited by IBM's practice of "endeavor-based work." This is yet another new concept that we haven't learned how to train people for. It means that you do not perform a single, specialized task and are not called upon over and over to perform one specialized kind of function. More and more people at IBM don't even have job descriptions in the conventional sense anymore. Instead, they contribute certain kinds of talents or even dispositions as needed to a project and stay on that team as long as they contribute to its success. He's confident that people know when the time comes to move on to the next project. They rarely need to be told. Why would they want to waste their time or anyone else's when their contribution is no longer needed? That's unfulfilling for everyone.

No supervisor tells you to move along? How is that possible? In this system, what happens to performance evaluations? Job reviews? Because endeavor-based organization doesn't depend on the amount of time you spend in the office or on the project, but on how you contribute to the success of the team, how do the usual statistical ways of measuring individual performance pertain? It's actually quite simple. Hamilton notes that "we are measured against business results, against each other's contribution, one's team contribution, one's credibility, and the deliverables (performed work)." That seems a more logical and thorough way of measuring an employee's complex contribution to the success of a company than the standard method at many businesses, where a supervisor is required to grade each employee and sometimes even to rank who in a unit is best, second best, and so forth. That system of rating employees against one another instead of according to the work they actually perform (either singly or as a group) is inimical to endeavor-based work.

From this perspective, one sees why John Seely Brown is convinced that the gamer mentality is well suited to the new world of business in the twenty-first century. Endeavor-based work is equivalent to your guild self-organizing for maximum effect and then pulling off an epic win together.

"Endeavor-based organization is a little like the movie industry," Hamilton says. You work hard together to make the movie, each contributing some aspect—whether it is acting or doing stunts or creating special effects or getting the financing together. When the film's in the can, you then go on to the next job. But you don't all go on to the same next job. The team disassembles, but everyone has learned who performed best at what, so for a future task, you know whom to call. Endeavor-based organization is structurally different from forms of work ultimately grounded in an assembly-line organizational model, with each person always contributing in the same specialized way to the same team to make the same product.

In endeavor-based organization, hierarchy must be lax and shifting. In one endeavor, someone who has official company experience or status might be the most expert and therefore be appointed the team leader. An hour later, as the workflow changes, that same person may sink back into the position of being a learner. Hierarchy isn't the guiding principle so much as trust is: depending on one another's capacities to work together. I think about Duncan Germain at Voyager Academy asking his sixth graders to write down who they will depend upon for a given task—themselves, their partner, or their toon. The answer varies per endeavor, per person.

Not much at IBM happens in the old model of one person, one job, one focus, one specialty, one expert, one manager. Because everyone is connected and because the teams shift constantly, everyone is a potential resource for everyone else. Proximity doesn't have to be a consideration. "I may need someone from India, China, Brazil, Canada, or the U.S.," says Hamilton. This is the "global teaming" on his list. That concept, too, evolved over the last decade. Global teaming requires an inherent humility, an intuitive and inquisitive gift for unlearning and learning, because one's patterns and expectations constantly come into productive collaboration with those of people schooled in other traditions, other cultures.

Hamilton underscores the fact that he's always learning too, and he tells me about the time colleagues in Brazil turned to him for some diversity training. No problem. IBM takes pride in its diversity and embraces the Internet crowdsourcing adage that you solve problems best by having as many eyeballs—as many different points of view—on any problem as you possibly can. You succeed by seeing differently. So the North American contingent of IBM had lots of company material they'd developed about managing diversity. But instead of simply trying to foist a prepackaged diversity program on the Brazilians, Hamilton asked his colleague to say a bit about what diversity signaled in Brazil. Something that emerged early was confusion over different norms of social office etiquette. In Brazil, coworkers often greeted one another with a hug or a kiss on the cheek. This was out of the comfort zone of many North American or Asian colleagues. Diversity as an ideal is different from the kinds of issues and tensions that might arise from a diverse work environment in actual practice, and he has learned that you need to ask, not assume, that you know what *different* means. You can't have one uniform set of materials on diversity that works everywhere.

It's that way for everything with global teaming, Hamilton says. "You cannot work in China without a deep understanding of how Chinese business works—and that means listening to the Chinese," he says. "It's not about telling them what *should* work there just because that's how it works here."

Learning only works in a corporation that embraces learning all the way down, Hamilton insists. Take corporate jamming. The term is borrowed from the improvisational sessions of jazz musicians and is IBM's way of drawing upon its worldwide network of employees. IBM turns to its global workforce not just to solve problems but to define what those problems are, using its trademarked protocol of Collaborative Innovation.[31] The company is convinced it learns more from these jam sessions than from hiring expensive "experts" to chart the future. The jam sessions also reinforce IBM's core value of cross-cultural collaboration.

The 2005 Habitat Jam invited IBM employees to join a virtual meeting. Tens of thousands of people from 158 countries participated and focused for three days on ways to improve the environment, health, safety, and the quality of life in the world's urban centers. Chongqing, China, is now thought to be the world's largest city, but it's growing by over half a million inhabitants a year, faster than anyone can count, so no one really knows for sure. It has nothing like the infrastructure and social supports an enormous megacity requires. Similar cities are springing up seemingly overnight in areas where IBM has a workforce, and no one yet has a handle on matters of infrastructure, ecology, or health and safety conditions in them. Habitat Jam addressed these new modes of urbanization, creating the agenda for the UN's World Urban Forum in 2006.

The seventy-two-hour Innovation Jam of 2008 brought together over 84,000 participants who contributed some 30,000 individual posts, divided into over 2,700 themes and 2,300 conversational threads on a diverse range of topics. The topics included geeky new statistical methods for crunching massive data samples gleaned from seemingly incompatible sources ("intelligent decisioning"), construction of Web sites where citizens can map local services

of benefit to new arrivals, a blueprint of best practices and protocols to institute IBM-wide carbon-free days, and creation of an electronic marketplace that
allows retired IBM employees to offer their services for special projects. In many
cases, the resulting ideas were almost instantly put into practice, instead of lan-
guishing in endless committees and subcommittees.[32]

IBM derives millions of dollars in entrepreneurial ideas from these corporate jams, as well as the equivalent in job satisfaction from those who offer
ways to make the company a better working environment. Because anyone can
contribute to an idea as it is developing, there is also a larger, companywide
buy-in for new ideas that IBM might want to roll out as a result of the jams, an
inestimable contribution to the bottom line because rapid employee adaptation
to new workplace mechanisms and strategies is key to corporate innovation.
Because an atmosphere is created in which anyone can contribute ideas or build
upon the ideas of others, there is also a better chance that the contributions
from the highly diverse, multinational workforce might help the corporation
to see its own blind spots. It counts on its own workforce, in other words, to
help chart the leading edge of its own innovation, looking to the corporate jams
to propose ideas outside the tried-and-true twentieth-century business school
methods of strategic planning, flowcharts, goal setting, organizational mission,
revenue optimization, targets, projections, benchmarks, market analysis, and
milestones.

IBM obviously finds the jams to be productive, for the company actually
implements many ideas that arise from them. The method is considered to be
part of the corporation's new digital-age trademark. The jams are efficient on
both the practical level (innovative ideas worth implementing) and the philosophical (setting a tone of global inclusion and collaboration). There is no way
to measure innovation, creativity, and cultural change by old metrics designed
to tally efficiency, yet all of these are necessary for success. If I happen to be the
person who tosses out the idea that thousands of others work to develop and
that turns into a company winner, it may not feel as if it counts as my *individual*
contribution according to a Taylorist scheme of things, but because the jams
are documented and public, my contribution is recognized in the new metric
of interconnection. You can be sure that the person who is throwing out great
ideas for the company is also being pinged with offers to join this or that project
team. That openness of contribution leading to subsequent participation, with
evolving collaborative roles and responsibilities based on one's participation and

actual performance (rather than on one's job description or title), is exactly how work flows in the new open, digital, interactive workplace. The corporate jam offers a new model of productivity and efficiency because each person and each office in the corporation is linked to every other one and any great idea can be taken up by anyone and implemented in a new environment. Wheelbarrows can't hold what I bring away from participating in such an event.

"Boardrooms are not big enough to handle the kind of learning we need now," Hamilton emphasizes. He notes that the "we versus me" model of mass collaboration works well with a corporation as big as IBM but that even to call it a model is misleading, because collaboration works differently, on a different scale and in different formats, in different contexts. Not all virtual meetings are alike. Even structure is something that has to be evolved in context. For example, when IBM calls meetings in virtual environments, in North America thirty people might show up. In China, it's typical for four hundred people to come to a virtual event. The flow of people is also different, with the Chinese group moving into a large virtual lecture hall but also, in the manner of face-to-face conferences, circulating a lot, breaking into small groups, and backchatting in order to pull away and interact even as the main presentation is going on. It's a very efficient process. Hamilton evokes business leader Don Tapscott's concept of "wikinomics," the new economics of collaborative enterprise, noting that "fifteen minutes of four hundred people's time is a lot of hours."

HAMILTON ALSO BELIEVES THAT PLAYFULNESS is part of creative, innovative, collaborative, productive work. One function of play in our lives is to learn how to enjoy working with others, and he is passionate about creating immersive virtual environments that renew our spirit of enjoyable, engaged learning. He believes virtual environments encourage relaxation, informal chatter, and friendly, happy relationships. This is where "@play" comes in. He insists that, when so much work is accomplished virtually, play is more crucial than ever. Through play, people learn to trust one another, to engage one another; they learn one another's skills and sensitivities, blind spots and potentials in an arena where the stakes aren't high. They learn respect. You need that in order to develop trust, and you need employees to be confident that, if they contribute an idea about carbon-free days to an innovation jam, they will be heard and valued, not ignored or punished. Believing in the importance of one's role is crucial in endeavor-based organization.

One of the drawbacks of the new workplace is that when people work remotely, there's no place to take a break and chat about sports or politics or celebrities and have a water-cooler conversation, which is that part of the work life that binds people together and shapes trust and commitment. "People often forget the social nature of why we work," Hamilton says. One way he helps fix this problem is by creating digital water coolers—places where people scattered halfway around the globe can come together and chat each other up.

Hamilton uses the virtual environment of Second Life a lot in his work. Second Life was developed by Linden Lab in 2003. Accessible on the Internet, Second Life provides a virtual space where its "residents" interact, socialize, collaborate as individuals or in groups, all by means of avatars designed by the individual resident. IBM has its own "islands," or virtual meeting spaces in Second Life, for its employees. And it's not the only company there. About thirteen hundred organizations and businesses now hold meetings in Second Life. For a modest monthly fee, employees can choose and customize an avatar, fly to an "in-world" meeting place, and hold either a private meeting or a public one, if they wish.[33]

If you are using Second Life for the first time, it all seems a bit bizarre—and Hamilton would say that's its charm as well as its effectiveness as a space for loosening people up, especially people from different cultures, to interact and collaborate together. You don't "look in" on Second Life. You can't just be an observer. You become a Resident, as users are called, which means you represent yourself in the virtual world with an avatar, an onscreen representation of yourself, typically with its own name and look, both designed by you. Even before you begin interacting with anyone else, you have to choose what kind of avatar you want to be. You can use default avatars, but even they require choices far beyond the normal ones before a business meeting of wearing or not wearing a tie today, choosing the flats or spike heels. Do I want my avatar to be male or female, Chinese or Caucasian American, young or old, thin or fat, full head of hair or bald? Even if I don't want to spend my Linden dollars "skinning" my avatar—customizing it with inventive features—I cannot be part of Second Life without thinking about how I represent myself in the world and entertaining the possibilities for imagination, whimsy, and playacting. You can buy yourself shirts, pants, hats, or dresses—or you can decide to be Tyrannosaurus rex, Sherlock Holmes, a Transformer, or a Disney character. The choices are endless. When you are conducting an IBM meeting in Second Life for thirty people or four hundred people who have all chosen and/or skinned their avatars, you are in a different kind of workplace. Hamilton thinks that is a good thing, a necessary thing, to relieve some of the tensions of a busy work life.

Chuck's own avatar wears a kilt, and his shaved head is dotted by impressive, punky spikes. He helps new SL'ers choose their avatars and then wanders around the SL spaces, including the IBM islands. Chuck works with "wranglers" (experienced Second Life Residents) who partner with first-time users until they get the hang of interacting in SL. The learning curve is steep at first, because you use your keyboard, keypad, mouse, and other features of your computer in specialized ways. For example, Ctrl-2 activates the "move" tool and pairs with other keystrokes to allow you to gesture, touch, pick things up, interact with others, and walk, run, jump, or fly.

Communicating in Second Life is a learned skill. Once in Second Life, you open a "chat bar," and when you meet another Resident, you type in what you want to say. Your conversation can be public to anyone else in your vicinity in SL. There's also a "shout" feature which allows you to broadcast across distances when addressing a large crowd. You type in ALL CAPS, and your avatar is automatically animated to look as if she's shouting. You can right-click and send a private IM (instant message) that can only be seen by the Resident you are addressing. You can address Friends in your contact list, and if they are online at the time, they can join you in SL. You can install Voice Chat when you set up your Second Life account so that you can talk to anyone else who has Voice Chat, again at close range or to a crowd. There are many other features that mimic those in RL (real life), but with a difference: All these communications might be happening not only with punky guys in kilts but also among robots, superheroes, animals, cartoon characters, or anything else. Contemporary digital campfires replace informal gatherings in boardrooms and conference rooms, as well as providing opportunities for digital events like U2 concerts or music in the park for symphony lovers.

IBM has figured out that the movement of people is costlier, more cumbersome, and frequently more stressful than bringing people together to meet in other ways that never require you to leave your office or even your den. In our post-9/11 world, travel is less convenient than it once was. In the recession, corporate travel bans to contain costs make it harder to communicate face-to-face, since fewer employees can see one another. And there are other issues too. When the H1N1 flu epidemic looked as though it might take over the world and possibly even result in nation-by-nation quarantines that would limit air travel drastically, many organizations began creating secure virtual meeting spaces where international delegates could meet without having to travel physically in order to be together.

"Geography is now history," Hamilton says, quoting one of his colleagues in India. Trained originally as a designer, then in systems management, Hamilton's work life now blends the two. He is the systematic designer of synchronous virtual spaces.

Sometimes IBM employees use Second Life simply for efficiency. In Beijing, traffic snarls can eat an entire day, but conference calls can sometimes be too impersonal for important meetings, so IBM China often chooses to meet in Second Life. It's possible to upload into SL virtually any documents you'd use at a standard business meeting, including PowerPoint slides and live or prerecorded video. There are also tools for virtual brainstorming, such as whiteboarding and 3-D mind mapping—interactive tools for a shared virtual space, even if one person is based in Tokyo and another in Moscow.

Of course, there are also the notorious "adult" spaces in Second Life, as well as a considerable amount of in-world gambling. Even antiglobalization rallies have been staged in Second Life, including one in 2007 when an Italian union staged a twelve hour protest against IBM's virtual campus there.

Hamilton notes that, if you want, you can replicate everyday business practices in SL, then asks, "But *why*?" If suddenly you have this new workplace tool that erases distance and builds in the capacity for imagination and fun, why would you want to replicate the material limits of the twentieth-century workplace that the virtual world transcends? For a company that wants to imagine the future, what better start than business meetings where the boundaries are neither architectural nor institutional, but mental. "If you let employees imagine a future without the usual limits," Hamilton asks, "why would you then limit what they come up with together?" That would be equivalent to forcing an avatar with the ability to fly anywhere at will to stand in long lines at a virtual airport, waste hours on a cramped virtual commercial jet, in order to journey to a virtual meeting she could have arrived at instantly by pushing the Fly button on the SL screen. When you have all the possibilities of your imagination at your disposal for interacting with colleagues around the globe, why would you choose to replicate practices where those matters were fixed by the walls and rules of the industrial-age workplace?

That's a central question of this entire book. Given the new options in our digital world, why, exactly, would we want to do things the way we did them before? Why would we choose to measure the new possibilities of the digital age against a standard invented to count productivity in the old industrial regime? Given the newly interconnected world we all now live, learn, and work in, given the new ways of connecting that our world affords, why would we *not* want to use our options? The question isn't which is better, the past or the present. The question is, given the present possibilities, how can we imagine and work toward a better future?

The term *affordance* is used by technology experts to signify those things we can do now, because of the way we are all connected to one another, that we could not do in the twentieth century. If the desktop in my office now can transport me instantly to Beijing where my Wonder Woman avatar can high-five your Monkey King avatar before a meet-up where a guy in a kilt leads us in brainstorming ideas for the next worldwide IBM corporate jam session, then where, exactly, is my workplace? Given these affordances and possibilities, does attention to task mean what it once did? Is fretting over multitasking even relevant anymore?

Hamilton questions the presumed equation, so fundamental to twentieth-century ideas of productivity, between a standardized workplace and efficiency. If the world is no longer uniform, is it really effective anymore for the workplace to be? And refuting another binary of the industrial workplace, he wonders if fun really is the enemy of work. It is faster and less stressful to bring a half dozen people together in a virtual space than to have them find a place to meet in central Beijing. But Hamilton argues that, unless that virtual space is inviting, it will fail. "Play can be an incentive to work," he notes, especially in virtual spaces. His comment reminds me of Aza Raskin's emphasis on the importance of "shiny things." Hamilton would agree that shiny things can motivate innovation and collaboration. The same passions and curiosities that motivate us to learn in our daily lives can be reignited, given the right, inspiring workplace.

In the right setting, we *want* to learn. That's true of preschoolers who want to know the "why" of everything until spontaneous curiosity is schooled out of them. It's also true of adults taking up the guitar or perfecting their golf swing or enrolling in Ancient Greek at the local community college in their leisure time. We shouldn't squelch that impulse to learn just because we happen to be at work. And we shouldn't have to if we can do a better job matching what a worker loves to do with what the workplace needs.

Hamilton believes curiosity and passion can be built into the contemporary workplace. Philosophers from Confucius to Freud have insisted that meaningful contribution (not laziness) is the human ideal. Hamilton thinks that because it is now possible to work at just about anything anywhere in the world—to find that niche where your talents and interests are needed—the twenty-first-century workplace holds greater possibility than ever before for more people to be able to do what they love and love what they do.

Our Taylorist legacy—whether in the factory or the modern office or in its educational system—separated work from fun, home from work. *Play* was not a good word in the twentieth-century workplace—or in the twentieth-century school, for that matter. Hamilton insists that, in ruling out play, we squander one of our greatest resources. From infancy on, enjoyment is, after all, one of our most powerful motivators. Hamilton believes pleasure is an underutilized affordance of the contemporary workplace.

If Web 1.0 was about democratizing access, and Web 2.0 was about democratizing participation, Hamilton is predicting that Web 3.0 will be the democratization of immersion. He thinks more of us will prefer meeting together and doing business together in virtual environments a decade from now. We will enjoy the playful creativity of having our avatars, in their kilts and their spiky hair, wander around in virtual realms.

Chuck Hamilton's optimism is not unrealistic. His job is to manage learning strategies for one of the world's leading corporations. Nor is his life entirely virtual. He spends a lot of time on planes, flying to give presentations fifteen or more times a year. Obviously, from his own experience, he knows that sometimes face-to-face is the more effective model. Yet he's convinced that we're still at the very beginning of exploring what we can do, what we can make happen, in virtual worlds. None of us knows where the virtual will lead us.

||||||||||||||||||||||||||||||

Thinking about the way Chuck Hamilton works and the ways he helps others to learn to work at IBM reminds us that it's time to reverse some of the assumptions motivating our debates about work and learning in the twenty-first century. Are the virtual worlds he works in "artificial"? Or instead, are *we* using the artificial props of an outmoded tradition to keep us from realizing the potential of the twenty-first century, a world that already *is* profoundly digitized and distributed?

For over a century we've been schooling ourselves and our children to take advantage of the affordances of the industrial-age workplace. It's time to rethink the affordances of the digital workplace.

There is nothing *natural* about either an educational system or a workplace. It's always about the best way of succeeding within a given context. By practicing certain forms over and over, they become habitual. We stop thinking about our assumptions and values, what we consider to be valuable, what counts and how we count. We see what we expect. When suddenly, abruptly, our context changes, we are forced then to pay attention to all the things we didn't see before. And the context of our work *has* changed in dramatic ways in the last decade. If we feel distracted, that is not a problem. It is an awakening. Distraction signals a serious realignment of our attention, a *necessary* realignment if we are going to flourish in the future. As with all changes in how we pay attention, we have to unlearn our past—our past practices, our past habits—in order that we can learn better ones to take advantage of future possibilities. The outcome isn't sustainability but *thrivability,* the potential to thrive in the conditions that comprise our collective future.

If Chuck Hamilton is right, then a lot of us have it backward. If there is stress in the workplace right now, maybe the problem is not that the workplace is changing too fast but that it isn't changing fast enough. The world has changed. *We* have changed.

It is stressful when we are forced to pretend that everything continues as it did before. We're like those kids waiting for recess so their normal lives can begin again.

Holding on to old distinctions—at school, at work—may well be the most anxious and unproductive component of our twenty-first-century lives. The vestiges of the twentieth century may be the make-believe that is harder and harder to sustain because it does not accord with the reality of our lives. A lot about our schools and our workplaces feels like the Amish grocery store I mentioned earlier, with villagers in homespun clothes transacting business using 3G wireless telecommunications systems. There is nothing wrong with anachronism, but one has to juggle one's principles ever faster to keep up with the inherent contradictions that this digital age pushes to the fore.

As Chuck Hamilton in his kilt and spiky hair would say, gathering other SL Residents around him, it's exciting to contemplate what lies ahead, but it's even better to enjoy what's already here. "Pull your blanket a little closer to the fire," he urges. "The best is yet to come."[34]

7

||||||||||||||||||||||

The Changing Worker

If a behemoth like IBM can transform itself in order to thrive in the changing workplace, then so can we. We can scale down its lessons and find ways to implement some of these practices in our own working lives. To be candid, most of us probably still work in ways that would be entirely familiar to a time traveler from the distant twentieth century. What are the tools and methods we can use to transform our own working lives in the future?

We can find some answers by looking at individuals who, in ways large and small, are changing how they work and tweaking the definition of what it means to be a good worker. Each has discovered something that others have overlooked. Each has found the right tools or the right partners to make a change that works. These are success stories, some modest, some global. They embody our principle of collaboration by difference—but difference is different in every context. Each of their unique stories has a lesson for us that is less about major corporate forms of reorganization than about what we can do, as individuals and collectively, to work with one another and, through that collaboration, learn to see new possibilities for working in unique, creative, and productive ways.

||||||||||||||||||||||||||

1. Seeing Ourselves as Others See Us

Today Tony O'Driscoll is a three-hundred-pound African American woman. Tomorrow he could be a paraplegic veteran adjusting to an office environment after Iraq or a middle manager from Ohio negotiating a complex transaction with a team in Mumbai. In Second Life, you can be anything—and that, O'Driscoll maintains, makes it the ideal tool for learning how to work together in the new global workplace, where we may not even know the partners on whom our success depends. "We're all learning this new way of working together," he says. "Right now, an immersive virtual environment like Second Life is one of the best tools we have."

Once upon a time, people fantasized that the digital world would be a utopia, free of the habits, routines, and prejudices that plague the real one. It turns out that most of the problems of the real world extend into the virtual too. Avatars, weird as they look to someone entering Second Life for the first time, bear an uncanny resemblance to their human counterparts, most people choosing hunky or gorgeous renditions of themselves—typically the same gender, race, religion, and nationality, even when the avatar happens to be a robot or a superhero or an animal. We carry our baggage with us even when given the opportunity to remake the future—we just don't know that we do. But here's where Second Life provides us with an important tool. By becoming an avatar—a character we make in a drama we shape with other avatars we encounter—we have a unique perspective from which we can witness our own behavior in action. As we know from attention-blindness experiments, this is something that, in the real world, it's difficult to do on our own.

O'Driscoll is coauthor of *Learning in 3D,* the definitive book on using virtual environments for workplace instruction.[1] He has been part of training sessions in immersive environments that help prepare people for any number of situations, including dire crises for which it would be impossible or prohibitively expensive to do realistic training simulations in real life. In this, he is kin to game boosters like Jane McGonigal, who see games as an ideal space for experimentation and collaboration. Money, time, space, danger, skills, and difference can all be overcome by the affordances of virtual worlds. You can simulate how you would respond in a nuclear-plant meltdown or enact on-the-spot protocols in response to the spread of an infectious disease through a hospital wing. But what makes O'Driscoll's eyes light up is the change he sees in individuals and companies when they find themselves stunned to discover their own unacknowledged assumptions. Like all great teachers, he knows that once people have caught themselves in an unproductive pattern, it is much easier to help them find ways to break that pattern and find a better one.

So today he's Lawanda, head of a human resources department in a midsize corporation. To become Lawanda, O'Driscoll picks from a number of default avatars that allow him to be an African American businesswoman of a certain age. However, to put on extra pounds, Lawanda is forced to customize. That means she has to take some time to figure out how to do this, and she has to spend a few extra Linden dollars because the default Second Life avatars all come with an idealized H-W (hip-to-waist) ratio. That has Lawanda annoyed.

She's even more annoyed when she steps into the group that is assembling for this session and finds herself ignored. They're all busy texting among themselves, joking, getting acquainted. Wonder Woman over there is getting a lot of attention. So is that ridiculous Iron Man. There's a lot of LOL-ing going on over his antics.

Lawanda is starting to feel invisible, like the ugly duckling at the school dance. As an HR expert, she *knows* why she's being ignored. She's read all the research on weight bias, especially against women—how college students would rather have a heroin addict for a roommate than an obese person, how being even fifteen pounds overweight means doctors take you less seriously and are more likely to misdiagnose and undertreat you, and how, in business, being overweight lowers your chances of being hired or promoted.[2] She also is familiar with the research on how women executives are paid less, promoted less, and given less credit for their ideas than their male counterparts.[3] And she knows the even more depressing studies about African Americans in the workplace, including one conducted by researchers at the University of Chicago and MIT. They sent out the exact same résumés, changing only the names—such as switching a name like Lisa to Lawanda—and found that applicants with black-sounding names were 50 percent less likely to be contacted for a job interview.[4]

As head of the HR division, Lawanda knows the research, but she expected more from Second Life. Feeling dismissed, she attempts to interject herself into the conversation. She tries to retain her professionalism but ups the assertiveness level in her texting, maybe more than she intended.

Now her colleagues respond! Their text messages come back to her in a burst, some apologetic, some defensive. The tone hovers somewhere between patronizing and defiant, between hostile and obsequious.

The facilitator of the session intervenes at this point, using the Shout feature in Voice Chat, and asks everyone to switch out of first-person perspective, what Second Life calls mouselook, a perspective from which you view the world as you would in real life, from your own point of view.[5] Looking down, you see your hands and feet; you see your own face only if you look in a virtual mirror. The facilitator suggests that everyone view in third person, what is sometimes called the God's-eye view, from which you can watch your own avatar in a dynamic with all the others around you.

Would that this were as easy in real life! Immediately, everyone sees what Lawanda has been feeling so keenly. Most of those in the group have been chatting with one another in friendly clusters, facing one another. Lawanda is at a remove, and some of those who texted her never even turned to face her. That's equivalent to talking to the back of a coworker's head. A silence falls. Everyone absorbs the meaning of the scene. Then Wonder Woman steps forward. "Hi, Lawanda. I'm Wonder Woman. Nice to meet you. Is this your first time here?" Lawanda doesn't hesitate, extending her hand too, texting back. Everyone has been given the opportunity to start fresh. The meet-up begins again.

"That moment," O'Driscoll says, "is almost impossible to duplicate in real life. The ability to see yourself as others see you." What's equally important, he says, is then deciding, if you don't like what you see, to make a correction on the spot and do it better the next time.

With the help of their Second Life facilitator, all of these executives were able to catch themselves in the act—or their avatars did—and then, equally important, they were able to see themselves make a successful correction. O'Driscoll believes experience is believing and that the added emotion of actually catching and then correcting yourself helps an insight carry over to future practices. In the Second Life training sessions he has participated in, he has seen people experiencing how they actually work together (and not how they think they do), and then trying to work together better. He thinks people will be more likely to succeed at identifying and correcting such matters in the future, in both real and virtual hallways.

Normally, O'Driscoll appears in Second Life not as Lawanda but as his avatar Wada Trip, a white guy in his forties with dark curly hair and a goatee. Wada Trip isn't a superhero, just a very friendly, decent man who happens to look an awful lot like the real-life Tony. He's a friend of mine, a colleague, and we've worked together on a number of occasions. We're having coffee together in real life and he's wearing his usual comfortable jeans and casual knit shirt, like his avatar. He's a fit, good-looking man with a pleasant voice that carries a hint of his native Ireland. His eye has a bit of the legendary Irish twinkle, too.

He can relate dozens of stories about individual workers and companies that thought they understood what was happening but didn't. They were stuck because they could not, in real life, remove themselves from the situation they were in, and that meant they couldn't really see it. That's one reason he changed careers after a dozen and a half years as an executive at Nortel and IBM, working in new-product development, sales productivity, and performance.

"One day I realized that most individuals and companies weren't understanding change until it smacked them in the face," he jokes. He saw so many people in the business world who thought workplace change happened always somewhere else, at Google or Apple, but not here. All they felt was the anxiety of a changed workplace, but because they were in the midst of it, they weren't seeing how much their own working situation had changed. Because they didn't see it, they didn't know how to engage with the change or prepare for it. They felt out of control, which led to panic, not decisive strategizing.

"They were looking so anxiously at the bottom line, working so hard to measure productivity in old ways, that they couldn't see the ground changing beneath their feet. They weren't ready for it. They weren't prepared at all. They could not see the change they were part of."

He decided to leave the corporate world to become a management consultant, helping individuals, groups, and companies to train for the changing global business environment. He also joined the faculty at the Fuqua School of Business at Duke, where he teaches strategic management, innovation, technology, and organizational learning and improvement.

When I ask if it is typical for profs in business schools to teach in Second Life, he just laughs. Educators, as we have seen, aren't any more likely than businesspeople to see how their institutions could be changing.

Collaboration and *context* are key words for him. Without those, we cannot succeed at work in the future. In the example with Lawanda, changing point of view allowed him and everyone else in the experience to see in a different way. Too often in the actual workplace, when there is cultural diversity or conflict, we simply ignore it. We *manage* difference rather than sorting it out and seeking to understand its roots. Without guidance, a bad experience can be destructive and can confirm and reinforce prejudices without revealing them. That's how interactions often happen; our annoyance is glossed over or attributed to some external cause. And no one learns a thing. Who hasn't been in a workplace scenario like that?

The whole point of collaboration by difference is that we cannot see our own gorillas. We need one another to help us, and we need a method that allows each of us to express our difference. If we don't feel comfortable offering an alternative point of view, we don't. And without such contribution, we continue to be limited or even endangered by our blind spots; we don't heed the warning signals until it's too late and an accident is inevitable.

To those just beginning to learn how to collaborate with others different from ourselves, virtual environments offer not theory but rapid feedback.[6] One of O'Driscoll's precepts is, "It's not about the technology, it's about the neurology."[7] We don't see the rules of our culture until something startling makes an impression on us and forces us to reconsider. In immersive environments, it is easy to create the surprising, unforgettable moment that re-forms our ideas of what's right and what's wrong and helps prepare us to see differently the next time.[8] Given all we know from the science of attention, what O'Driscoll says about immersion makes perfect sense. Seeing our own blind spots in real life often requires disruption.

"Sometimes failure is necessary before people or businesses change," O'Driscoll notes. The bottom line, he says, is that how we work and who we work with is changing fast, maybe too fast for us to comprehend the deep nature of the change. Being a great collaborator has replaced the old icon of success, the "self-made man." In preschool, we learned to play well with others, but for most of our education, we were rewarded for our individual achievement scores, for leaving those others in the dust. That means we need special help in recapturing the old preschool skills, especially how to work with people different from ourselves. Real creativity comes from difference, not overlap or mirroring. We may think we respect difference, but we don't even see the ways we have been schooled in disrespect for what doesn't measure up, what doesn't meet our standards, what we have been taught doesn't count.

"It's when we have a problem we're not able to solve with the usual methods that we learn how much we need others," O'Driscoll insists. "And sometimes what those others help us to learn is ourselves."

2. Seeing Talents Where Others See Limits

If Tony O'Driscoll gets high marks for teaching us to see the value in different colleagues, Thorkil Sonne deserves an A+ for his reassessment of who makes the ideal worker and is worthy of our attention. He has a knack for spotting talent that everyone else misses. As the founder and CEO of Specialisterne, this Danish entrepreneur has succeeded in hiring some of the best software-performance testers in the business. Although this sector of the computer industry sees the highest employee turnover, Sonne's consulting firm is stable, reporting annual revenues around $2 million since its founding in 2004.[9] Specialisterne's mission is to guarantee that your new software package with the bells and whistles doesn't come bundled with bugs and glitches. Finding the right testers to make that happen is no small feat. Testing is grueling work; it requires the close inspection of interminable sequences of code for the one wrong digit that can bring down a system. Testers must work extended periods without losing focus, and for this reason the burnout rate approaches that of air traffic controllers or simultaneous language interpreters, occupations notorious for causing cognitive overload and subsequent meltdown. Yet at Specialisterne, testers are not only eight times more accurate than the industry average but also love their jobs and are three to five times more likely than testers at competitor firms to stay in their positions for over a year.

Sonne's secret is that, out of his fifty employees, not one has been diagnosed with the cognitive condition known as NT. NT is an abbreviation that stands for neurotypical. It is a slang term, slightly derogatory, that members of the Autie (autistic) and Aspie (Asperger's syndrome) community use for what are elsewhere known as "normal people." Sonne has found that NTs are "disabled" and even "handicapped" when it comes to the demanding work of software-quality testing. In the words of Thomas Jacobsen, an autistic employee at Specialisterne, "Going through a program looking at every detail, testing the same function over and over again in different situations," is neither difficult nor boring for him or his colleagues. He knows it's regarded as such by most NTs, but he says it "doesn't disturb those of us with autism. That's our strength."[10]

Thorkil Sonne understands the software industry, and he is also the father of a son who is a high-functioning autistic. He is a conduit between those worlds. Like many parents of autistic children, Sonne worried about what would happen to his son when he became an adult. How would he become independent, finding meaningful and gainful employment? He realized one day that his son's great gift happened to fit precisely into a sector of his computing industry where those skills were needed most, where there was a shortage of talent. That's when Sonne quit his former job and started Specialisterne.

Although no one knows for sure if the actual number is rising or if we have a different method of diagnosis, it seems that more people who are born now will eventually be diagnosed with autism than a decade ago. The current number is estimated at twenty-six thousand children born each year in the United States, according to statistics from the U.S. Centers for Disease Control and Prevention. There aren't hard statistics, but Sonne estimates that currently about 85 percent of the children diagnosed with autism are either unemployed or underemployed. The fear of institutionalization haunts parents and autistic children alike. So it is no surprise that Sonne's model is attracting attention and starting to be replicated by other firms around the world. Specialisterne is neither make-work nor a charity. It solves a legitimate and even pressing need within the software industry, while also addressing a major social issue: independence for persons with autistic-spectrum disorders.

For Sonne's fifty employees, performance testing software is also a great reminder of how "disability" is not a fixed term but one relative to a historical moment. Within the realm of their profession as software analysts, autistics are not disabled. It is the NTs who come up short. Viewed against the requirements of software performance testing, NTs are characterized by an inferior ability to detect errors in numerical sequences, a high level of distractibility, and a tendency to pay undue attention to social roles and social status. NTs have an especially difficult time adapting to the social norms of those who set the standards in the performance testing of software. To put the matter bluntly, NTs are *handicapped* at software performance testing.

Being able to see the other side of disability is what makes Thorkil Sonne a model CEO for the workplace of the future. Rather than seeing his son as entirely defined by (and therefore limited to) his diagnosis, he was able to understand what gifts his son might have and where those particular gifts might be useful. This is a model of work very different from the conventional one and indeed a different model of personhood. What defines Sonne's employees as autistic also defines their skill at software performance testing. In that realm, they have not just average but extraordinary abilities.

I hope that somewhere my Autie and Aspie readers are cheering. The language of deficit and disorder is usually directed at them, not at NTs. In the world of software performance testing, most of us are disabled. We tend to be less efficient, to waste more time, to lose attention more easily. We'd rather indulge in office gossip or Web surfing than try to find the one wrong number in a string of code. We miss more work, take too many breaks, and can't be counted on to stay in the job. In short, we NTs are inferior workers.

At the same time, we are in the majority. We set the norm and the pace. So workplace issues do arise. Many of the autistics who excel at debugging work find it difficult to deal with the social habits of NTs, which necessitates finding ways to manage work relationships such that both NTs and testers can be as efficient as possible. That's why Sonne requires that his potential employees at Specialisterne enter a rigorous training program that includes several months of screening as well as training in how to interact with NTs or, in times of crisis, how to back out of a situation in order to collect oneself. Sonne also assures future employers that his company will take responsibility for the employees if any personnel problems arise on the job. A supervisor always has someone to call at Specialisterne who will automatically step in and serve as a "translator" between the NT employer and the autistic employee. This arrangement provides a backup system for everyone, ensuring that a trusted "normal person" who understands the particular issues of the autistic employee can help solve the problem. This person's role is analogous to that of a mediator helping to negotiate the right truce or trade agreement between nations that do not share the same language or cultural values.

Sonne notes that in placing consultants from Specialisterne in an office setting, his company also makes sure some conditions are conducive to his workers' success. Autistic employees often prefer a cut-off, windowless, austere cubicle to an open workspace. Most NTs would gladly change offices. The autistic employees tend to prefer a solitary, concentration-intensive task that they perform on their own over one that requires teamwork and give-and-take. They also prefer specific instructions, as precise and unambiguous as possible, and don't particularly want to be able to "do it their own way." Nor do they want an open-ended instruction to "figure it out." Once established in such an environment, they thrive, enjoying the work and exceeding expectations for quality and quantity of productive output. The key to Specialisterne's success, Sonne observes, is facilitating "situations that fit employees' personalities and ambitions and don't force everybody into one mold. That just causes stress, and workplaces already produce too much of that."[11]

It is useful to think about Specialisterne as a metaphor for work in the future. What stresses NTs may not bother the autistic employees. And vice versa. And that's the point. A workplace is never "*a* workplace." Everyone thrives in different situations and brings different assets. Ron Brix, a fifty-four-year-old computer systems developer for Wrigley Corporation, applauds the work Sonne is doing, both at his company and as a public relations man for those with autism-spectrum disorders. Brix has Asperger's syndrome himself—and he's convinced that it makes him good at what he does. Asperger's is responsible, he says, for his single-minded focus and attention to detail. Brix insists that an "autistic person has both great gifts and deficits." He continues, "My whole career was based on skills that came as a result of, not despite, my autism."[12]

|||||||||||||||||||||||||||||||

Sonne is a matchmaker, a broker, a conduit, and a node, a switching point between multiple worlds that, without him, would be invisible to one another. Like Tony O'Driscoll, he, too, has mastered the lesson of collaboration by difference: Valuing others who do not mirror our talents can help *us* succeed. It's a role more and more of us will be performing in the future. Sonne translates between, on the one hand, the needs of the software industry for a certain kind of employee who can do a certain kind of work and, on the other, the autistic community with certain talents, interests, predispositions, and social talents. To accomplish this, he must assume multiple perspectives at once—businessman, software developer, amateur expert on autism-spectrum disorders, and father. This is true for all of us. To make collaboration by difference work, we have to understand how our own multifarious talents might come into play in new ways.

The lesson of twenty-first-century work on display in Thorkil Sonne's thriving company is that we need to reassess the value of what we have to offer. In a digital world where arrangements in the workplace are being altered profoundly, where virtually any job can be outsourced to anyone anywhere in the world, what is the skill set of the normal worker? What constitutes a "valuable" and "valued" repertoire of skills? If the job requires scrupulously and continuously reading software code for irregularities that computers cannot diagnose, then those who have that skill are invaluable—no matter how they previously have been diagnosed, no matter that they previously have been labeled as disabled. What matters is the unique talents that they contribute to solve a problem—to do the work—better than anyone else can. That these workers are also autistic isn't relevant to the particular, specialized skills that they contribute and that make them uniquely capable of collaborating in the solving of a software-industry problem.

If Sonne has found value in these individuals in whom most people previously saw none, it is because he was able to see how the digital world has changed the expectations of what a worker is supposed to be. Online, there are increasingly precise ways to match jobs that need doing with workers who can do them, and since one might not need to be physically near one's employer, there can be literally a world of possibility. The "long tail" metaphor applies not just to specialized interests of consumers that can be met by obscure products available online but also to the range of skills any one person may possess and the specific jobs to be done.

For managers, this lesson is essential, but even as individuals we would do well to become our own version of Thorkil Sonne and reassess our own skill sets in the context of a changed workplace. We are so used to valuing ourselves by the standards of twentieth-century work that we often fail to see skills that we've undervalued because the world didn't seem to value them. Or we think of ourselves as a category—dyslexic or "a C student"—without appreciating the ways that we might mine our own idiosyncratic talents. It's the kind of step-back moment that Tony O'Driscoll is trying to use Second Life to help workers discover.

What is relevant in a new, decentralized world of work may not even be a skill for which we know how to measure or test. When I think about the new worker in the digital, global workplace, I am reminded of that old study of hockey players, which attempted to determine what made one player great and another simply average.[13] The testers were surprised when legend Wayne Gretzky ("The Great One") turned out not to be any faster, any more accurate, or any *one thing* better than anyone else. But where he was off the charts was in his astonishing ability to anticipate where the puck would be after a certain kind of stroke and then respond to it almost before it landed there. He was able to intuit how to get there before anyone else. And he was making the same intuitive projection forward for the position of the other players that he was making for the puck. Interactivity personified! It was as if he factored into his instant calculations not just speed and vectors but the reaction times and patterns of his teammates in order to anticipate where he needed to be when.

How do you keep stats on that kind of integrated, holistic, physical, perceptual, cognitive, athletic ability? As we have seen, just because we have not come up with a way of measuring an ability doesn't mean that that skill does not exist. Choreographers, CEOs, conductors, and "web weavers" like Internet guru Tim O'Reilly all have these complex interactive skills to bring together for some larger purpose. The manager combines collective charisma with roll-up-the-sleeves, do-it-yourself instincts; resource management skills; and creative vision. At the same time, we, as individual workers, can assess ourselves on a new scale too, finding ways to mine the different attributes that, in their totality and in relationship with one another, make us unique. We can take our own audit of what counts and learn to appreciate ourselves differently, as desirable workers in the future.

|||||||||||||||||||||||||||||

3. Seeing Work as Part of Life

If Thorkil Sonne is able to take advantage of the skills of his autistic employees by also being responsive to their needs (providing breaks or isolated working conditions), what about the needs that any of us have, needs that, if met, might help us to be exemplary workers? Simply put: What's the key to unlocking what's best in us when it comes to our jobs? That's where Margaret Regan comes in. Located in a sprawling, fourteen-room brownstone in Brooklyn, her FutureWork Institute looks more like someone's living room than a consulting firm and think tank for redefining work in the future. There's a four-year-old lying on the floor in the play space, scribbling in her coloring book while one of FutureWork's clients paces nearby.[14] He's thinking out loud with Regan, president and CEO of FutureWork, about how to rethink and prepare his workplace and marketplace for a more competitive future. Take a good look at this scene, Margaret Regan suggests, gesturing toward the little girl, the daughter of one of the FutureWork consultants. What you see in this brownstone *is* the cutting-edge future of work.

Regan is a dynamic entrepreneur who spent most of her work life at Towers

Perrin (now Towers Watson), the global risk-management and financial consulting firm, fourteen thousand strong, that has been an industry leader since its founding in 1934. Regan loved working at Towers Perrin but decided to go out on her own as a consultant for a very pragmatic reason. Her clients said that, if she owned her own business, they could garner the tax credit available for contracting with a woman-owned business. They also challenged her to experiment with what work might be like in the future. She presented the idea to her colleagues at Towers Perrin. They saw the opportunity and offered to back her during the crucial startup year.

But she had another motivation. She wanted to see for herself if you really could run a profitable business entirely on the principles that she was espousing to others. Most people who came to her at Towers Perrin listened to her advice and then made only some of the strategic modifications she suggested. No one risked everything for her vision of a revamped workforce. She decided that if she didn't try, who would? So she embarked on an experiment for which she would be both the principal scientist and the chief guinea pig.

Regan was operating from a simple principle: The workplace of the future had to start taking into account the life desires, not just the work ambitions, of workers. She was convinced that the best, most creative workers in the future might not be the workaholics with the eighty-hour workweeks, but people who had figured out what way they love to work and how they work best. For some, that really is eighty hours, full throttle. For others it might be twenty-five hours. For others still, it could be twenty hours a week for a while, such as while young children are at home or an aging parent needs care, and then, when life circumstances change, a move to a more intensive time commitment again.

It's all about match, seeing the entire workplace as a collective, collaborative whole, and understanding what each member can contribute and in what way. At FutureWork, the key was to rethink the basic principles of uniformity and standardization. What if an entire workplace could be reshaped to the rhythms of its workers? Why is it necessary or natural that everyone works best on the same schedule, with the same time commitment, regardless of other life commitments? In preindustrial times, people managed their work around their lives. In a postindustrial digital age, with so much of our lives online and so many opportunities for flexibility over time and space, why are we holding on to such an arbitrary value? The standardized categories, such as working hours or full time and part time, may have worked for the twentieth century, but do they always and necessarily make sense for the twenty-first?

So when Regan started up, her first order of business was to get rid of anything that remotely resembled a punch clock. No badging-in at FutureWork. She got rid of office business attire next, then threw the conventional org chart out the window. By the time she was done, there was such a good match between what people wanted to be doing and what they were actually doing on the job that you wouldn't know they were at work except for the fact that clients were walking into FutureWork and leaving with a bill for the services they were rendered.

This loose, nonhierarchical approach is more than just a way to make the office more informal; it's an attempt to release people from the straitjacketed idea of the twentieth-century worker. Rather than have people squeeze themselves into the definition of what a worker should be, Regan wants people to bring the best parts of themselves to the idea of working. Her lesson is that we are losing talent by not allowing workers to contribute productively, but differently, depending on other factors in their lives. She argues that there is a looming workplace crisis that people are not seeing because the current recession obscures the real "gorilla" of twenty-first-century U.S. work. Between 1999 and 2025, the annual U.S. labor-force growth rate will shrink from 11.9 percent a year to 0.2 percent a year, a figure that is below that needed to replenish the labor supply. While others think about slimming down their workforce in a recession, Regan, counterintuitively, is pioneering better ways to retain the best workers. Change and instability, she underscores, are double-edged. Those who feel insecure in their positions are also willing to take risks to find another one.

Regan finds ways that those with pressing life demands outside the workplace can still be fantastically productive and creative workers. Workers in the economy of the future need to have work structured around their lives, not the other way around. Now *that* is a novel principle!

She must be doing something right. FutureWork has been profitable from its inception. She ended up not even needing the backing of her old firm, Towers Perrin. Her business is thriving, and, proof of her own claims, the company is not losing any employees.

So how do you do it? How do you make a business work without a punch clock or an expectation that everyone works full-time? The overall principle is to understand how each worker contributes differently to the larger goals of the company, to make overt the collective energy everyone invests, and to create systems of rewards that allow some people who are more committed to Future-Work's success to reap more of those profits. Salary is pegged partly to contribution, in different directions that might change over time. While some employees want shorter hours, one team member asked to work the equivalent of two jobs in order to have enough money to renovate his dream house. Later, after those jobs were finished, he arranged reduced hours in order to enjoy it. As long as everyone together can figure out the right mix and balance for FutureWork, why not also make arrangements that work best for workers?[15]

Given the flexible structures at FutureWork, it is quite easy to drop in and out of a full-time core-staff position as circumstances or preferences warrant. FutureWork is reimagining a nonexploitative alternative to the idea of freelance work. FutureWork maintains a circle of "affiliates," consultants who work on major projects, as well as a larger circle of "training facilitators," who work on a per-job basis, as they are needed and as they are free to take on a particular job. Numerous companies have been fined for exploiting workers in these situations, most notably Microsoft in the "permatemp" suit that unfolded over the course of nearly fifteen years: Microsoft was hiring full-time workers through temp agencies in order not to pay them benefits or offer them job security. Future-Work offers benefits to all of its core employees, including those who work highly unconventional part-time hours. It also has a group of other affiliates who either do not wish to have full-time jobs or who have other positions but appreciate FutureWork's methods and stand ready to be asked when a particularly exciting or relevant job comes up. In pure business terms, by knowing what size workforce it consistently can support and then offering additional work to its affiliates as it arises, FutureWork protects itself and its employees. Even in the leanest times, it has been able to support its core staff with its work outlay and then add additional consultants as times improve. This allows the institute, a relatively small operation, to take on larger jobs when they are available and then include a cadre of loyal affiliates who appreciate FutureWork's mission.

"We never lose talent," Regan says definitively.

When I ask Regan about the biggest surprises in running FutureWork, she mentions that, at first, she was sure her corporate clients, many of whom are highly traditional in their approach to business, would want something that looked and felt familiar—core staff, a single building, clear titles and ranks, and expertise. She quickly found that her clients could not have cared less, so long as the job they contracted for was being done and done well. Indeed, she points to an exceptionally high rate of return customers, with referrals often coming from within the same large corporation. Designing company policy based on false assumptions about what others will or won't approve, she notes, is one of the biggest innovation killers.

"Workers have changed more than workplaces," Regan notes. She can list all the different ways that workers' lives have changed—children coming at a later age or to those who are single, aging parents, environmental and other illnesses, or just dreams that don't conform to normal success at work. She's proud of the team member who reduced her hours in order to devote more of her life to volunteer work. Experiences like that, Regan insists, make FutureWork more productive in a way that cannot be measured—except by the firm's success. Her company is succeeding on its own and having a tremendous impact on others by modeling its methods for them and advising others on how to implement them too.

"We act as though workers have to fit the workplace. Why isn't it the other way around?" She believes that is the key question for workers in the twenty-first century. I ask her what she considers will be the next big thing. "Educating leaders about the shortage of talent looming in the future workforce," she answers definitively. "And treating talent like a scarce and precious resource, not one to be squandered."

She believes that "younger workers—so-called Gen X and Gen Y—insist on meaningful work, and they want it anytime they choose, anywhere they want. They work hard—but they aren't 'chained to their desks.' Time is the new currency—and many young people will gladly trade money to get more time. Most of them don't even have desks anymore. We're beyond the era of the wired desktop. We're entering the era of the wireless network. That changes everything about the potential for work in the future. That's why, if we're going to stay productive, we have to reshape the way we work."

She is adamant that the future of work is already here. Most of us just have not recognized how profound the change is. "We all need to work together to co-create the flexible, inclusive workplace of the future."

When I ask Margaret Regan if she misses her old life, if she would consider returning to the stability and security of her senior corporate job at Towers Perrin, the answer comes in a flash: "Not on your life!"

4. Seeing How and When We Can Work Best Together

When management consultants like Tony O'Driscoll and Margaret Regan talk about transforming organizations and translating future work trends, I am reminded of one of the most revered of Duke's recent alumni, Shane Battier, who now plays professional basketball with the Houston Rockets.

Spoiler alert: This is not going to be an objective accounting. If you teach at Duke, how can you not admire someone who led your university to the NCAA championship while earning a 3.96 grade-point average? He majored in comparative religion, studying Middle Eastern Studies, among other subjects, with some of the toughest professors at the university. If he hadn't chosen the NBA as a career path, he could have gone on to become a scholar. He was as polite, thoughtful, and decent a student as you could find anywhere, never mind being the Big Man on Campus. The guy even married his high school sweetheart. (I warned you that I wasn't going to be objective.)

Here's my off-the-court Shane Battier story. I had the privilege, in 1999, to help create a large center for some of the most innovative international and interdisciplinary programs, all in a building dedicated to John Hope Franklin, a faculty member emeritus at Duke who was the foremost historian of African American history. He published *From Slavery to Freedom* in 1947, and it went on to sell 3 million copies and mark a field and a life as exemplary as any I have ever known. He was a hero, locally and nationally, and when we opened the John Hope Franklin Center, several hundred people from North Carolina and all over the country turned up, including John Hope's old friend, Bill Cosby.

As we were busy preparing, my office received a call from Shane. He was a sophomore at the time, already a hero, but a different kind than John Hope. He was calling to say how much he admired Dr. Franklin and wanted to be there for his talk that day, but he didn't want his own local celebrity to detract from John Hope's celebration. I called John Hope and relayed the message, and he said of course he wanted the "young man" there. We came up with a plan for how to handle this, and I hung up the phone full of admiration for the tact, modesty, and maturity of this nineteen-year-old scholar-athlete.

He plays that way on the court too, always aware of who is around him and what influence anyone is having on anyone else in a given situation. He's a genius at figuring how and when we work best together in fluid and ever-changing situations, who shines when, and how to encourage that stellar contribution in the actual moment. He's the glue that makes the whole team better. Even more remarkable, there are no real stats that keep track of, quantify, or even make obvious the fact that he's doing anything special himself. Whether points, blocks, rebounds, or even assists, the *team* does better when he is on the court. That's measurable. Why is not.

A few years ago, this remarkable collaborative ability was even dubbed the Shane Battier effect by sportswriter Michael Lewis.[16] According to Lewis, the Battier effect is the ability to lead your team to a win not by your own prowess but by arranging a situation in which each participant plays to his very best ability in relationship to the opponents. It is also a remarkably modest form of leadership that lets others shine and at the same time empowers others to take responsibility and to take charge. Collaboration by difference becomes the quintessence of court sense.

Many people have noted that basketball is the sport most like life. I would add that it is the sport most like work in the digital age. Stats aren't the thing. Neither is individual achievement. It is learning to work in a way in which one is always aware of context and competition, in which one leverages one's own abilities in a given situation with others in that situation in order to succeed. As the situation changes, other abilities are needed—yours, those of your coworkers—and what also changes is whom you work with and how you work together. It is *always* situational. It is *always* context. And it is *always* about moving, sometimes with the ball, sometimes without.

Shane Battier's nickname is Lego, because when he comes on the court, he makes all the pieces fit together.

That's exactly what defines the manager of the twenty-first century, exactly what Tony O'Driscoll, Thorkil Sonne, and Margaret Regan exemplify, too. Legos. Putting the pieces together.

〰〰〰〰〰〰〰〰〰〰〰〰

5. Seeing the Possibilities of Mass Collaboration

Jimmy "Jimbo" Wales is the Shane Battier of the Internet. In the same way that Battier facilitates others on the court, Wales is the Lego of Wikipedia, making the biggest collaborative intellectual venture in human history as good as we can all, together, make it. That's the victory: making Wikipedia as strong as it can be. Wikipedia gets more reliable and better written every year.

On the day I visit him, I find Wales sitting alone in a small, glass-walled office. Famous for his modest demeanor, understated and almost humble, he is dressed casually, not even close to looking like a jet-setter, from his logo-free cotton knit shirt down to his scuffed brown soft-leather shoes. He doesn't notice at first when I approach the small office where he sits amid some boxes at an unadorned table. He is absorbed by his Mac with a look of frustration on his face that we've all had at one time or another: *computer trouble.* The leader and figurehead of the world's most ambitious and impressive voluntary, nonprofit, collaboratively created and edited intellectual project in human history—a project with the modest goal of imagining "a world in which every single person on the planet has free access to the sum of all human knowledge"—has been thwarted by a two-year-old MacBook that refuses to reboot. Somehow, that's comforting.

The project began in 2000 as Nupedia, then morphed into Wikipedia in 2001. In interviews, going back to the early days, when Wikipedia started having some traction, Wales showed himself remarkably undefensive, even when interviewers were skeptical, snide, or even hostile that amateurs could produce anything like a reliable encyclopedia and that it would be useful to have a compendium that had entries on everything from aa (a kind of lava) to Z (an industry term for trashy, low-budget movies). No one believed then that Wikipedia had a chance to produce anything but a laughable mess worthy of humanity's lowest common denominators. Except Wales wasn't laughing. He had faith in the power of mass collaboration, governed by standards created and enforced by volunteer editors. He took in the criticisms and brought suggestions back to his team of volunteers. They would go up on their public wikis, where anyone could bat around the suggestions. A new practice would evolve. Is that any way to make a *reference* book? Most people didn't think so, but in less than a decade, Wikipedia went from being an impossible dream to being pretty much taken for granted as an indispensable reference, logging half a billion readers per month.

Like Battier, Wales was seeing not the Lego pieces but the overall design, even as that design was constantly changing. His basketball court was the wiki, a perfect tool for handling public documents as they evolve, for accepting any contribution and any edit to that contribution while preserving the original.

A wiki allows you to create an interactive playbook for a digital age, this one masterminded by an extraordinary player-coach. For nearly a decade, Wales has been anticipating, moving, cutting, and pivoting, too, depending on what is the best way to succeed given everyone who is on the court. It's no small responsibility to run a worldwide nonprofit, to keep it as censorship-free and as government-free as possible, given all the world's despots and all the world's regulations, and to continue to make it a charitable public good even now, when it has been valued at over $3 billion.

Unlike the warm Shane Battier, Jimmy Wales is famously taciturn. When he talks about Wikipedia, he seems dwarfed by its magnitude, almost intimidated. He's modest, the opposite of a self-promoter, but a tireless booster of the Cause of Wikipedia. While there may be arguments about whether he or Larry Sanger had a bigger role in creating Wikipedia and how that fraught early collaboration morphed into the present-day Wikipedia, Wales is now its chief evangelist. He is the most important of all the Wikipedia network weavers in a vast web as big as the world.

It costs about $10 million a year to make it all work. Most of the money goes for the hardware and software that keep Wikipedia secure, functional, and safe for those billions of annual visitors, plus all those nuisance types who would like to despoil it just for the heck of it. Only a modest amount goes to pay an exceptionally small group of actual employees, the core staff of about thirty-five who work hard to keep Wikipedia's infrastructure running smoothly. The core Wikipedia staff of employees is housed in one modest room, elbow to elbow without walls, in a cement office building across from the train tracks in the SoMa area of San Francisco. This room could be in any graduate school in the country, the "bullpen" where the math teaching assistants hang out.

On my recent visit to the offices of Wikimedia Foundation, there wasn't even a receptionist to greet me. One of the technology specialists separated himself from the group to chat for half an hour or so as I waited for Jimmy to arrive. The technology specialist showed me around the office, which basically meant waving his arm toward the collection of people, mostly men, bent toward computer screens. I waved, they waved back. End of tour.

When Wales phones to say that he's tied up with computer woes over at Wikia, Inc., the for-profit company that he founded and that has headquarters a couple blocks away, my guide offers to take me to meet with Wales there. On our walk, he notes that his previous employer was a typical, hierarchical nonprofit where all instructions came from the top down and no one did anything without explicit approval, deliberation, sign-off, and notification from above. At Wikimedia Foundation, the opposite is the case. Everyone pitches in. They each have assigned tasks, but the hierarchy is practically invisible, the hand light, and they're always jumping in to help solve problems together.

A lot of what happens at Wikimedia Foundation is either DIY or Do It Together rather than outsourced or delegated. The tech director explains that he loves the freedom and the responsibility of working at Wikipedia, plus the excitement of pioneering a new method. To be one of a few dozen people in the room making sure the largest encyclopedia ever compiled in human history is running as efficiently as possible has to be a head-turning and inspiring job—or at least this educator happens to think so!

I thank him when he deposits me at the entrance of Wikia, Inc. I'm in the elevator before I realize I never even got his name. Like everything else about Wikipedia, our communication turned out to be productive but anonymous.

The terrain at the for-profit Wikia, Inc., is a bit plusher, although it is still a far cry from Trump Tower. There's no receptionist here, either. The hallway off the elevator has some quieter floor covering, and there's fresher, brighter paint on the walls. There is again one very large room, with some smaller meeting spaces around the periphery, but the room is better lit, more immediately pleasant, with nicer desks and more space between them. It's less like the TA lounge and more like the room of assistants for the CEO of some large corporation, except this is all there is.

It takes me a few moments before I spot Jimmy in the small glass box of an office, focusing on his troublesome laptop. I met him once before, in Newark, New Jersey, where we both landed because of interrupted travel plans, I at a conference that had to be displaced there, he because LaGuardia was socked in with fog. It was a good conversation then—he's not one for small talk—and when I contacted him, he readily agreed to meet with me again, something he does not often do. We greet each other, and I mention Newark, New Jersey, and he recalls the drink we shared with mutual friends. Given that he is on the road constantly for Wikipedia, that's quite an astonishing detail to remember.

I know his history. He became wealthy quite young as a commodities trader in Chicago and then with a "men's interest" Web site dedicated to sports, technology, and women. Like a number of the Internet's most prominent open-source advocates, he's a libertarian, an avowed devotee of Ayn Rand, although he's denounced the American Libertarian Party as "lunatics." He was influenced by the Austrian economist Friedrich von Hayek, particularly the essay "The Use of Knowledge in Society," which argues that information needs to be decentralized, that each of us knows only a fraction of what all of us know collectively, and that decisions are best made by combining local knowledge rather than by ceding knowledge to a central authority.

Another person who has influenced his thinking is Eric S. Raymond, whose essay "The Cathedral and the Bazaar," an impassioned paean to an unregulated, open-source Internet, might just be the geek Internet Bible.[17] Raymond wrote that the Internet thrives not by regulation, control, or hierarchy but by difference and by magnitude: "Given enough eyeballs, all bugs are shallow" is his famous dictum. He means that if you have enough people looking at open-source computer code, then you will find solutions and be able, collectively, to program out any bugs and glitches. That's how Mozilla runs. But it's not just eyeballs. The key is to have many eyeballs and different kinds, even opposite and contentious ways of seeing. Without that calculated difference, you'll have a wearying consensus. There has to be the freedom to balk and dissent and snark—the Internet way. If contribution lacks bazaarlike cacophony, you end up with flaws, buggy proprietary software. You end up with something like *Microsoft.*

Microsoft, of course, is the antithesis of the open-source collaborative, contributive, participatory model of working together. Wales insists there is no contradiction in serving as the de facto leader of the free-knowledge world as a philosophical libertarian. Leadership in one of the world's grandest philanthropic projects, in the global sharing of knowledge, he notes, does not in any way compromise his libertarian principles or sacrifice his self-interest.[18]

I look around, instinctively waiting for that IT guy who is always there at the CEO's shoulder, making sure the fearless leader doesn't have to spare a moment's frustration over something as trivial as a laptop, but there's no one to come. During the hour I'm there, Wales politely apologizes for needing to get this laptop fixed before he hops a plane to Bulgaria, or maybe today it's Bucharest. I insist it's fine if he keeps trying, and we time our conversation between the blips and beeps that the laptop emits, warning that this or that download hasn't worked either.

He turns away from the laptop and mumbles an embarrassed apology for being so distracted. What comes next? I ask him. Are there lessons we can draw from Wikipedia for everything else we do, for the ways we might work in the future?

"Huge ones," he says without hesitation. "It's about working with accountability—everything is public—and about transparency and visibility. That's the beauty of a wiki. You know who's doing what. That's a benefit to business and to society. There are so many problems to fix in the world. Why waste time having people all working on the same thing when they don't even know about it? I visit big corporations and I hear all the time about people spending a year or two on a project, and then they find out someone else is working on exactly the same thing. It makes no sense. It's a waste of valuable time. There are too many problems to solve."

I wonder if he is implying that the next step might be a gigantic social wiki, where everyone in the world could contribute insights and solve world problems, editing one another, contributing, always knowing who said what, insisting on transparency, on clarity, and on the facts—not on chest-thumping or bragging but on solutions that have plausible precedents to build upon and research behind them, focusing on real problems that can be solved, that are doable.

"It could work," he says. "Why not?" His voice is a little defiant this time. "We've learned from Wikipedia that the crowd is smarter than any individual. We have a lot to learn from the crowd. The only problem is that collaboration is messy. It looks bad at first and then, over time, it looks better than it would have if you'd done it yourself, gone the straight route."

I laugh, saying that I've heard him compare the collaborative efforts of Wikipedia to sausage making: You might like the product when it's done, but you really don't want to see how it's made. He chuckles. "It *is* a little like that."

None of the work done on Wikipedia fits conventional formulas of work. On the Community Portal, one finds a long to-do list: "Welcome to the community portal. This is the place to find out what is happening on Wikipedia. Learn what tasks need to be done, what groups there are to join, and share news about recent events or current activities taking place on Wikipedia."[19] If this were a commercial publisher, all of the tasks would be managed by editors, but here you choose what you want to do and just let others know you are doing it. You can decide to offer "peer review or give feedback on some articles," or you can help develop the next stage of Wikipedia's GIS (geographic information system) tools. Or you can translate articles into English.[20] The list recalls Tapscott's idea of "uniquely qualified minds," meaning you yourself are uniquely qualified in a certain situation. In my office or yours, that might be equivalent to coming in each morning and perusing a menu of tasks that need attention, choosing which one you might wish to take on during a given day, and simply signing up, assuming the responsibility, with all of your cowatchers keeping an eye on how you were doing.

I'm thinking about Legos. And sausage makers, too. Jimmy Wales fits both descriptions. Wikipedia could never have worked, everyone says, if it hadn't been for his leadership, both steady and willing to bend, to change directions. There's definitely genius in leading hundreds of millions of people in creating knowledge together, in letting a collaboration of this magnitude evolve by creating its own community rules even as you keep a constant focus on what those rules mean, including in the tensest and direst of situations, such as with breaking news where a false story or too much information released at the wrong time could mean someone's life.

That was the case in November 2008 when Bill Keller, executive editor of the *New York Times,* phoned Wales and asked him, human to human, editor to editor, if he could follow the lead of the *Times* and not publish the news that reporter David Rohde had been kidnapped by the Taliban. Someone already had posted the news to Wikipedia, but Keller argued that giving the Taliban publicity would increase the chance that Rohde would never be released alive. Wales agreed. For seven months, he kept the news off the site, as did the *Times*. Keller called Wales before news of Rohde's release broke, and Wales finally gave the OK for the entry reporting the kidnapping to be published. "I was right. You were WRONG," the angry user wrote after his post was finally admitted as passing community standards.[21] Everyone expected an outcry from the Wikipedia community, but there was mostly sympathy.

It was a complex moment, and a coming of age—old media reaching out to new media, new media exercising restraint—a benchmark in the history of new forms of online mass collaboration.

"Public collaboration is still in its infancy," Wales says, "All your mistakes are out there in the open, correcting one another, setting one another straight. It's democratic. We've never seen anything like this before, and we don't really know how much we can do, how much we can change, how far we can go with it—but it's looking like we can go very far. It takes work. Collaboration isn't easy. But we're just beginning, and I think we're going to go a long way."

There it is, the largest encyclopedia in human history, a lot of term papers tendered voluntarily because people, we now can see, have an innate (whatever that means) desire to share knowledge on the World Wide Web—so long as all partners know their contribution is valued, all are working toward a larger vision, and the contribution then "sticks" and has a public face.

Plus, one other thing: No one is exploiting anyone else.

Here's a reversal of rational choice theory, the notion based on the idea that people act in ways that maximize their benefits and reduce their costs. Writing a Wikipedia entry can cost a lot of one's time; it can take days. You receive nothing in return, not even your name on the entry. Anyone can edit what you write. If there is a benefit, it's not one that we've calculated before. And it certainly isn't monetary. Wikipedia works only because it is *not* monetized. No one is profiting from the unpaid labor of others. That's the central point of Jimmy Wales fixing his own laptop.

There are many ways to profit from social networking. Just ask Mark Zuckerberg, who has no trouble data-mining Facebook's user-generated content to the tune of several billion dollars. Jimmy Wales protests that no one should ever make a profit from Wikipedia. Ruefully, he chuckles that if he or anyone else ever tried to, someone would probably kill him, and that's probably true. Certainly Zuckerberg has not made fans by his constant redefinitions of privacy and private content on Facebook, and it's not clear how pending lawsuits will determine who owns Facebook's content in the end. Like a lot about collaboration in the digital age, it's by no means a clear matter when judged by the old standards of intellectual property or ownership—and we haven't yet formulated the new ones. Projects like Creative Commons are pioneering hybrid ideas of intellectual property, but it's not always a simple matter in a collaborative endeavor to agree to "share alike."

Who *does* own the content on Wikipedia? Wales would say we all do. He's probably right that there'd be a price on his head if ever he tried to retroactively claim ownership for himself.

Jimmy Wales's computer makes a sickening grinding noise that has us both jump to attention. "That doesn't sound good," I say.

"Excuse me a minute," he says politely and turns back to the screen. I suspect he's more comfortable working on the computer than talking about himself anyway. He wrinkles his forehead again, trying to figure out what could be going so wrong.

That indelible image of Jimmy Wales in a glass box at Wikia, Inc., down the street from the even lower-rent Wikimedia Foundation, trying over and over to make a temporary fix on a laptop that is destined for the Mac Genius Bar Recycle Bin, is touching. It's a glimpse of a new mode of collaborative leadership that is counterintuitive, one that by principle, practice, and prior experience should not be working at all. Except, indisputably, it does.

||||||||||||||||||||||||||||||

Tony O'Driscoll, Margaret Regan, Shane Battier, Thorkil Sonne, and Jimmy Wales are all player-coaches. All are experts at collaboration by difference and all have learned the Lego lesson that difference itself varies depending on context. And so does collaboration. There's no one way to do it; you just have to get into the thick of things and make it work.[22] Yet when the principles of difference are honored, when we gear our interaction to the unique skills and styles of our collaborators, something great happens that is far larger than the sum of the parts. We are all differently abled. And putting those different abilities together can add up to something grand, something that makes us winners.

The principles are simple and can be summarized in three points:

1. All partners know that their contribution is valued by the group and/or by whoever is evaluating performance, including the group itself.
2. All individuals see that their contribution is crucial to the success of the larger vision or mission of the group.
3. Everyone knows that if their contribution passes muster, it will be acted upon and acknowledged as a contribution by the group and will be rewarded in a meaningful way as part of the success of the group.

This is all well and good when it comes to remaking a technology company or a consulting firm, but we have to ask ourselves whether the new workplace of the twenty-first century is also reshaping the most twentieth-century of jobs. Can this fast-moving, shifting, on-the-fly form of collaboration by difference work in a really traditional workplace? I don't mean in Second Life or in Wikipedia, but in the most Taylorist of occupations. Let's say, for example, a construction site. Even if you could allow that you could build the largest encyclopedia the world has ever known using this method, even if you could

admit that the Internet itself is unregulated and collaborative, that's still somehow different from building a building in this loosey-goosey Collaboration 2.0
way. Right?

|||||||||||||||||||||||||||||||

6. Seeing the Future of Work by Refusing Its Past

Nothing else in Greensboro, North Carolina, looks like the lobby of the
Proximity Hotel. Once the hub of North Carolina's textile and furniture trades,
both of which moved offshore over a decade ago, Greensboro is now a typical Southern town abandoned by its former industries. It began to woo new
businesses in the insurance, health, and pharmaceutical industries in the late
1990s, but the recession has set back that rebuilding effort. With a population
of 250,000 and an unemployment rate around 12 percent, it's an incongruous
place to find a spanking new and opulent hotel. Open your eyes in the lobby
of the Proximity and you'd be sure you were in a major urban center, maybe
L.A. or Tokyo. You could imagine music fans whispering in this lobby about a
sighting of R.E.M., camped out in the penthouse suite before a concert. When
Greensboro aspires to luxury, it's typically chintz or brocade, yet at Proximity,
the soaring pillars are square and unadorned, made from some dark material
striated in rich bronze, with amber tones worked to a lacquerlike patina that
defies categorization. An iron reception desk cantilevers from the wall. Behind
it is stunning dark silk Fortuny drapery, almost Art Nouveau.

This doesn't look like Greensboro, but that's not the real surprise. What's
most interesting about the Proximity is that it's the "greenest" hotel in America,
the only one to have received a Platinum LEED designation, the top certification given by the U.S. Green Building Council. LEED stands for Leadership in
Energy and Environmental Design, and the platinum designation is the most
coveted award for environmental sustainability that any building can receive.
To earn it, a building has to meet a very long list of standards and benchmarks,
both for process and product. I know. Duke takes great pride whenever one of
its new buildings receives a LEED award. It requires effort and commitment
to receive even a Silver one. But Platinum? The highest level? For a hotel? It's
a curious occurrence, and practically in my backyard, and that's what's drawn
me here.

Looking around the lobby, I feel as if I'm in a dream. It seems utterly implausible, the whole thing. It doesn't even *look* eco-friendly, whatever that means, but even acknowledging that thought makes me realize I've come with my own preconceptions. I was expecting something that looked earthier, homespun, homemade. But this hotel is stunning, even by my snobbiest cosmopolitan standards. Nothing in my expectations about Greensboro led me to believe this hotel could exist.

But that's not even the best part. Scratch the surface and you find out that the Proximity was constructed almost entirely by local construction workers, builders, craftsmen, plumbers, and electricians, which is all but revolutionary, given the building's feature-forward attitude. This isn't Marin County, and workers here aren't typically trained in environmentally sustainable building techniques any more than they are in urban sophistication and design. The more you learn about the hotel, the more you get the feeling that a great amount of thought and energy was put into its development, and it's clear that this involved a conscious effort to break with past practices.

So how did the Proximity end up putting a new environmentally friendly face on Greensboro? Its story is largely that of Dennis Quaintance, CEO, entrepreneur, developer, businessman *cum* environmentalist. Greeting me in the lobby wearing a sports jacket, loose-fitting trousers, and sport shoes, he doesn't fit my image of a slick developer and CEO. Small and wiry with a thick head of dusty blond hair and a ready smile, Quaintance talks with an accent that blurs the distance between his childhood home of Montana and North Carolina, where he has lived for thirty years. That he was in charge of this cutting-edge design defies another stereotype. That he is an environmentalist fits his healthy, outdoorsy look—but not his portfolio. A successful businessman before he developed Proximity, he previously had made conventional buildings with conventional materials.

If disruption is the key to changing how we see the world, then it shouldn't be so surprising that the event that changed Quaintance's life—and his thinking when it came to development—was of a fairly common kind: the birth of his twins in 1998. Their arrival got him focused on the world in a way he'd never thought of before. During the long nightly walks he and his wife, Nancy, would take together, they began trying to imagine twenty or thirty or forty years into the future, when the twins, by that point adults and parents themselves, would look at them and give them a thumbs-up or a thumbs-down for how they had lived their lives. They might be proud of their parents' success, but wouldn't they be furious that, knowing all the facts in an environmentally fragile world, Dad had gone on building project after project that took from the earth without giving back? Wouldn't they feel that their parents, and a whole generation that knew better, had betrayed them, bartering their future for a little luxury in the moment, choking the air, robbing the earth, polluting the atmosphere?

"We decided right there we had to do it better," Quaintance says. "But we didn't know how. No one did. There weren't good examples. We had to learn it all, we had to learn it together." Quaintance called together about sixty local workers in various aspects of the construction trade, people he had worked with on his previous projects, "tobacco-spitting types," he jokes, not environmentalists by a long stretch. He set them the challenge. He wanted people on the Proximity Hotel project who were willing to learn together, which meant unlearning just about everything they had ever done in the past. From the point of view of present business, those practices may have been successful before. But from the point of view of *future* business—creating a world where our children and grandchildren could live and work together productively—everything they'd been doing up to this point had been a dismal failure. Working with him on this project meant starting over.

"We realized that we wouldn't get any different outcomes with all the same inputs," he says, and so he set them all a unique challenge, to invent a new business model for innovation based on a conviction, on what he called a "sincerity" about wanting to build an eco-friendly hotel. "We were asking people to put our sustainable sincerity first in every decision we made," even before the bottom line. So he made them a proposition. He insisted that he had to have absolutely hard, fixed, nonnegotiable prices for any conventional materials and designs but, for anything on which they took a chance, going for something environmentally sustainable, "all we needed was a good guess." He budgeted uncertainty and speculation into the project, but only for those aspects of the project that met their ecological ambitions. He wanted the builders to know that "we were in this together. We didn't expect them to be the only people taking the risk. They had to trust me. If they thrive, we thrive. If they fail, we fail. You can't ask people to go into uncharted territory while you wait safely on the shore."

It worked. On a project they estimated would cost $30 million to build, they had about $2.2 million in overages. However, because the hotel was designed to be more efficient to cool and heat, the heating, ventilation, and air conditioning system (HVAC) required smaller components than they originally budgeted for. These sorts of integrated design advantages ended up saving over a $1 million, bringing the total project within about $1 million of their original budget. More to the point, they were using 39 percent less energy than anticipated, saving money on bills and earning tax credits.

And in the process they ended up with over *seventy* sustainable practices, a valuable cache of new methods that could be profitably employed in future projects. Most of these they developed themselves, but they kept careful records and made sure to make a list of these practices on the hotel's Web site so anyone else could learn from them. Quaintance is happy to discuss them with anyone who wants to know. One hundred solar panels heat 60 percent of the water for the hotel and restaurant. They restored a stream and planted it with local species, and they used salvaged local walnut trees for the bar in the bistro and bamboo plywood for service trays in the hotel rooms. Sensors on the kitchen exhaust hoods detect heat and smoke and adjust the fan speed automatically. That saves a lot of energy and makes the kitchen cooler and quieter.

Some of these solutions are simple, and one wonders why they aren't already standard practices. But they aren't. Quaintance and his builders invented, researched, improvised, and tracked down the best systems they could find. In typical hotel restaurants, the refrigeration system is air-cooled and, while effective, is often highly energy inefficient. So when Quaintance and his contractor thought about how they might do it better, they hit upon the idea of using geothermal energy, a way of distributing excess heat caused by the appliances into the earth rather than into the room where additional cooling would be necessary to keep down the temperature. Even the elevator contributes to the energy solutions in the hotel. It uses North America's first regenerative-drive system, which means that it captures energy on the way down and feeds it back into the building's electrical system.

As many of the materials as possible were manufactured locally, which saves on shipping costs. This practice provides a training ground for local craftsmen to learn sustainable design and building techniques that will be useful on future jobs and will continue sustainability in other projects. The builders not only used recycled materials wherever possible, but also recycled the construction waste into the materials in the building: 87 percent of construction waste was recycled, diverting 1,535 tons of debris from landfills.

As a guest, I can read about these features, but what's clear without any reading is that I've never stayed in a more pleasant hotel room. There is none of the stifling hotel air or the terrible chemical smells abundant in so many hotel rooms, usually due to the use of volatile organic compound (VOC) paints, adhesives, and carpeting. At Proximity, everything is low-VOC and filtered. Circulating air makes it feel almost as if one is outdoors. The temperature is perfectly controlled, none of those standard hotel blasts of overheated or frigid stale air. Seven-foot-square windows that can actually open flood the guest rooms with natural air and natural light.

The hotel opened at the crisis moment in the recession, but despite the economy, it is doing well. Quaintance and his team garnered good publicity from earning the nation's only Platinum LEED designation for a hotel, and they are exceeding the industry average in the luxury hotel market. Quaintance now knows that they will have made back the extra investment added for the sustainable features within four years, even in the recession.

And because everyone on the project had researched and learned together, others on the project also felt positive effects in their businesses. When he began the project, Quaintance told people that he was sure they'd curse him along the way, but in the end, they'd thank him. And that started to happen even before Proximity Hotel was built.

Steve Johnson, an electrical contractor who had worked with Quaintance on several previous projects, was skeptical about the whole thing when he began. Johnson had liked working with Quaintance in the past but said up front that he expected this project to cost too much. Johnson wasn't sure the project would succeed. Still, he did as assigned, taking it as his personal charge to learn everything he could about green electrical materials. He researched online. He informed himself. He met with his suppliers. He asked them about green building materials and challenged them to come up with new ways of thinking about sustainability. At some point, skepticism turned to conviction, hesitancy to excitement. By the time the Proximity was erected, Johnson says he was thinking of it as "his" building and "our" building," not Dennis Quaintance's building. "Every one of us felt that way. We *all* began to take it personally," he said.

One day he called Quaintance to tell him that he just had received the largest contract in his company's history, awarded because he had demonstrated that he knew how to build green. "We were the first green electrical contractor in the state, the first to build a Platinum LEED building, and people found us, they came to us," he says with unmistakable pride. "People who visit our buildings, who stay in the hotel, feel the difference immediately. They don't even know what good, healthy air quality is in a building, but when they experience it, they don't want to go back. Word spread that we were doing something different, we were doing it better."

Sixty five years old, with five grandchildren, Johnson agrees with Dennis Quaintance that the real issue is the legacy he's passing on. Johnson says he talked to his grandchildren the whole time Proximity Hotel was going up. Like Quaintance, he acknowledges that, although he always works hard, he worked harder on this project than on anything else—and felt better about it, prouder. He wasn't learning alone and he wasn't learning just for himself. He knew he was gaining knowledge that he could share and pass on, including to those grandkids. "I was learning and I wanted them to learn. We all have to be better stewards of the atmosphere and of the earth. Someday it will be their turn."

He had promised Quaintance that the next time he built an office building, he was going to go green. When he built his own new six-thousand-square-foot office building, it received its own LEED rating. "We all love to come to work here." He smiles. "I figure productivity is up twenty-five percent. It's healthy here. There's light. The air is cleaner in our office than in most of our homes. It makes me mad that the new building code says that when you build a new home, you have to put in a carbon monoxide detector. You shouldn't be detecting toxic chemicals. You should be eliminating them. If you build green, you don't need a *detector* for toxic chemicals."

He knows he wouldn't even have had that thought a decade ago. Working on Proximity Hotel changed him. Like Quaintance, Steve Johnson has had a revolution in the way he thinks about building, the earth, his own ability to change his practices, his own importance in the world, and his potential to make a difference.

What surprises him the most is that virtually all the materials he used came from within a one-hundred-mile radius of the project. It was all there; he just hadn't seen it before. Even in the catalogs from his most trusted suppliers, there were badges for "Green"—he'd read right over them, as if they were for someone else, not for him. He hadn't even really noticed them before. Now buying local was not only another way of cutting the carbon footprint but also a way of building a local network of people who could share practices and references, people you could count on if you needed them. It was bigger than any one of them, bigger than Dennis Quaintance or even the Proximity Hotel. There were ripples from their success. There still are, and whenever people come to them with problems, he says, the solutions are here. "They are right before our eyes. All we have to do is change our focus. Suddenly we see what we were missing before.

"We were a team. We were dedicated. We made a difference," he says. "Now, I have people calling me all the time. And I always say, Go green. The main thing I'll tell anybody is it's the *sensible* thing to do. It makes sense."

For Dennis Quaintance, the teamwork, the sense of collective purpose, the connection, and the ability to pass on what they learned so others can learn too were the most gratifying parts of the whole endeavor. "That's the thing when you do this together, when everyone contributes something unique and important. You build not just a hotel but a community," he says.

What changed for Dennis Quaintance, Steve Johnson, and the other members of the Proximity team was their way of viewing the future. Instead of seeing only through the lens of profit, they were determined to view the future through a dual lens of good business and good environment. Quaintance's job as the project's leader was to challenge everyone to look through that second lens. He challenged everyone to rethink the project from scratch, this time through a sustainability lens, reexamining everything they thought they knew about best building practices with that new way of seeing.

"It's not what you don't know that gets you in trouble," Quaintance insists. "It's what you *do* know. What gets you in trouble is the thing you think you know, what you've done forty-four times so you think you know how to do it the forty-fifth."

If you see only through the lens you've been wearing all along, innovation is invisible. You'll never see it. "We had to recognize that process is more important than outcome." And the process was to rethink everything they thought they knew in order to solve the puzzle together. What works? What doesn't work? What actually conserves energy? What looks energy-efficient but turns out to be bogus, window dressing? Dennis Quaintance didn't know he was going to earn Platinum certification. All he knew was that, because everyone was learning together, doing the research together, and exploring together, it was important to document everything they did so others could learn from their experience. The twins joked that it was "toilet of the week" club at the

Quaintance house because each week they would try out another environmental design at home, testing it on the family until they found something that worked, that used less water, and that was beautiful and did the job well. It was that way with everything. They knew they were doing their best together; they knew they were sincere. They were shocked not only to receive the highest Platinum rating, but also to be the first hotel in America to earn it.

"It's embarrassing that no one else had done it," Quaintance says. "All we really knew when we started is that we wanted something better. Everyone pitched in. Everyone figured out a solution to a problem we didn't even know we had. Everyone contributed. We were in it together because we knew it was important to do this. The crazy thing? It wasn't even that hard."

||||||||||||||||||||||||||||||||

Each of these examples of the changing worker is unique, but eventually there will be far more like these. In the best version of our future, all bosses will be like Thorkil Sonne, all workers like Steve Johnson, all companies will run like FutureWork or like Quaintance-Weaver Restaurants and Hotels. And while we're at it, let's dream a future in which all the world's problems will be solved by massive collaborative efforts like Wikipedia and all basketball games won by player-coaches like Shane Battier.

The lessons offered by these individuals are essential for millions of businesses, for millions of workers as we move forward. It is still the rare workplace that is structured to take advantage of the fluidity of what we offer, as individuals, as partners, in the collaborative venture that more and more is how we need to work. Crowdsourcing talent is a skill we're all just learning, a configuration in which each of us is a node connected to peers and extending outward, so our solutions can help others find solutions. We are only beginning to access, cultivate, reward, and even teach in these new ways.

The digital age hasn't yet found its Frederick Winslow Taylor, but even before we have our theory of these new ways of working in a digital age, more and more of us embody the practices of a changing world of work. A friend who works at one of the world's largest pharmaceutical companies recently told me that she has never met the two colleagues with whom she works most closely, one in the United Kingdom and one in Bangalore. She writes software algorithms for global delivery systems in tandem with these partners. They work together every day and have done so for over a year in the endless day of global business, wherein her workday ends when the ones in India and then in the UK begin. She said it took a while for them to synchronize their workflow, to find a comfortable way of working together, but now that they have, they are so effective that they don't want their team to be separated. *Separated* is an interesting word in this context of what we used to call strangers, whose livelihoods depend on one another and whose trust in one another might well be greater than it is among next-door neighbors. She states that the reason their team is so successful is that they have similar communication styles and utterly different talents and perspectives. They look forward to meeting one another someday.

WE'VE HAD OVER A HUNDRED years to nurture individualist labor and achievement, to learn to pay attention to task in order to stay on track, to think of productivity as meeting a goal. We're still very new at thinking of productivity as pushing the imagination in such new ways that one doesn't even know the result of one's own experiments, rather like Tim Berners-Lee or Jimmy Wales watching in wonder as thousands and thousands of anonymous individuals contribute to an open-source project—the World Wide Web or Wikipedia—that no one really knew would work and without either compensation or guidelines for how to do what they are doing. The guidelines—or community rules—evolved not top-down but with everyone contributing as they went along and got deeper and deeper into the project. Success wasn't meeting a quota but finding ways to facilitate through collaboration a never-ending, ongoing way to explore possibilities of a collective nature on a global scale that no one had imagined even as it was happening. Twitter? Really? Twitter's haphazard, nonlinear, yet meteoric rise is the digital-age equivalent of a business plan, a twenty-first-century version of the blueprints and logbooks and stopwatches and punch clocks and assembly lines of Taylorization or the modern office.

It's hardly surprising that our collaborative skills still have rough edges. Yet more and more of us, in ways huge or modest, are honing our collective talents already. The accomplishments may not be grand in an individualistic superhero way, but they are in ways that are enormously satisfying because they're based on respect for the unique contributions of others and ourselves and on the conviction that, by working together, we all have the potential to win.

The individuals we've seen in this chapter inspire us to think about the usefulness of our old assumptions as we go forward. That practical introspection about what habits and practices do or do not serve us is, basically, what is required right now. It's not possible to do this on our own, but with the right tools and the right partners, we can change. We can change our schools and our workplaces, but even more important, we can change our expectations, in what we value, in what we attend to, in what commands our attention. We can start to believe that these are not natural and fixed but that they truly *can* change.

And so can we. That's where the lessons of attention take us next. Biology is not destiny, not any more than current formal education equals learning or than the workplace we've inherited from the industrial era has to shape our dreams for our own productive lives in the future. Even aging is fixed less by biology than by our expectations of what that biology means. That's another of the surprises of this chapter. Not a single one of these cutting-edge thinkers I found in the work world is a "millennial." All were born considerably before 1985. They may have mastered many of the principles of the digital world, but not because they were born at a particular historical moment. Instead, they saw a need to change. They wanted to change. And they took advantage not just of the Internet as a tool, but of the Internet as an opportunity to think about working together differently and better.

That's why I resist any argument that says "the Internet makes us smarter" or "the Internet makes us dumber." These pronouncements too readily deny our will, our resilience, our ability to change the filter, to see in a different way. We know from the science of attention that it is not easy to change on our own, precisely because we cannot see ourselves as others see us. But others *do* see us, and if we create the right conditions, they can help us see what we're missing— and we can help them.

The science of attention teaches us that we tend to pay attention to what we have been taught to value and that we tend to be astonishingly blind to change until something disrupts our pattern and makes us see what has been invisible before. The same science also teaches us that we don't have to depend on surprise, crisis, or trauma to help us see better. Rather, we can seek out the best partners, tools, and methods to help us see. We can internalize the message of the basic brain science of attention. We can assume there are things that we must be missing, and then we can build that knowledge into the ways that we work with others. Through that process, we can find more fruitful, productive, and invigorating ways to organize our lives. By building upon what others see, we can learn to take advantage of what we've been missing, important things that have been there right in front of us for the taking all along. The six individuals we've seen in this chapter already have succeeded in making this change. They have put their ideas into practice, and so can we, not only at school and at work, but *everywhere else.*

As we shall see next, that unlearning and relearning includes changing how we think about ourselves, how we view our future at any age, and how we imagine aging in the future. If that sounds unrealistic, keep in mind what Dennis Quaintance had to tell us. With the right tools, partners, and methods, it's not only possible: *It isn't even that hard.*

Part Four

||||||||||||||||||||||||

The Brain You Change Yourself

8

IIIIIIIIIIIIIIIIIIIIIIII

You, Too, Can Program Your VCR
(and Probably Should)

Instead of answering my question, the physical therapist at the Duke Sports Medicine Center wrote a series of numbers on the whiteboard and then drew a line down the middle, forming two columns. It looked something like this:

29	31
39	41
49	51
59	61
69	71

I was only halfway through physical rehab at that time. I'd spent three months after surgery trying to get my right arm and hand functioning again, following one of those freak accidents that changes your life in a matter of minutes. My husband, Ken, and I were visiting our friend Diego at a summer place he was renting on the incomparably beautiful island of Capri, off Naples, Italy. We decided to go swimming in the famous opalescent waters near the Blue Grotto one day. Beaches on Capri aren't long stretches of sand. They are flat places on stern cliffs that jut from the Mediterranean. Diego and Ken had already cannonballed into the water. They bobbed there, smiling, waving me in. It wasn't a long jump but I got cold feet and decided to enter via the ladder embedded in the cliff face.

Apparently my foot slipped. I don't remember the interval between touching one toe to the cool metal rung and, sometime later, being untangled and lifted to safety by a stranger. I remember Diego and Ken swimming toward me, a look of horror having replaced the earlier smiles. It was my good fortune that the Italian gentleman waiting to take the ladder into the water after me happened to be a doctor. Apparently, when my foot slid, instead of falling into the water, I got my arm tangled in two directions, between the top rungs and also between the cliff face and the ladder. When the doctor described what he'd witnessed, he made an abrupt gesture, like someone killing a chicken by wringing its neck.

On the advice of the doctor, Ken and Diego bundled me into an island taxi and rushed me to the tiny emergency room on Capri. I was obviously not the first person to arrive after a collision with the rock face. The colossal nurse who presided there looked like someone out of a Fellini film, but I was soon happy for her size and her dominatrix disposition. Even before the X-rays came back, my arm was starting to swell, and she said it was best to fix a dislocation immediately or it would be weeks before I'd have another possibility to set it in its socket. Nerve damage could result. With one expert chop to my shoulder, she re-placed my arm. With similarly precise power moves, she restored first my elbow and then my thumb. I don't actually remember any of these, for I seem to have blacked out on impact each time.

The X-rays showed that nothing was broken, but we all could see I was in bad shape. The nurse put my arm in a sling, supplied me with meds, and sent me on my way. She didn't even take my name, because in Italy all health care is free, even for tourists. We took the ferry back to Naples, and I pretty much stayed drugged during the flights to Rome, then to Dallas, then to Raleigh-Durham, where I went to the emergency room at Duke and eventually they figured out what to do with me.

Claude T. Moorman III, an athletic man who lettered in football for Duke, was the brilliant surgeon into whose hands I was delivered. The complexity, extent, and unsystematic irrationality of my damage resembled that of an athlete leaping high to catch a football only to have his legs tackled out from under him, his arm entangled in tons of human flesh. I didn't look mangled the way a crash victim with similar damage would, but basically nothing was working; a lot of tendons and muscles and nerves had been severed. A good deal of the surgery had to be improvised. X-rays of my postsurgical arm look alarming or comical, depending on your sense of humor, with titanium gizmos that look as though they'd come from Home Depot. I've had follow-up tests in the seven years since the accident, and apparently I'm the only person left, anywhere, with a cup hook holding up her biceps.

Besides actual screws, what was drilled into me in the weeks after the surgery by Dr. Moorman and everyone else was the importance of physical therapy. I was fortunate again to be at a place like Duke, with its splendid medical facility and athletics, and was soon passed on to the wise and patient Robert Bruzga and his staff at Duke Sports Medicine, for what would turn out to be six months of mind-numbingly boring exercises and surprising glimpses into the human psyche in multiple forms, including my own. Bob and his staff were great about giving me small challenges to meet—almost like the boss-level challenges in game mechanics. As a teacher and a researcher on the science of attention, I was fascinated by how creative they were at coming up with new goals, new modes of testing. Although progress was glacial, the methods kept me mentally alert, focused and refocused, and physically reaching—another lesson useful for both the classroom and the workplace.

The day I could move my right arm one inch forward from plumb position was a triumph. That took about a month.

Many factors conspired toward my recovery. First, I was given a tip by one of my most inspiring rehab buddies at the center, a man I knew only as Bill, a seventy-three-year-old runner who was getting his knee back in shape, after surgery, for an upcoming triathlon. He told me that Duke athletes who were rehabbing injuries or working on skills tended to arrive between four and five o'clock. I started coming to rehab at three, would work for an hour, and then get a second boost of energy from watching these young men and women work at a level of intensity that both put me to shame and inspired me.

On some afternoons, a young star on our basketball team, a first-year student, would also be there working with Bob Bruzga and others to improve his jump by trying to snatch a small flag dangling above his head from a very long stick. As the student doggedly repeated the exercise over and over, a team of professionals analyzed his jump to make recommendations about his performance, but also recalibrated the height of the flag to his most recent successes or failures, putting it just out of reach one time, within the grasp of his fingertips the next, and then, as his back was turned, just a little too high to grasp. It is a method I'd seen used to train competition-level working dogs, a marvelous psychological dance of reward and challenge, adjusted by the trainer to the trainee's desires, energy, success, or frustration, and all designed to enhance the trainee's ability to conceptualize achievement just beyond his grasp. It is also a time-honored method used by many great teachers. Set the bar too high and it is frustratingly counterproductive; set it too low and dulled expectations lead to underachievement. *Dogged* is the right word. That kid just would not give up. I would watch him and try again to inch my arm forward.

Another inspiration came from reading the work of Vilayanur S. Ramachandran, one of the most inventive neuroscientists of our age, who also supervises actual physical rehab work with patients, putting his research to immediate and transformative use. One of his most famous "cures" is for phantom limbs. A strange phenomenon no one has really figured out, phantom limbs are experienced as real by the amputee, as if they'd never been severed at all. Sometimes they cause pain, sometimes they itch, sometimes they are just plain spooky in their simulation of the missing limb. Ramachandran was one of the first people to actually ask patients to describe these limbs. In his marvelous book *Phantoms in the Brain,* Ramachandran offers a vivid description of one such encounter:

> I placed a coffee cup in front of John and asked him to grab it [with his phantom limb]. Just as he said he was reaching out, I yanked the cup away.
>
> "Ow!" he yelled. "Don't do that!"
> "What's the matter?"
> "Don't do that," he repeated. "I had just got my fingers around the cup handle when you pulled it. That really hurts!"
> . . . The fingers were illusory, but the pain was real—indeed, so intense that I dared not repeat the experiment.[1]

I especially appreciate the way Ramachandran, in this account, respects the intensity of the real pain experienced in a phantom limb and strives not to inflict it again.

Although my own arm was real, the nerve damage was extensive, and I often experienced odd sensations and even abrupt movements, though it still dangled limp at my side. It used to be thought that a phantom limb resulted because the nerve endings had been irritated during amputation and were still sending signals back to the brain. Ramachandran theorizes that certain areas of the cortex near the ones that would normally control the missing arm continue to respond as if the arm were still there. He came up with a novel treatment using a mirror box originally devised by thieving street magicians in India. The thief would trick gullible believers into thinking they were seeing one hand in the box when they were actually seeing a mirror image of the other hand. The free hand was actually under the table picking their pockets. Ramachandran used one of these mirrored boxes so that patients could think they were looking at their severed hand and give it instructions. Open, close: What looked like the phantom limb would open and close. The brain, in this way, could be convinced that its signals were being dutifully responded to so it could calm down and stop giving them. The mirror trick didn't work in all cases but was successful in several. The disturbing phantom limb went away.

Since the nerves in my arm had all been traumatized in one way or another and often signaled to me in unpredictable and inexplicable ways, I began to try variations on Ramachandran's mirror experiments during my physical therapy. I would often lift my good (left) arm in front of a mirror, but I would stand so that I could not see my injured arm. I would lift my left arm in the correct movement, and visualize in exact detail my right arm doing the same. It probably was only going up an inch or two, but I would stare at the image of the working arm in the mirror, imagine it as the injured one, and lift, partly with my arm but partly with my imagination.

It took about a year for me to be able to successfully use a computer mouse again, my thumb being difficult to rehabilitate. I would mix it up, mousing left-handed with a big mirror leaning against my computer screen so I could visualize it was my right hand mousing. I tried lots of other tricks with mirrors too.

Did all these exercises inspired by Ramachandran's neuroscience do the trick? I have no idea. Probably not, from a purely medical point of view (if one believes in such a distinction). But psychologically, it seemed to be working, and if you were to look at me now, you would not know this right arm of mine, blazing away at the keyboard, once hung like a twelve-pound anchor at my side, nor would you suspect that my busy mouse thumb had ever been compromised.

The third huge help came from Sadiik (as I'll call him), the Somali taxi driver who would pick me up at my home, drive me to physical therapy three miles away at the Sports Medicine Center, and then drive me home again. The first time he picked me up, he came around to open the taxi door for me, and I could see that he limped and carried one shoulder higher than another. He told me a little about himself over the course of the next several months, and although he looked to be in his late thirties or early forties, I learned he was actually a decade younger than that. He'd been living in the United States for five years and had brought his entire family here. Many of the taxis in my area of North Carolina are owned by Somalis. Sadiik was cautious about describing how he was injured, but *war* and *camp* were two words I heard in the course of his very circumspect admissions. He was extremely wary of saying anything that might hurt his chances of earning his green card, and I, too, learned to divert the conversation if it seemed to be heading in a direction that would cause him anxiety later on.

Every day as he dropped me off at rehab he would say, "You are so lucky," and I would say, "Yes, I am very lucky." Every day when he picked me up after two hours of rehab, he would ask, "So what did you do *better* today?" Because I knew he would ask that question, I felt inspired to do something, each day, that would give me something to answer. If I failed to lift my arm an inch, I would try to ride the exercise bike an extra twenty minutes. If I was too tired for aerobics, I would work on stretches. I reported something better every day, until that day, in March, with the leaves finally coming onto the trees again and the first cherry blossoms beginning to bloom, when Sadiik watched me approach his taxi, and said quietly, "This is the last time I'll drive you, isn't it?"

It was both exciting and sad to tell him that I'd been given permission to drive myself to physical therapy. It was on that last ride home that I thanked him, again, for helping to inspire my recovery and he told me, for the first time, that my persistence had also inspired him do to some stretching exercises he was supposed to do, but that he'd given up years ago because of the pain they caused. Sometimes, when there wasn't another fare, he would park his taxi, wait for me, and say his Salaah, his evening prayers. If he had time, he said, he would then do the arm exercises while he waited for me. The stretching was starting to ease the pain in his shoulder.

I hadn't realized before that my daily response to his inspiring question had also been motivating him. This, I began to see, was collaboration by difference of the tenderest kind.

So what about those numbers the physical therapist wrote on the board in answer to my question? On the days when I went to the Sports Medicine clinic, I would end my therapy session doing water exercises in the luxurious therapy pool. At first I'd been angry at how awful the changing room was—freezing showers that were hard to turn off and on, uncomfortable seating, no one to help you get a swimsuit on or off (no easy task wet and one-armed). But I soon realized that the unpleasant locker room was intended to remind people that the pool was for therapy. It wasn't intended to be a spa. This was important because I discovered there was a substantial group of people who would have preferred to use physical therapy as a time to loll in the pool and chitchat, mostly to complain. I'm convinced that if the shower room had been any more pleasant, they would have been there taking up limited space in the pool every single day.

As you can see, I was irritated by these pool slackers, as I called them. I resented their bad attitude. Unlike the septuagenarian triathlete, unlike the basketball player perfecting his jump, unlike Sadik, and unlike the dedicated therapists at the center, those pool folks drained my energy. If I focused on them (including my irritation at them), I felt defeated that day. So mostly I pretended they weren't there. I loved the pool when I was there by myself or with someone really dedicated to physical therapy. Parts of the pool were eight feet deep, allowing me to do all manner of exercises with the various flotation devices. The water was a luscious 85 degrees and limpid turquoise as soothing as the Caribbean. The two-story windows looked out on a beautiful patch of forest land. The warm water felt fabulous on my dangling, near-dead arm, and I'm sure the water made those other folks in the pool feel just as good. I was unkind not to appreciate that—but I didn't. They'd be there almost every day, the same cluster of "old people," many of them younger than me. The most common topic of conversation was their surgery. Typically, they were on a second or third or even fourth surgery for the same injury, often to the rotator cuff. The water felt so good, and I was able to perform exercises there with the benefit of the buoyancy that I could not do without, so I loved that part, but I dreaded how often people would come into the pool, meet one another for the first time, and then immediately join a chorus of complaint about "getting old," how the "body doesn't heal fast anymore," often pointing over at some young athlete walking laps with water-weights over and over and over, "Look how the young ones heal!" someone would say, floating there doing nothing.

So that was my question, the one that elicited the chart on the whiteboard with the two columns of numbers. Was there any hope for me, realistically speaking, of regaining the full use of my arm *at my age?*

The therapist said there was absolute, irrefutable evidence that, within the same family of injuries, the more time you spent in serious and constructive rehab, the greater the extent of your recovery. There actually was no real evidence that, given the same amount of careful rehab effort, a young person healed more quickly or better than an older person. It was impossible to measure individual by individual, of course, because no two injuries are alike and no two bodies are either. But in the rehab studies, controlling for as many body types as possible, the results were rather like Timothy Salthouse's studies of IQ in which he gave a large group of people the different types of IQ tests under different conditions, and found greater variations within one individual than between individuals. Physically, as well as mentally, we are each our own bell curve.

That's when the physical therapist wrote the numbers on the whiteboard. It turned out those were ages. On the left were the ages before the turn of a new decade, and on the right the age one year after turning thirty, forty, fifty, or sixty. She called it a "decade event." She said she had no rigorous test for it, but she and the other therapists had noticed that if people were a year away from a decade event, they often felt as if they had lucked out, that time was still on their side. These patients often worked hard, did the exercises, got better, and felt exhilarated at the "near miss" of recovering, while still in their twenties, thirties, forties, or fifties.

However, she said that, anecdotally she found the opposite also to be true. If people had just gone through the ritual of passing into a new decade, it was hard to convince them that the birthday was not the kiss of death for future recovery. They would often begin physical therapy with a lament, half joking but also real: "I knew my thirties would be downhill." Her point was that if you begin with a negative rhetoric and mind-set, it is hard to inspire yourself to do what are, in the end, excruciatingly tedious repetitions of the same exercise toward recovery so incremental it hardly counts as positive feedback. She and her colleagues had seen enough evidence in the training areas to convince them that those who'd just had a major birthday tended to exercise less, had a rotten attitude, and never made up the lost ground. Their poor recovery confirmed what a bummer entering this new decade was, with all its harbingers of approaching decrepitude. If the patient was thirty-three, he or she seemed to be on track again; by thirty-five the decade event was irrelevant, but then the pattern would start again, the ticking clock of aging making him determined as he approached the new decade marker and resigned to failure on the other side of it. The therapist was convinced that a twenty-nine-year-old had more in common with a thirty-nine-year-old, in terms of attitude toward and success at recovery, than with a thirty-one-year-old. Attitude was all. Old age didn't seem to have nearly as detrimental an effect on healing as one's prejudices about age and the body's ability to heal.

That rhetoric about "I'm too old to be cured" is something I heard over and over during my six months of intense rehab. I suspect that physical therapist's intuition was correct, that the self-fulfilling prophesies of middle age hurt us incomparably more than the creaking of joints or the forgetting of proper names.

I see the same thing now, in my weekly or twice-weekly Pilates classes with Radonna Patterson, who stands five feet tall but is as strong and as tough as any other Texan you'll find. A professional dancer, she suffered a catastrophic back injury that effectively ended one career and brought her to another as a physical therapist. Convinced from her dance background that every part of the body is linked to every other, Radonna now combines her training in traditional physical therapy with Pilates, Gyrokinesis, yoga, breathing techniques from Eastern meditation, and deep-tissue massage. She even partners with an elderly doctor of Chinese medicine, who comes once a week to work with various machines and techniques. Radonna also uses a lot of hard-eyed determination. She doesn't tolerate whiners. After getting my arm functioning about halfway, Bob Bruzga delivered me to Radonna. She sized me up, put me through the paces on her machines, and assessed what I could do. Then she began challenging me to do more.

Given the mechanics of my arm now, there are some things I'll never do again, but most of the time, I am unaware of those limits. As with attention blindness, we can rarely see our own body patterns, and I can't tell that my arm "limps" except in a few very specific situations. Radonna, though, delights in finding weird movements that I'm not able to perform with my "bionic" arm, and when I admit I can't do something, she'll come up with some pretzely exercise to work on that. She's the kind of teacher who stays up at night thinking about such things and then tests them the next day. It's not surprising that most of her clients are recommended by doctors or that many are medical professionals themselves. She's not big on self-pity, so almost everyone in Radonna's studio looks and acts considerably younger than their years. It's not unusual to see someone seventy-five suspended upside down from the towering piece of Pilates equipment known as the Cadillac.

No one is required to go to Radonna's. People are there to work, at any age. What the physical therapists confirmed as the significant correlation between healing and the determination to do what it takes to heal—even mindless, repetitive, constant, dogged exercise—is alive and well at Radonna's studio.

THE SAME PATTERNS OF EFFORT my physical therapist found also hold true for the brain. Some biological differences among elderly, middle-aged, and young brains exist, of course. I'm not saying they don't, any more than I'm saying that physical challenges don't exist. But *attitude* plays a tremendous role, at any age, in one's cognitive and physical health and in one's ability to make a change. Other factors that I experienced during rehab and that we've seen throughout this book also play their roles: inspiration from the young, indefatigable athletes working on improving their game, as well as from the seventy-three-year-old triathlete who told me about them; the daily challenges of situational, strategic learning geared to my successes (rather than a standardized lesson plan for progress imagined for a "mean" that does not exist); the kind reticence of Sadiik, whose life and example disallowed my wallowing; and, of course, reading neuroscience and being convinced (whether it was "true" or not) that my mental attitude played as big a role in my rehab as my physical exercises—and that the two were connected.

Avoiding the downers, the people who were convinced the exercises did not work, was also key to the cure. What the chart epitomized was unforgettable: At any age, if we believe our age controls us, it does. If we believe we control it, we have a fighting chance.

|||||||||||||||||||||||||||||||||

Back in the distant twentieth century, the ultimate symbol of our lack of control over our body and over technology in a changing world came together in the device known as the VCR. Younger readers might not remember what those are, so for their benefit, I will explain: VCR stands for video cassette recorder. You hooked up this device the size of a large briefcase to your television set, you programmed it to record a TV show you wanted saved, and then, later, you could play back a wobbly, grainy, seasick-colored tape of the show. A modern miracle! That's how TV was "revolutionized" back in the day, kids. This was before you could watch streaming video via Hulu on your laptop or download from TV.com on your iPhone. In fact, it was before Hulu, TV.com, iPhones, or the Internet even existed.

The VCR was clunky, and it quickly became the late-twentieth-century symbol of the technological generation gap. You could divide the world into "old people," who couldn't figure out how to program a VCR, and "young people," who would program it for you if you asked nicely. Gender exacerbated the situation. "I'm too old to learn how to program the VCR," someone, invariably female, would say, and a teenager would step up to do it for her.

"Age," Radonna would say, "gets blamed a lot of times when the operative word should be *lazy*."

Lazy would be an odd word to use for the brawny industrial age, but it certainly is the case that much of our conditioning and many of the institutions designed to prepare us for work in the twentieth century were also about delegating expertise and oversight to some people and therefore delegating the status of the amateur to others. But go back before the twentieth century, and you find a different system. In preindustrial society, you figured out solutions, worked in communities, turned to the person you knew who was "good at" fixing oxcart wheels, and paid him with the Speckled Sussex hen you bred. It's not that he was the expert and you weren't, but that, as a community, you exchanged expertise all the time—collaboration by difference! And the next week, when the guy who repaired the wheel on your oxcart needed help mending fences, he might drop by to ask you to pitch in on that job. By contrast, if the blinking light on my VCR is driving me nuts and no teenager is handy to fix it, I do exactly what my century trained me to do: I go to the Yellow Pages, look up "TV repair," and call an expert to come to my house and reprogram my VCR. I don't pay him in chickens. I hand over forty bucks.

Expertise existed long before the industrial age, as well as after, but the hallmark of the twentieth century is that it *formalizes* who is and who is not an expert, who will and who won't be charged with paying attention to certain problems but not to others. Expertise in the twentieth century gets fully credentialed, institutionalized, formalized, and, perhaps most important, put into a strict and even rigid hierarchy of importance: What is most valued, respected, and/or remunerative is, implicitly, the standard against which others who lack these skills are judged. The parsing of humanity according to who is delegated to be the official judge of quality in each realm is exactly what the industrial age is about. The division of labor into white-, blue-, and pink-collar workers; the division of home and work; the division of fields; the division of the "sciences and technology" from the "human and social sciences"; the division of talents ("good at art," "good at math," "good at school"); the division of abilities (and therefore disabilities); the division of the young, the adult, and the aged; and even the division of the brain (executive functions in the prefrontal cortex, emotional reactions to music in the cerebellum): All of the different ways the world and human nature are parsed and segregated and *valued* that may seem natural to us have evolved into carefully stratified levels, separate spheres of influence, and rigidly credentialized and institutionalized roles and functions in the last one hundred years of human history.

And then comes the World Wide Web—all the world's documents in all the world's media—riding atop the Internet, which links each to all and allows anyone with access the ability to connect to anything and anyone at any level at any time without intervention by an expert. In the twenty-first century, if your oxcart wheel needs fixing, you don't go to the Yellow Pages to find someone you can pay to fix it. You go to Oxcart.com and post a photograph of your broken wheel and ask for advice on the best way to fix it. Within minutes, someone in Thailand will be giving you point-by-point instructions, and then someone in Bali might be offering a different opinion. If the twentieth century was all about training experts and then not paying attention to certain things because the experts would take care of the matter for you, the twenty-first is about crowdsourcing that expertise, contributing to one another's fund of knowledge, and learning how to work together toward solutions to problems. We're just at the beginning of that process, but a first step is unlearning our assumptions, not only about who is or isn't the best expert, but also about our own ability to *become* expert in a given situation. That means getting rid of a lot of tired ideas about ability and disability, prowess and infirmity, and even about the inevitabilities of getting old. Life is a constantly evolving, changing, customizing process of remixing and mashup, learning and unlearning and relearning, over and over. There is no pinnacle to which we ascend and from which we fall. Those "decade events" don't spell our doom—as much as we might like the excuse that, now that we've turned thirty or forty or fifty, we are "getting old" and therefore can delegate a lot—like programming the VCR—to those still young and vigorous.

The more immediate point is that we are not getting away with anything, in the long run, by blaming age and avoiding change. All we do when we use aging as our excuse is reinforce our sense of ourselves as losing control. And feeling in control, it turns out, also is a factor in our success. A study led by Margie

Lachman, director of the Lifespan Lab at Brandeis University, underscores this point. In 2006, Lachman's team studied 335 adults whose ages ranged between twenty-one and eighty-three. The study was funded by the National Institute of Aging (NIA) and specifically looked at the sense of control that middle-aged and older adults perceived themselves to have over their own cognitive functioning. All of the participants were set the task of recalling a list of thirty "categorizable" words—fruits, trees, flowers, etc.

What the study revealed is not that younger participants were able to recall more nouns than older participants. The study showed that those who were *confident* in their cognitive abilities did better on the test than those who were not.

That is very good news. In purely biological terms, there are differences between twenty-one-year-old brains and sixty-year-old brains. Myelin is a chief difference, and those who want to assert the biological determinism of aging often emphasize it. A twenty-one-year-old is at the very height of myelin production, neurally fired up and ready to go. Myelin is the fatty substance that surrounds the power source—the axons—within each neuron in the brain. Myelinated axons transmit information a hundred times faster than unmyelinated axons, and myelination keeps impulses flowing smoothly along the axons.

We're back to the Hebbian principle we saw in chapter 2: Neurons that fire together wire together. Well-myelinated neurons fire best. From twenty-five years on, myelination begins decreasing slowly and then declines more rapidly after sixty or so.[2] Demyelinated axons in the middle-aged and elderly are thought to contribute to diminished short-term memory and working memory (the kind of memory needed to remember a sequence of operations, such as programming a VCR).

Demyelination is a fact of brain biology, but a bit of a squirmy one. No one knows how much demyelination occurs within any one individual and at what rate, nor how much the actual diminishing of myelin changes that person's life. There is no way to measure individual change, biological or behavioral, in anything like a sustained way. There's a leap of inference here that the biological change *necessitates* a corresponding change in performance or behavior because, as we saw from the chart on the whiteboard, will and attitude can outweigh diminished capacities to such a great extent that chronological age becomes a relatively minor part of the equation.

The president of the United States, the so-called leader of the Free World, was 48 when he took the oath of office, and he was considered a young man.

Similarly, the average starting age of a new CEO of a Fortune 500 company in America is 48.8. Even in today's supposedly youthful digital information industries, the average starting age for a CEO is 45.2 years.[3] Their middle-aged neurons are fraying as we speak! But what does that mean? Our definition of *young* is variable, depending on the circumstances. Forty-eight is "young" to lead a nation or a corporation but "old" when it comes to physical rehabilitation or the VCR or the occasional memory lapse.

Lachman's study tells us a lot about the importance of attitude relative to biology. Confidence in one's sense of cognitive control might even outweigh the supposed debilitating effects of demyelination. She found that, statistically, there is a greater correlation between confidence and memory than between age and memory. As Lachman summarizes, "One's sense of control is both a precursor and a consequence of age-related losses in memory."

Anyone over the age of twenty-five needs to go back and parse that sentence. Lachman is saying that feeling confident and in control helps your memory; having a sense that your memory is good helps you have a good memory.

And the reverse is also true. If you are facing middle age and finding yourself at a loss for the occasional proper name, you can exaggerate the loss and soon be undermining your own confidence. If you are positive that anytime you forget a word you are seeing a harbinger of worse things to come—dementia or Alzheimer's, just on the horizon there, right around the next cognitive corner—you shake your confidence. Lachman's study suggests that loss of a sense of control makes things worse, and pretty soon the process becomes self-fulfilling, as any new lapse looms large while successes are invisible. It's the self-reinforcing selectivity of attention, yet again.

My eighteen-year-old students don't care a bit if they or their classmates forget a noun. The last time I taught This Is Your Brain on the Internet, I conducted a little experiment. There was nothing rigorous or official about it; I simply made a mark in my notebook whenever I caught one of them forgetting a proper name or a date. The discussion in class was always energetic and roamed widely, covering anything from parallel GPU (graphics processing unit) 3-D rendering to the brain waves of Buddhist monks during meditation. The students ranged in age from eighteen to twenty-two, with one "elderly" graduate student (he was twenty-eight) the exception to this undergrad demographic. If any student forgot a word or name, I made a little tick and watched to see how they recovered from the lapse. This happened a lot more often than you would think: on average about five to seven times per class.

How did they react? They themselves barely noticed and no one else cared at all. Their memory lapses didn't disrupt the flow of the conversation and seldom diverted them off course from the points they were making. If it didn't matter, they'd just skip over it with a phrase like "I'm blanking on his name." End of matter, moving on. If it was something important, another student often pitched in to help, suggesting the forgotten name. On other occasions, the student who couldn't remember in class would post the information later to our class blog, often with relevant links to Web sites or to an interesting video of the same phenomenon on YouTube. No embarrassment, no apology.

What I never witnessed once in my fourteen weeks of keeping track was the kind of self-conscious stumbling around that characterizes middle-aged forgetfulness. I have a coworker who stammers and blushes and apologizes and stops any conversation dead in its tracks whenever she forgets a name. We all feel embarrassed (and annoyed) as she hems and haws and apologizes instead of finishing what she had to say in the first place. Of course, that failure and humiliation make her feel less confident, and that exacerbates the potential for forgetfulness next time.

You know what I'm talking about—the way middle-aged people break into self-consciously apologetic jokes about "senior moments" or lament that "old age sucks." My friend, novelist Michael Malone, calls this tiresome midlife game Name That Noun! It's a self-preoccupied game. It brings everything else to a halt and calls humiliating attention to the forgetful moment and the forgetter. Given what we know about attention, we know that every time we remind ourselves of a painful failure, we are setting up the anxieties that contribute to the next one.

I admire people who find ways to cope easily and graciously with inevitable stumbling blocks. No one does it better than Nannerl Keohane, who was president of Duke University from 1993 to 2004. My friend and also a mentor, Nan engages affably and generously with everyone she meets. She's an extraordinary human being by any standard, and she was always gracious as president. No matter how many dozens of faculty, students, alums, and interested others would come to greet her at our university functions, she found ways to make a personal touch with almost everyone.

"You have an incredible memory for names," I once said after observing her at a very long public event. She demurred, noting that her ability to summon up the exact proper name in a circumstance was good enough for her purposes, but she didn't consider it exceptional. She said what she always did try to do was make some meaningful connection with every person she met, so there was something significant in any exchange. "That's always more important than remembering names," she said.

I observed her more carefully after that. "So good to see you," she might say upon warmly greeting someone at a party. And then, shortly after, "How's your son liking college?" She'd refer to a past encounter or simply connect with a comment offered by the other person. Sometimes she said the person's name, sometimes not. I had no way of telling if she didn't say a name because she didn't remember it or because the conversation was so engrossing that I hadn't even noticed whether she'd said the name or not. The point is that the name was not the point. A conversation began, a connection was made, a human contact reaffirmed. If she had played Name That Noun! and made a big deal about either remembering or forgetting the person's name, the flow would have stopped and the focus would have turned from her warm, human interactions to her own accomplishment of (or failure at) being able to remember a proper name.

The ability to remember names is an achievement, but being able to connect to the person you are addressing—*whether or not you can recall that person's name*—is a far greater one. It's also a better way, neurologically speaking, of remembering names in the future. The more you stand and stammer, the more you retread the tracks of forgetfulness. If you move on to another association with the person, you have a better chance of triggering the memory of the name you misplaced.

We're back to the old VCR routine of the 1990s. If you program it, you become confident about programming it the next time. If you palm it off on your teenage son, you create a bad habit, a pattern of avoiding what you don't want to do and blaming it on your age. If Radonna were around, she'd be chiding you and urging you to climb on the Cadillac and give yourself a good stretch, upside down.

THE CONFIDENCE OF YOUTH, EVEN in the face of forgetfulness, is a good enough reason to hang out with young people. James R. Gaines, formerly a top editor at *Time, Life,* and *People* magazines, recently started at FLYP, a multimedia online publication that combines text with motion graphics, cool Flash animation, and streaming audio and video. At sixty-one, Gaines was suddenly the oldest person in every room, often twice as old as anyone else. He quickly noticed that he was the only one who cared about his age. His young colleagues wanted his new ideas, not his past experiences. He learned to think in the present and the future, not so much in the past. He's the boss of the operation and yet daily he finds himself relying on much younger people to help him with things like video codecs and MySQL databases. He's learned that his reliance on his young coworkers changes not only the hierarchy of the office but changes him. He has the wisdom of the boss but also the energy and enthusiasm of the learner. Collaboration by difference has the collateral effect of turning one into a student again, even reversing the typical position of student and teacher.

This isn't like palming off the programming of the VCR onto the youngsters. This is a collaboration of those uniquely qualified minds, cross-generationally, but with the same qualities of trust and asymmetrical exchange that we saw in the last chapter. There is even a Shane Battier–like facilitated leadership at work here, for when it comes to the latest of the latest technologies, Gaines finds himself deferring leadership to someone else.

He enjoys being on the flip side of their youthful wisdom. He was especially pleased when a young coworker, who had been working for FLYP longer than he had, commented, "Seeing someone like Jim find what we're doing exciting has made me see it in a new way, sort of like when I started out. His enthusiasm for it reminds me why I went this way." Gaines finds it charming (and only ever so slightly galling) when someone half his age compliments him in an avuncular fashion, "Fine young man, that Jim! Going places, that boy!"

What Gaines adds to FLYP is a certain calm and an ability to put small upsets and victories into a larger perspective. That skill is appreciated by his young coworkers as much as his expertise and insights. The asymmetry of the contribution does not diminish respect in either direction but adds to it. Gaines finds his new role inspiring. "The young people I work with now will be the settlers of that frontier, and I can't think of anything I would rather do than help them get there," he says.[4]

Complementary skills help everyone see differently and better. That is emphatically true cross-generationally. We *are* susceptible to peer pressure. Hang out with middle-aged people all the time and it can be like the people in the therapy pool. You start reinforcing one another's worst habits.

If schools and the workplace operate on a twentieth-century industrial model, think of the institution of the nursing home. As an outgrowth of what used to be called the sanatorium, the development of a segregated facility for the elderly is yet another part of the division of labor and productivity we've seen was characteristic of the twentieth century. We now know that the elderly do better when their peers are not limited to other old people, and studies have shown vivid improvements if you pair older people with the young, infants, the disabled, orphans, prisoners, dogs, hamsters, or even goldfish. In fact, giving a very infirm person a pet to take care of, a pet for which she is responsible, is probably the single best thing you can do for both of them. As one friend likes to joke, it is equally good for both "the petter and the pettee."

The key is restoring a sense of vitality, purpose, difference, and control. Of course, control is mostly an illusion—so why not cultivate the illusion to one's benefit? It's a bit of a trick with mirrors and phantom limbs sometimes, but if the trick works, go for it. Here's Margie Lachman again: "The more you believe there are things you can do to remember information, the more likely you will be to use effort and adaptive strategies and to allocate resources effectively, and the less you will worry about forgetting."[5]

As my students in This Is Your Brain on the Internet know, if you forget something, why worry? What's the big deal? You can always google it later on.

||||||||||||||||||||||||||||||

I am optimistic about the most recent discoveries in the neuroscience of aging. I believe in the power of a can-do spirit and the rejuvenating energy that comes from working with others who don't use their age as an explanation for their shortcomings. That said, I want to say outright and explicitly that I am skeptical about many grand claims about rejuvenation. Some are overstated and overly optimistic. Others smack of hucksterism. Amazing feats of neurogenesis—neural plasticity and recovery—are reported all the time, but aging still exists, and so does brain damage and disease. In the late 1990s, when stem cells were discovered in the hippocampus—a key area for memory—some predicted an imminent cure for Alzheimer's, brain-stem strokes, and other forms of damage. That miracle hasn't happened yet.[6] Every year or two some new health evangelist promotes (and some get rich by advocating) a new substance or practice: green tea, blueberries, cocoa, ibuprofen, ballroom dancing, aerobic exercise. Others tout the rejuvenating properties of extension courses, learning a new language, plenty of sleep, or positive leisure activities. Daniel G. Amen, a popular motivational speaker on aging, advocates antioxidants such as ginkgo biloba and vitamin E (in fish oil), natural vitamin supplements, aerobics, psychological exercises ranging from meditation to goal-oriented positive thinking, music therapy, good nutrition, lots of sleep. Plus ample sex. There are worse ways to rev up those neurons.

But lots of people are getting old right now, worldwide. Aging is big business. There's a geriatric industrial complex. Claims of a fountain of youth flowing from this or that substance or practice need to be viewed with critical distance.

I have another, more personal reason for being aware that not all damage to the brain can be cured, all aging reversed, all IQs improved. Tragically, around the same time that I fell off the ladder on Capri, my younger brother suffered what his doctors called a series of "neurological events," several small strokes. He'd had brain damage for many years following the removal of a benign tumor in his early thirties. The inexact nature of surgery at the time and what was called "overzealous use of radiation" left him with substantial hearing impairment, some memory problems, a diminished sense of time, and an increased tendency to lose his temper unexpectedly and for no reason. Still, he was able to hold down a job, has a beautiful home, and has a wonderful wife and son, and he remained very intelligent, an avid reader, and conversant on many topics, so much so that people overlooked his impediments. With the more recent strokes, however, he slipped beneath some cognitive threshold. His high intelligence could no longer mask the impairments. He was put on disability at work and is no longer allowed to drive a car. He finds this frustrating when he is awake, but increasingly, much of his day is spent asleep.

Through my research and contacts in the neurobiology community, I was able to help my brother and his wife find and get appointments with some of the best doctors in his city. My brother would say they caused him only misery. They were the ones who recommended he no longer be allowed to work or drive. His quality of life since then has diminished. No one is optimistic of seeing a cure-all in his future.

Whenever I hear of miraculous possibilities ahead or read of astonishing incidents of neural plasticity, I am thrilled for the hope they offer. But I also remain aware, on an intensely personal level, that not all brains change for the better. Not all damaged brains heal themselves. My decade of research on the science of the brain cannot alleviate my beloved brother's condition. I never forget that.

|||||||||||||||||||||||||||||||||

Rather than seek some specious fountain of youth, I prefer to think about what the aging brain does well and to imagine ways those skills can be maximized. Roberto Cabeza, professor of psychology and neuroscience at Duke University, has found that the older brain excels at, in technical terms, "cross-functional complementarity." In lay person's language, that means being skilled at sharing knowledge with all kinds of different people, drawn from many different sources with distinctively different levels of credibility, and being able, from the brew of thought, to come up with creative solutions. Collaboration by difference! Who knew? Midlifers turn out to be even better at it than young people.

Cabeza's lab has also found that midlifers do well in a wide array of problem-solving situations and excel at episodic memory—that is, relating a sequence of autobiographical events in order and with proper causality. Midlifers are also excellent at perceiving accurately and paying prolonged and sustained attention—especially if they are given an opportunity to control and shut out sources of distraction. Midlifers are good at figuring out (all that problem solving, synthesis, generalization, and episodic memory) what will cause them the most distraction and compensating accordingly.

One of the most remarkable findings of all from Cabeza's intriguing tests is an increase in what's called midlife brain bilaterality. The neurons flying along paths from one to the other hemisphere of the brain seem to help one another. If one part of the brain isn't working up to speed, the senior brain finds another part of the brain that can pitch in.[7] This cross-brain assistance also suggests a complexity of thinking, almost an in-brain collaboration by difference!

No single test measures this complex synthesis of abilities, but it is probably more important in everyday life than knowing the missing number or letter in a sequence, anticipating what the three-dimensional shape will look like if turned upside down, figuring out which word goes with which other word, and other features of conventional IQ and cognitive testing.

Indeed, Harvard educator Howard Gardner has spent the last three decades arguing that there are multiple intelligences and that all of them are important. Here's his list of the major different forms of intelligence that any one individual can possess in different combinations and with different strengths and weaknesses:

1. Linguistic intelligence (as in a poet)
2. Logical-mathematical intelligence (as in a scientist)
3. Musical intelligence (as in a composer)
4. Spatial intelligence (as in a sculptor or airplane pilot)
5. Bodily kinesthetic intelligence (as in an athlete or dancer)
6. Interpersonal intelligence (as in a salesman or teacher)
7. Intrapersonal intelligence (exhibited by individuals with accurate views of themselves)[8]

When we speak of declining, age-related mental capacities, are we referring to these capacities? We know, for example, that, on average, mathematicians tend to be young and historians do their best work later in life. Is one smarter than the other? Does one have more cognitive resources than the other? Is the pure abstract thinking of the mathematician really superior, cognitively, to the associational searching and reading and analyzing and interpreting and synthesizing and then focusing and narrating that are the historian's gift and trade?[9] One could also note that it is mathematicians and statisticians, not historians, who tend to make up the statistical measures by which we judge cognitive excellence, achievement, and decline. On the other hand, it was two historians who created H-Bot, that robot who can whup most of us on standardized tests.

|||||||||||||||||||||||||||||||||

It may seem hokey to equate just trying harder or coming up with alternative methods of thinking with a rejuvenated brain, but these methods work in surprising ways, psychologically and—if we can even make such a distinction—biologically. Dr. Yaakov Stern, professor of clinical neuropsychology at the College of Physicians and Surgeons of Columbia University, has spent much of his career studying the most frightening of age-related mental deteriorations, Alzheimer's disease. From years of study of degenerative age-related disorders, Stern has concluded that the single most important thing anyone can do to prepare for the possibility of brain disability is to have a backup plan. Seriously.

I love this idea and readily admit it appeals to me largely because I'm the kind of person who goes through life with a backup plan and often a backup plan to the backup plan. You probably figured that out from my account of the various physical therapies I undertook to rehab my arm.

For someone like me, Stern's theory is reassuring. It is based on the remarkable results of autopsies performed on 137 people who had been diagnosed with Alzheimer's disease in the 1990s. In clinical terms, Stern notes, Alzheimer's disease is more than its well-known symptoms of confusion, memory loss, and cognitive decline. It is also a brain pathology characterized by plaques and tangles that cause deterioration, starting in the hippocampus (affecting memory) and moving to the cerebral cortex, where it influences reasoning and language abilities. Because official diagnosis of Alzheimer's cannot be confirmed before autopsy, the diagnosis of a living patient is usually "probable Alzheimer's disease" and is based on the symptoms the patient manifests while alive.

Stern's research team at Columbia University assumed that there would be a great correlation between the severity of the symptoms manifested in the living individuals and the degree of deterioration found in the autopsied brain. That's a perfectly logical theory if you believe that aging and cognition are purely biological. But while the team found that there were correspondences, there were also surprises in both directions. They found some severely disabled AD patients with brains that turned out to be less diseased than expected and patients with few AD symptoms whose brains were badly ravaged. What could account for this difference, Stern and his research team wondered?

Stern's idea is that, in the same way that the visual cortex of certain people afflicted with blindness can be adapted by the brain for sound and music, something similar happened in some AD patients. Complex operations of working, procedural, long-term, and short-term memory were taken over by the undiseased parts of the patient's brain. In other patients, other parts of the brain filled in and averted the usual variety of emotional and behavioral conditions associated with Alzheimer's.

Why this cognitive boost in some patients but not others? Stern attributes the discrepancy to what he calls cognitive reserves—a backup plan. Some people have a reserve of intellectual, cognitive, physical, and emotional experiences that allow them to create complex and interconnected neural pathways that can be called into service throughout their lives. In some cases, these new routes are useful in the case of brain damage or disease. If a neural pathway is blocked, it is as if the brain finds another way that isn't. If you have plenty of cognitive reserves, you can go around even a major blockage, taking a back road or a circuitous route toward your goal. According to Stern's theory, those patients who had cognitive reserves had seemingly been able to tolerate progressive brain deterioration without manifesting catastrophic behavioral or cognitive symptoms.[10]

So how do we get our cognitive reserves? Based on this research, Stern has created a list of things you and I can do to pump up our own cognitive reserves well in advance of any neurological decline or disaster: continuing education, meaningful and enjoyable work, pleasurable leisure activities, physical exercise, social interaction, learning new skills, learning a new language, and—of course!—taking up new computer skills, playing video games, and interacting on social networks. All those mentally stimulating activities build reserves of neurons, synapses, and neuromotor skills that we might be able to call upon if we need them. If we are lucky enough that we don't need them, they are there to be enjoyed anyway.

I was able to learn firsthand of the amazing ability of these cognitive reserves to take over after a friend of mine, Ichiro, collapsed with an aneurysm several years ago and was rushed to the hospital in Osaka, Japan. While on the way to the hospital, the ambulance driver was sure he was hearing Ichiro speaking gibberish. "No, it's French!" his son said. Ichiro is Japanese. He learned English as a second language, then fell in love with a French woman and learned French. Because new languages are acquired and remembered in different areas of the brain than first languages, the parts of Ichiro's brain that remained intact after his stroke remained able to function even at the height of his distress. When I saw him a few years later, he had relearned Japanese, with notable gaps and lapses, but his English was better than it had been when I first met him over a decade before.

What Stern calls cognitive reserves are what, throughout this book, I've been calling learning. In fact, the concept of cognitive reserves corresponds nicely to the digital model of learning. That is, the more actively you are pursuing Toffler's idea of learning, unlearning, and relearning, the more likely you are to be working to ensure that your brain is not only active but stimulated in ways that are always new. And if we buy Stern's ideas, then the positive outcome is that, the more your brain has been rewiring, then the more likely that the symptoms and signs of degenerative brain conditions will be less when you are alive than they will be when you are dead. That's not a fountain of youth, precisely, but it's not a bad bargain, either.

LET'S RETURN TO THAT OLD VCR again. The kind of mental workout one gets from mastering any new electronic device and incorporating it into one's daily life is beneficial to rewiring the brain. Think of it as building up those cognitive reserves.

The VCR is also a great symbol in other ways. First, it really was a crappy technology. VCRs broke a lot. They were always on the blink. It was a makeshift and conceptually user-unfriendly interface that seemed designed to separate young and old, female and male. It was almost the opposite of the user-powered Internet and the World Wide Web. If I seem a slap-happy technology utopian, it may well be because I've seen before and after—and the VCR was the epitome of before. The technological device that was designed to bring you pleasure brought more than its share of pain. It was a little like my physical rehab. To get it going, you had to go through a lot of procedures, and the final product was never as good as the original.

The Internet and all the mobile devices that came after may seem like an extension of the VCR, but I think of them as its antithesis. The digital age, and the World Wide Web itself, is based on an idea that it works only if *everyone* participates. Rather than reinforcing distinctions, all the new kinds of devices depend on inclusion. And, amazingly, they *are* inclusive. Over 80 percent of Americans between fifty and fifty-four are now online. We can look at the numbers another way and note that older Boomers (those born between 1946 and 1954) make up 13 percent of the total adult population of the United States and exactly the same percent of the Internet-using population.[11]

Contrary to the myths about our digital era, the average blogger in the United States is not fourteen-year-old Amber but a thirty-seven-year-old white, middle-class male. Social networking media are similarly popular across demographics, with approximately 40 percent of Facebook and MySpace users over thirty-five.[12] "Digital immigrants," in other words, are streaming through the technological equivalent of Ellis Island. Including *old* digital immigrants. In the seventy-six-plus age range, a surprising 27 percent of Americans are using the Internet, and where the Internet is easily and inexpensively available to them, the elderly are, statistically, the fastest-growing and most active demographic of new adopters.[13]

Seniors are taking advantage of the brain tonic that is the Internet, turning into practice all the research going back to the 1980s and 1990s demonstrating the cognitive benefits of interactive digital media. The latest research suggests that engaged use of the Internet provides better mental calisthenics than the more conventional senior pastimes of reading, television or movie watching, or playing cards, board games, or word games like Scrabble. Coupled with bilateral exercise (i.e., walking, running, or dancing) for those who are able-bodied, "silver surfing" is the single most beneficial senior activity we know.

|||||||||||||||||||||||||||||||

I met him on the Internet. He called himself ToughLoveforX. He had a way with winning phrases. Soon, I was one of his followers.

This story isn't going where you think it is.

ToughLoveforX is Michael Josefowicz, a retired printer who has an active life on Twitter. I began to follow him, which means, for those who don't use Twitter, that I selected to receive Josefowicz's tweets. They roll past me every time I check out my Twitter page, several times a day. Twitter is called an asymmetrical technology in that you can follow anyone you want and read what they tweet. You follow people because you are interested in what they have to say. Because he seemed always to be on top of the most exciting news and research on education, social media, publishing, civil society, the brain, and other areas that interest me, I began following ToughLoveforX—or TLX, as I now call him for short—and found that every other day or so, he would be sending out a link to something invaluable for my own thinking and research.

I'm not the only one who thinks TLX has a bead on something worth seeing. He currently has 3,202 followers. That's a pretty substantial audience, given that, of the 75 million people with accounts on Twitter at the end of 2009, only the top 1 percent had more than 500 followers. The average Twitter user has exactly 27 followers.[14] TLX describes himself as: "Retired printer. Lots of time for blabla. If I can help fix high school. So much the better." That candid, offhanded, unpretentious style is why a lot of us follow him. Another reason is that, in addition to having so many followers, he himself is avid at following others, 3,410 of them to be exact. He has made himself one of the most informed people out there, anywhere, on all of the research in neuroscience and learning.

There's a silly status game that celebrity wannabes play wherein they measure the ratio of who follows them to whom they follow. The actor Ashton Kutcher, for example, follows about 400 people while over 4 million follow him. But the beauty of TLX is that he is a true information node. He reads voraciously, he sifts and sorts through all those Twitter feeds, and he distills it all, publishing what he thinks is the very best of his own personalized, constant stream of information. For the 3,202 of us who follow him, he is a filter on educational reforms happening anywhere in the world. Like Dennis Quaintance's "sustainability filter," TLX is an "education filter" who helps many of us— thousands, actually—to see the complex and changing landscape of education with more clarity and depth.

That is why I sent him a DM (direct message) and said I'd like to interview him to find out how he got into this tweet-filled life. We DM'd a few times, then we tried something we called Twitter Tennis—an open-to-anyone exchange of ideas on Twitter. From there, we went to e-mail and had a long phone conversation. He told me his story. He'd owned his own printing business, retired, and then could have dwindled into a solitary old age. Instead, Michael Josefowicz of Brooklyn, New York, became ToughLoveforX, a vocal champion of education reform for kids, someone who turned his passion for kids and education into an information resource that reaches far more people, several times a day, than most people ever reach in a lifetime. It costs him nothing. It turns leisure time into productive time, and now that he is followed by some pretty prominent educators, he has an impact on the world. Or as TLX says in his inimitable style, in 140 characters or less per tweet, it's the butterfly effect: "initial condition of a dynamical system may produce large variations in the long term behavior of the system." Not bad for a retired guy.

The son of immigrants, he'd gone to Columbia University, was a student there during the famous protest strikes against the Vietnam War in the 1960s. Josefowicz maintains an independent Boomer spirit, but he confines his activism to his online research and communications. He is amazed that so many people look to him for insight, that so many people follow him on Twitter— and pleased. He says it took him about six months to get into the swing of it, and once he did, he was astonished that he, one guy in Brooklyn, could have an impact. In fact, his slightly sinister Twitter name came from a simple inefficiency. Before Twitter, he had a blog called Tough Love for Xerox, begun as a protest against a policy by the company that denied some retired employees benefits they were owed. A lifelong admirer of Xerox because it made the digital photocopier that transformed his printing business, he was disappointed at the company's shabby treatment of these retirees and wrote about it in his blog. When he transferred the same name to Twitter, it cut off his name after the X, and he liked the mysteriousness and kept it.

He's frank about the pleasure it gives him to be such an information conduit on a topic close to his heart. He doesn't feel like the isolated, classically useless retiree but, thanks to his online life, knows that he has a purpose. By twentieth-century rules, when one is not gainfully employed, one is not important. Thanks to the low participation barrier and the excellent possibilities for communication online, there are now many new ways to be gainful, to be important. I think back once again to that kindergarten class at Forest View Elementary. "We're kindergartners." In the end, that's really what it is all about, taking pride in who we are and in how we inhabit our world.

When I called Josefowicz, he was as delighted as I was about the way a random connection on Twitter had turned out to be anything but random, actually a linkage of strangers who were paying attention to the same issues, the same world. When I asked him if he could do one thing to improve the world, what would it be? he said without hesitation, "Scale John Seely Brown." I had to laugh. JSB is one of the most profound thinkers of the information age, formerly the legendary "out of the box" director of Xerox's PARC, the think tank of all think tanks. He also happens to be my friend. Scaling JSB is a great idea.

"He'll like that," I told Josefowicz, "I'll tell him when I talk to him tomorrow." Josefowicz was momentarily speechless. He never dreamed his tweeting would bring him one degree of separation from someone who had been an intellectual hero responsible for shaping his vision. This was a twist on collaboration by difference—almost communication by difference, with a certain loop of karmic "what goes around comes around" thrown into the mix.

Within a week, I'd written a "Scaling John Seely Brown" blog. JSB tweeted it. TLX retweeted. So it goes. That's not exactly a circle, but it's a connection forged by a shared vision, not by sameness but by contribution. For TLX and many others, it is what growing older happily in the twenty-first century is about.

Having vigorous links to the world and to others goes a long way toward keeping TLX young, and the statistics are on his side. Health insurers have found Internet use rivaling or, in some studies, exceeding the success rates of antidepressants in managing geriatric mood disorders. A research team at the Centre for Mental Health Research at Australian National University has even demonstrated the surprising result that Internet human contact can be as effective in mitigating depression as face-to-face human contact, especially in the long run, replacing the familiar despair of losing friends and relatives one by one with a sense of being part of a vibrant and growing community of caring and attentive acquaintances.[15] Better geriatric health has also been attributed to Internet-assisted self-help, because the elderly become their own health advocates, finding reliable sources of medical information online, taking greater control of their own health care, resulting in fewer doctor bills, less medication, and a greater sense of well-being and security.[16] The punch line of all of this is: Don't retire. Rewire! Don't retreat. Retweet!

In Holland some years ago, an inventive group of medical professionals at an institute in Amsterdam called the Waag Society came up with an idea to use the Internet to pair elderly shut-ins in their nursing home with young partners, some who volunteered and some who were paid a small stipend for participating. In most cases, the senior citizens had to learn how to use a computer to talk to their young partners. Some of the kids were jobless, with too much time on their hands, hanging out in Internet cafés. Some of the elderly were virtually senile, also with too much time on their hands. It seemed like a crazy idea, but no one had much to lose. Why not try? There were low expectations on all sides.

The first surprise came when the old people jumped at the chance to do this. Their excitement wasn't at learning computers; that was just a means to an end. The excitement was in learning something "youthful" to be doing, something that connected them to actual young people. The second surprise was that, as soon as the connection was made, everyone started reporting that they felt happier. This wasn't an official experiment. There were no scientific tests. There were simple self-reports of feeling better. The next surprise was that these reports came not just from the oldsters but from the kids, too. As the old and the young began writing to one another, sharing their stories, they began feeling a sense of optimism.

That would have been a successful enough experiment. But soon the nurses were noticing that the seniors weren't requesting as much medication for pain, anxiety, depression, memory loss, or even their physical ailments. Cognitive signs seemed to be reversing too, and people were doing things that, a few months before, seemed beyond their mental grasp. There was only one problem with the experiment. New beds had to be added. Why? Because the oldsters were sticking around the nursing home longer than expected. Put plainly, they weren't dying as fast as they used to.

The results of this early use of the Internet to make human connections have been so positive that organizations dedicated to digital innovation, like the remarkable Waag Society, are now busy creating multimedia interfaces for computers specifically to facilitate online storytelling by the elderly and the physically impaired.[17] This being Holland, there is also now what amounts to a technological bill of rights guaranteeing elderly Dutch citizens their inalienable right to an Internet connection.[18]

‖‖‖‖‖‖‖‖‖‖‖‖‖‖‖‖‖‖‖‖‖‖‖‖‖‖‖

Until her untimely death in 2008, Australian blogger Olive Riley was banned from two of the world's most popular social networking sites. A blogger who also regularly posted vlogs on YouTube, Riley loved the Internet and kept up an e-mail correspondence with friends the world over, despite living in a medical facility. Yet she was unable to join MySpace and Facebook. Like all social networking sites, these two set their own community rules for who can or cannot belong; ninety thousand registered sex offenders in the United States, for example, were barred from MySpace in 2009.[19] However, Olive Riley was prohibited membership not for criminal or even antisocial activity, but because of her age. You must be 13 years old to join Facebook and 14 to have an account on MySpace, but few people know that there is also an upper limit on these sites as well.[20] On its application page, the Facebook birth-date box does not accept anyone whose birthday is before 1900. MySpace's registration system excludes those over the age of 100. When she passed away, Olive Riley was an intrepid, game, feisty, tech-savvy 108 years young.[21]

Dubbed the World's Oldest Blogger by the media who came to feel affection for her, Olive Riley embraced the digital age, but her passing didn't substantially decrease the average age of those participating in a Web 2.0 world.[22] Although her age made the charming centenarian blogger an international sensation, her story is representative of the many older Americans now piling onto the Internet. This is a remarkable change from the VCR era. And the difference is *interaction*. It is not the technology that is good for us, it is this volition—the sense of control—coupled with the connection with anything and anyone of the billions of others connecting online at any time. The best affordance the Internet offers is a chance to move beyond ourselves, even if we have limited capacity to move in a physical sense. And the best news is that this isn't the cod-liver-oil theory of learning, that you have to do it because it is good for you. It is the enjoyment that makes it so, and no one has to force you. Despite the stereotypes, older people are online as frequently as the young.

We need to apply the lesson of attention blindness to our midlife anxieties about the Internet and rethink where those anxieties come from. If we each were assigned our own private sociologist to follow us around, marking off instances of adaptation and successful change, we would be able to chart how much we learn and adjust every day. Because we are not "blessed" with this external analysis of our actions, we see only what we focus on, and we tend to focus on what frustrates, eludes, annoys, or shames us by making us feel inflexible, incompetent, and superannuated—a term that means, literally, being disqualified or incapacitated for active duty because of advanced age. We are hypersensitive when we hesitate, cringe, or stumble; we tend not to notice all the ways we have successfully integrated the new technologies into our lives.

Despite the fact that we adapt to enormous changes all the time, what we attend to most are those instances when lack of success brings a sense of inadequacy and shame. That sense especially accompanies middle age because, in modern society, especially in the United States, middle and old age are treated as shameful conditions, almost diseases or disabilities, in need of help or repair. Multi-billion-dollar youth industries—from plastic surgery to pharmaceuticals like Viagra—prey on our vulnerability. Rejuvenation is big business, especially as the world's population ages.

If the patterns of learning predict what we see, then it is past the time to unlearn our preconceptions of aging. Like Olive Riley, that indefatigable 108-year-old blogger, we can begin to find more exciting, open, inclusive ways of relearning suitable to this astonishingly rich and bewildering information age. In Olive Riley's simple lesson plan for the future, "We're never too old to learn."

Conclusion

||||||||||||||||||||||||||

Now You See It

Social networking pioneer Howard Rheingold begins his digital journalism course each year at Stanford with a participatory experiment. Shut off your cell phone, he tells his students. Shut your laptop. Now, shut your eyes. Then, the visionary author of *Smart Mobs*, who penned his first book, *Virtual Reality*, back in the medieval era (i.e., 1992), spends the next five full minutes in silence with his students, in the darkness behind their own eyelids, completely unplugged.[1]

Why? Rheingold wants his students to experience the cacophony of their own silence. We are so distracted by our feelings of distraction in the digital age that it's easy to forget that, with nothing commanding our attention at all, with no e-mail bombarding us and no Tweets pouring in and not a single Facebook status to update, our brain is still a very busy and chaotic place. The brain never was an empty vessel, but it's hard for us to appreciate all that's going on in there when we're not paying attention to anything else but our own attention.

Rheingold prepares his students for their five-minute self-guided tour through their own minds with a few words of advice. He's discovered that many of his students have never been in this space before, alone with their own mental processes, charged with tracking their own attention. To help them focus their introspection, he advises them to pay attention not only to *what* they are thinking but also to *how*. He asks them to chart their mind's way, to notice how one thought connects to another, to observe how their own mind operates when it is not being interrupted by e-mail, text messages, cell phones, or anything else. He asks them to note their patterns of attention and inattention. He has them keep track of how long they can stay on one, linear, single path. He challenges them to stay focused.

After they open their eyes again, he asks his students to report on what they observed. Typically, they express shock at how incapable they were of staying on a single track for even a few seconds, let alone for five interminable minutes. There's a constant bombardment, one idea or sensation or memory or impulse or desire tumbling into another, with each distraction short-circuiting the next and each interruption interrupted. They find it almost impossible to narrate what they observed in their own unquiet minds. They stumble as they try to reconstruct the sequence of their own mental processing. Five minutes with no interruptions, yet what captured their attention in that time was as fleeting and as difficult to recall as the figments of dreams.

WE WORRY ABOUT WHAT IT means to be surrounded by a ceaseless assault of information, the constant distractions of our digital age. "Drop that Black-Berry! Multitasking may be harmful" shouts a recent headline on CNN Health. com, reporting on a study of attention problems among multitaskers.[2] What the Rheingold experiment demonstrates with its simple eloquence is that to be human is to be distractable. We are always interrupting ourselves. That's what those 100 billion neurons do. Until we die, they are operating. They are never still. They are never off. Our brain is always wired—even when we think we're unplugged.

The lesson of attention is that, even when we are sitting in the dark behind our eyelids, unfocused, our mind wanders anywhere and everywhere it wants to go. It might notice the itchy nose, the annoying buzzing sound, a truck rumbling by outside, or the cricket noises generated inside our head by our head. This is the world of distraction that our powers of attention shut out from view most of the time; that diverting other world is nonetheless always there. We're simply blind to it. More than that, our brain expends an inordinate amount of energy at any given time working so that we do not notice our own inner distractions. It's selecting what to attend to, and that selection requires work. We've been doing it so long, as part of the process of learning to pay attention, that we aren't even aware of it until someone or something makes us aware. When we're not distracted by the normal world that occupies us, there is suddenly plenty to pay attention to that we never noticed before. Simply staying on track with any thought, unwired and eyes closed, requires selection and rejection.

It also requires *narrative*. When Howard Rheingold's students try to describe what they were thinking, they inevitably turn it into some kind of story, with a recognizable beginning and ending—that is, with clear categories. Like the Cymbalta ad, we use our mind's main story line to screen out the upsetting side effects. Like the infant learning what counts in his world, we have to master our culture's main plot, and we unconsciously repeat and reinforce it often, over and over, to make sense of our lives.

For well over a hundred years, our culture's story line has been about focused, measurable productivity. Productivity has been defined by artificial and manufactured separations dividing the world of work from the rest of life. Within the workforce, twentieth-century productivity has been based on a philosophy of the division of labor—into hierarchies, job descriptions, responsibilities, and tasks. Badging in, punching a time clock, workdays and weekends, lunch hours and coffee breaks, have all been prescribed, sorting our day into the "on" times for our attentive labor and the "off" times for our leisure. School has been organized to prepare us for this world of work, dividing one age group from another, one subject from another, with grades demarcating who is or is not academically gifted, and with hierarchies of what knowledge is or is not important (science on top, arts on the bottom, physical education and shop gone). Intellectual work, in general, is divided up too, with facts separated from interpretation, logic from imagination, rationality from creativity, and knowledge from entertainment. In the end, there are no clear boundaries separating any of these things from the other, but we've arranged our institutions to make it all seem as discrete, fixed, and hierarchical as possible. And though categories are necessary and useful to sort through what would otherwise be chaos, we run into trouble when we start to forget that categories are arbitrary. They define what we want them to, not the other way around.

We've sorted our life cycles in a similar way, with developmental theories of how children learn and geriatric theories of how we begin to forget. If there is any word that defines the twentieth century, it might be *normative:* a defining and enforcing of standards of what counts as correct. We've divided the "norm" from the non-normal, we've created tests to measure where we are on that scale, and we have elaborated forms of statistical analysis rooted in a theory that the mean is good and that it's important to measure how far we do or do not deviate from the mean.

The old idea of the brain mirrored, supported, and justified a certain ideal of human labor, too. Attention is, definitionally, *the concentration of the mind on a single thought, object, or task.* In the Taylorized model of attention, it is assumed that if we can control the amount of external stimuli we experience, we can become more attentive and productive. If we can minimize interruption, control the task, narrow the focus, eliminate input from the other senses, and blot out disruption, then our attention will be undivided and our thoughts will flow, one useful idea leading to the next, as neatly and efficiently as on any assembly line Taylor ever conceived.

Except that that's not how the mind works. As Howard Rheingold's experiment shows us, the mind is messy if left to its own devices. Try it yourself and you'll see: Shut off your cell phone, shut your laptop, shut your eyes. Even the most frustratingly disruptive workplace multitasking is sane compared to all the activity among all those neurons at any one given time, each making jumbled and associational connections that you would not, in your logical narrative, ever have devised.

If the brain were focused by nature, it would be easy not to be distracted when we had no external distractions. But it turns out that staying on track when there is nothing capturing our attention is intrinsically difficult to do. Neurologist Marcus Raichle's research at Washington University has recently determined that an astonishing *80 percent* of our neural energy is taken up not by external distractions at all but by the mind talking to itself.[3] Raichle hooked up neuroimaging machines to dozens of volunteers and found that the brain lights up the screen in all kinds of patterns whether a person is actually doing a task or seemingly at rest. Quiet, uninterrupted thinking turns out to be quite active neurologically, even when, later, the volunteer insists that he wasn't really thinking about anything. Brain chatter is so much a feature of our brains that we are rarely aware of it at the time or subsequently—unless we have psychologists to tell us that our brain was doing something when we were sure it wasn't.

Raichle has found that more of the areas of the brain light up when the volunteer is daydreaming than when the same person is engaged in a concentrated task. Remote parts of the brain are "talking" to one another in those down times, and the entire brain is about twenty times more active than when it's being stimulated from outside. Switching from one specific task to another also turns out to be energy efficient. It uses less than 5 percent of the brain's normal "at rest" energy. We use less brain energy multitasking than we use "sitting there at rest, thinking random thoughts."[4]

The brain is not built to take one task from start to completion without interruption, then the next task, and then the next one. Our schools and our offices may be built that way, but the brain loves its back channels and is constantly seeking ways to rebel against the order we try to impose upon it. Malia Mason, working first at the Neurocognition Lab at Harvard Medical School and then at Columbia University's School of Business, has begun to pinpoint how various cortical areas generate daydreams. She has suggested that mind-wandering, not focus, may turn out to be good for the brain and beneficial for the ideas it generates.[5] Connecting one part of the brain to another, all the time, is as crucial to brain health as aerobic activity is to the body. Mason's team has been finding that even when we're engaged in task-specific mental processing, parts of the brain we once thought were irrelevant to the specific task light up on fMRIs. Areas of the brain that are distant from one another are working that back channel, so even when we think we are focused, there's another conversation going on among the medial prefrontal cortex (primarily responsible for so-called executive functions), premotor cortex (coordination of body movement), and cingulate gyrus (limbic system and, by extension, memory and learning). Mason's preliminary theory is that the brain performs cognitive cross-training. Those interconnected neural training activities help us minimize our attention blindness, assist in multitasking, and give us a boost as we practice collaboration by difference. Taken together, they are what prepare us for unlearning one set of habits so we can learn something new.[6]

||||||||||||||||||||||||||||||

Our era may be obsessed with attention, distraction, and multitasking, but I'm convinced that these age-old concerns always emerge with new force whenever a major new technology makes us aware of habits, patterns, and processes that had been invisible before. I'm not convinced that our own era is intrinsically any more distracting than any other time. Because we're in a transitional moment, in the midst of tremendous social and technological change, we're experiencing the tension between the old and the new, a tension that exposes us to our own patterns of attention, patterns that are usually hidden behind closed eyelids.

The present conversation about multitasking raises implicit questions: Was there ever such a thing as monotasking? Were we ever good at it? Or did we just come to think we were decent monotaskers after a hundred years of school and workplaces reinforcing the ideal of focused attention to a specific task?

Are we experiencing something different than we did before? And a further question: Are we all experiencing the same thing? To return to an issue we've addressed in many ways in this book, how great are the differences among us at experiencing a phenomenon like multitasking? How great are our own differences from one task to another? Do we experience attention, distraction, and multitasking in different ways in different circumstances?

The answer is "all of the above"—more or less. Yes, information overload and the extension of the workday are real. As a society, we have not yet grappled with or conquered (and eliminated) the eighty-hour workweek, and clearly we must. That kind of work stress *is* bad for our health and is unproductive in the long run. Similarly, some of the extra time we're spending at work and some of the tension we're feeling happens because we have not yet routinized certain aspects of the new technologies in our lives, such as developing accepted e-mail protocols that can make lower-order decisions for us. It's like etiquette. There is no intrinsic reason why you use one fork for one course and a different one for another, but having a set of rules means you don't have to rethink the issue every time you're presented with a salad. Some of our moment's agony over multitasking is that we haven't yet figured out which digital fork to use at which times. The Internet is still in its adolescence. We will work out those rules. But until we do, we can expect to feel taxed.

The upshot, though, is that there is nothing *intrinsically, biologically* harmful about multitasking. Our brain science is there but we really haven't gotten our minds around that point yet. Because of the new information about the neurology of attention, a few psychologists have taken out the Taylorist stopwatch again, this time to measure not task completion but multitasking and other new attention issues. Psychologist Jonathan Schooler of the University of California–Santa Barbara asks volunteers to push a button anytime their minds wander while they are reading from lengthy books, such as *War and Peace.*[7] Even when they know their attention is being tested, volunteers report their focus deviating from the fates of Prince Andre and other members of the Bolkonsky family about 25 percent of the time.[8] Schooler is currently researching what he calls meta-awareness, the gap between how much our mind wanders and how much we think it does. Our widespread worries about multitasking *create* the problem; they don't just name it. It's the gorilla experiment again: Focus only on the horrors of multitasking and that's all you see.

It's interesting that even people who worry about the harmfulness of multitasking often rate themselves as being quite productive, despite being required to multitask. On the very same tests, though, they worry about the distractions everyone else has to encounter in the workplace. This is hardly surprising, given that we always have difficulty assessing ourselves. On the other hand, when these tests are set up to reveal to the subjects that they themselves are distracted, they then tend to *overestimate* the negative effect of the disruption, exaggerating both its impact and duration.[9]

Another recent study, this one led by sociologist Clifford Nass, whose office is down the hall from Howard Rheingold's at Stanford, made headlines by pronouncing that even die-hard multitaskers are worse at multitasking than those who concentrate on only one thing at a time. That was the gleeful sound bite ("See!") that hit the media. But the study itself has a very interesting conclusion that leaves the door open on that exact question. The authors conclude that "heavy media multitaskers . . . performed worse on a test of task switching ability, likely due to reduced ability to filter out interference from the irrelevant task set." Translated from psychologese to English, that means that the multitaskers transferred some knowledge from one task (the task the testers deemed irrelevant) to the other task they were performing. Does that make them bad multitaskers—or good ones? Don't we want to be applying information from one situation to another? Perhaps the "reduced ability to filter out interference" is actually a new mashup style of mental blending that helps one be a good multitasker. Outside a testing situation, in real life, it is precisely your susceptibility to distraction that makes you inclined to switch to another task and then, after a time, to switch back or to another one. But what if in all that switching we're not just forgetting but we are creating a new, blended cognitive map of what we're learning, task to task, from the process? In the end, isn't remixing what all learning should be?

Especially with a computer screen, that's certainly what surfing is. *You* do the switching. In a psychology test, you don't have volition. Something new flashes to distract your attention, then something else does. But what relationship does that kind of structured, experimental distraction have to the way we operate, now, on the Internet? The choice of when to stay on one page or when to click on an enticing URL and go elsewhere is exactly what defines the adept surfer. The experimenters acknowledge (not in the headline but in the fine print of their scientific study) that they were testing young multitaskers using an old-school test for what is known in the field of psychology as "stimulus-dependent attention," or interruption from outside. But multitasking, as we know it in our offices or even in an evening's leisurely Web surfing, is more various than that. Sometimes attention is derailed from the project at hand by an incoming text message, but other times we're just following our curiosity where it takes us, and at still other times, we are talking and texting together as a practice and a skill, as we saw with the adept back-channeling IBMers on their multinational conference calls. The IBMers' "reduced ability to filter out interference from the irrelevant task" is what allows them to text-chat while they are listening and speaking. Successful multitaskers know how to carry over information across tasks when they need to do so. More to the point, multitasking takes many forms, in many situations, and operates differently for each of us, depending on a multitude of factors. The headline-grabbing "multitaskers aren't even good at multitasking" is a nonsensical statement, when you break it down and try to think about it seriously.

The authors of the Stanford study suspect as much and worry that they might be using old tools and concepts to test something new, a new way of interacting in the world. They continue: "It remains possible that future tests of higher-order cognition will uncover benefits, other than cognitive control, of heavy media multitasking, or will uncover skills specifically exhibited by [multitaskers] not involving cognitive control."[10] The media didn't pick up that there were possible alternative ways of viewing the test results, because the media were blinded by their old view of attention.

I'm suggesting that the most important finding from this test is not that multitaskers are paying attention *worse*, but that they are paying attention *differently*. If frequent multitaskers have difficulty ignoring information from other tasks, is that information really irrelevant, or might they be creating a different, interconnected cognitive map, a mashup or synthesis from some trace of all of the tasks at hand? And could this mashup be a new way of paying attention, one very different from the cognitive sequencing of discrete tasks we tend, still, to label "productivity"?

If what we are seeing isn't discrete switching from one task to another but a form of attention that merges and remixes different strands of information almost seamlessly, then one ally we have in this new form of attention is the brain's own associational, interconnecting energies. We are only just now beginning to see experiments designed to test attention in this creative, self-generating, multifaceted way. It is as if we thought we could evaluate running ability by seeing how well we do on a treadmill, using uniform metrics and standardized scales. We haven't yet come up with ways to measure running and winning when the test is more like the parkour freerunning path over the tops of buildings and up walls, where the key to winning is also knowing how to tuck and roll so you can pick yourself up when you fall.

The new brain science helps us to re-ask the old questions about attention in a new ways. What are we *learning* by being open to multitasking? What new muscles are we exercising? What new neural pathways are we shaping, what old ones are we shearing, what new unexpected patterns are we creating? And how can we help one another out by collaborative multitasking?

We know that in dreams, as in virtual worlds and digital spaces, physical and even traditional linear narrative rules do not apply. It is possible that, during boundless wandering thinking, we open ourselves to possibilities for innovative solutions that, in more focused thinking, we might prematurely preclude as unrealistic. The Latin word for "inspiration" is *inspirare,* to inflame or breathe into. What if we thought of new digital ways of thinking not as multitasking but *multi-inspiring,* as potentially creative disruption of usual thought patterns. Look at the account of just about any enormous intellectual breakthrough and you'll find that some seemingly random connection, some associational side thought, some distraction preceded the revelation. Distraction, we may discover, is as central to innovation as, say, an apple falling on Newton's head.

Two neuroscientists at Cambridge University, Alexa M. Morcom and Paul C. Fletcher, have already argued along these lines to suggest that it is time to get rid of the metaphor of a baseline of focused attention from which the mind is diverted. By examining neural processing, they show that the brain does not perform one simple maneuver over and over and over again. Even if repetition is what your job or your teacher requires, your brain will often find playful ways of diverting itself. The brain is almost like a little kid figuring out a new way to walk the boring old route to school: Let's walk the whole way on the curb this time! Let's not step on a single crack in the sidewalk! Let's try it backward! Why not walk the same old route in the same old way every day? It's more efficient that way, but sometimes you want to mix it up a little. The brain is inquisitive by design. It is constantly and productively self-exploratory, especially at times when one's experiences are rich and new.[11]

Morcom and Fletcher conclude that, on the level of neural networks, there is no such thing as monotasking. They argue that there is no unique function that happens in any one separate part of a brain, no process that is singular in its focus, and no form of attention that is seamless or constant. Apparently, neurons just can't abide being bored. The mind always wanders off task because the mind's task is to wander.

Mind-wandering might turn out to be exactly what we need to encourage more of in order to accomplish the best work in a global, multimedia digital age. In the twenty-first-century workplace dominated increasingly by the so-called creative class, it is not self-evident that focused, undeviating, undistracted attention really produces better, more innovative solutions. More and more, researchers are taking what's been called a nonsense approach or even a Dadaist approach to attention—you know, the playful artistic toying with reality evidenced in the fur-lined teacup, a urinal displayed as art and labeled "Fountain," or an image of a pipe with a label "This is not a pipe." What confuses the brain delights the brain. What confounds the brain enlivens the brain. What mixes up categories energizes the brain. Or to sum it all up, as we have seen, what surprises the brain is what allows for learning. Incongruity, disruption, and disorientation may well turn out to be the most inspiring, creative, and productive forces one can add to the workplace.

Neuroscientist Daniel Levitin has said recently that "surprise is an adaptive strategy for the brain, indicating a gap in our knowledge of the world. Things are surprising only if we failed to predict them. Surprise gives us an opportunity to improve our brain's predictive system."[12] He's right on all counts. What I am suggesting is that our Taylorist history and traditions make us nervous about what we are missing in an era when work is now *structurally* changing, when surprise is inevitable.

As with all neural shaping, there's shearing too, so it is quite possible that because we are paying attention in a new way, some of our old ways of paying attention will be lost. The issue is, which of those matter? If they matter enough, what institutional structures and personal practices can we cultivate to make sure they survive? Some people worry that this "younger generation" will never read long novels again, but then I think about all those kids playing video games in line at midnight, waiting for the local bookstore to open so they can buy the latest installment of *Harry Potter*.

We need to sort out value from nostalgia, separating the time-honored tradition of one generation castigating the next from real concerns about what might be lost or gained for humanity as technologies change. If some younger people aren't bothered by multitasking, it may well be simply a matter of practice. The more you do something, the more automatic it becomes. Once automatic, it no longer seems like a task, but a platform you can build upon. But there is tremendous individual variation in our ability to do this. Most people would not consider driving a car while listening to a radio multitasking, but it certainly was when Motorola put the first radio in a dashboard in 1930. Media expert Linda Stone coined the phrase "continuous partial attention" to describe the way we surf, looking in multiple directions at once, rather than being fully absorbed in one task only.[13] Rather than think of continuous partial attention as a problem or a lack, we may need to reconsider it as a digital survival skill.

In most of life, our attention is continuous and partial until we're so forcefully grabbed by something that we shut out everything else. Those blissful episodes of concentrated, undistracted, continuous absorption are delicious—and dangerous. That's when we miss the gorilla—and everything else. The lesson of attention blindness is that sole, concentrated, direct, centralized attention to one task—the ideal of twentieth-century productivity—is efficient for the task on which you are concentrating but it shuts out other important things we also need to be seeing.

In our global, diverse, interactive world, where everything seems to have another side, continuous partial attention may not only be a condition of life but a useful tool for navigating a complex world. Especially if we can compensate for our own partial attention by teaming with others who see what we miss, we have a chance to succeed and the possibility of seeing the other side—and then the other side of that other side. There's a Japanese proverb that I love. It translates, literally, as "The reverse side itself also has a reverse side" ("*Monogoto niwa taitei ura no ura ga aru mono da*"). In a diverse, interconnected, and changing world, being distracted from our business as usual may not be a downside but the key to success.

||||||||||||||||||||||||||||

This past November I visited South Korea for the first time. I was a guest of KAIST University,[14] the equivalent of MIT, keynoting a conference on Digital Youth organized by the Brain Science Research Center. The other organizers of the conference were mostly from the air forces and aerospace industries of the United States and countries around the Pacific Rim.

It was a fascinating event, and I will never forget one exchange at the opening dinner when a senior flight surgeon in the U.S. Air Force sparred with his equivalent in the Korean Air Force. "So Korean medicine still believes in that hocus pocus about meridians?" the U.S. medical officer chided about the traditional Chinese medicine theory that the energy of the body flows through channels or meridians. "And you still believe in 'right brain' and 'left brain' divisions?" shot back the Korean flight surgeon, who has degrees in both Chinese and Western medicine. After some seriously self-deprecating chuckling between these two brilliant men, the Korean doctor added, "Sometimes excellent research can happen, even when the principles themselves may be simplistic." The rest of the conference was like that, with many conversations inspired by orchestrated collisions of different principles.

After a few days in Daejeon, where KAIST is located, we were then bused the three hours up to Seoul, entering like Dorothy into the Emerald City that is contemporary Seoul, dwarfed by the gleaming towering glass-and-steel skyscrapers. We spent an afternoon at the futuristic Samsung D'Lite interactive showcase and then held a "Bar Camp" (an unconference spontaneously organized by the participants) at a private university owned by Samsung, whose motto is "Beyond Learning." Seoul was more than living up to its reputation as the "bandwidth capital of the world." I felt that I was in the dynamic heart of the digital future. Orderly. Scientific. Innovative. New.

By accident, I discovered there was another Seoul coexisting with the gleaming digital city. I would never have seen it, were it not for the helpful concierge at the Seoul Hilton. I'd spent four days with the aerospace scientists and then three touring Seoul with multimedia artists, and as I was packing for home, it became clear that all the beautiful books and catalogs, plus the goodies and the souvenirs I'd bought, were not going to fit into my suitcase. It was around eleven P.M. when I made my way to the concierge desk and asked if there would be any place where I could purchase a suitcase before my bus left for the airport first thing in the morning.

"Sure! Go to the night market," the young woman offered brightly.

She told me to walk out of the Hilton, past the gigantic glitzy "foreigners' casino" (Koreans are not allowed entry), and toward Seoul Central Station, the main subway and train station in the city. I had been in Seoul Station several times and had walked this ultramodern urban street, with its multiple lanes, its overpasses and underpasses, a dozen times. She said about a block before the station I'd find a Mini Stop, the Korean version of 7-Eleven. That's where I would find a narrow alley leading down the hill behind the store. If I looked closely, I'd see light coming from the alley. That was Dongdaemun, the night market.

I knew that no concierge at the Hilton would be sending an American woman alone into a dark alley if it weren't safe. And from my googling before the trip, I knew Korea had virtually no unemployment and one of the lowest crime rates on earth. So I didn't hesitate to go out on an adventure. I had money in my jeans pocket, my room key, and nothing else—no identification, as I didn't have my wallet or my passport. I was on my own without even a phrase book on my final night in the bandwidth capital of the world.

I walked down the familiar thoroughfare outside the Hilton, paused at the Mini Stop, saw the dark alley winding down the hill, and yes, I could see some light coming from there, somewhere. No one else was on the street. I looked left, right, then took my first step into the alley.

In fairy tales, this is where you see the ogre. In horror movies, here's where girl meets knife blade.

I started down the hill. I saw the glowing light ahead. I turned. Suddenly, everything changed. It was unreal. I felt as though I'd stumbled onto the movie set of a bustling nineteenth-century Korean bazaar. Removed from the glistening glass skyscrapers on the highway, I was now walking amid a warren of stalls and stores and food wagons and fortune tellers and palm readers and Go players and barkers hawking fish and fish products, vegetables fresh and cured, tofu, pickles, sweets, and kimchi in all colors and textures. At some stalls, elderly men squatted on little stools, their heads tipped over their spicy noodle bowls, the steam swirling into the night air. Other stalls sold silks or leather goods, Samsung smartphones or knockoff Dior bags, stickers and pens and purses featuring Choo-Choo, Korea's slightly demented feline answer to Japan's Hello Kitty. There were electronics, plumbing fixtures, electrical parts, and yes, luggage, too, and all manner of men and women, young and old, all beneath the red awnings, carved signs, and dangling lights of the Dongdaemun night market, a different swath of life through the back alleys of the most modern, wired urban landscape imaginable.

I found my suitcase within two minutes and strolled for another hour, wheeling it behind me, filling it with new treats as I stopped at stall after stall. It seemed impossible that I had walked within twenty yards of this scene every night for three nights and hadn't even suspected it was here. Dongdae-mun existed alongside the high-tech digital Seoul. They were parallel worlds, although, no doubt, neither would have existed in quite the same way without the other.

After a while, I reluctantly reversed my steps, locating the exact alley through which I'd entered, across from the luggage store. Those fairy tales were in my head again as I crossed the threshold back into the ultramodern Seoul I had come to see, and now saw differently. I would finish packing. In the morning I would take the airport bus, and then, on the long flight home, I'd sit among people from all over the world, people like and not like me at all.

My world had changed. I'd glimpsed a new city, front side and back. In a guidebook on the plane home, I read that the Dongdaemun was founded in 1905. Its name originally meant, Market for Learning.

Whatever you see means there is something you do not see. And then you are startled or distracted, lost or simply out for an adventure, and you see something else. If you are lucky, everything changes, in a good way.

But the key factor here is that "everything changes" has more to do with the way you see than with what exists. The gorilla *and* the basketball players coexist in the same space. So does the physical world and the World Wide Web. And so does the twentieth century's legacy in the twenty-first.

To believe that the new totally and positively puts an end to the old is a mistaken idea that gets us nowhere, neither out of old habits nor finding new ones better suited to the demands of that which has changed. John Seely Brown calls the apocalyptic view of change *endism*. Endism overstates what is gone. It blinds us to the fact that worlds not only exist in parallel all the time but that, if we learn how, we can dip in and out of one stream and enter another and enjoy the benefits and reminders of each as we are in the other. Like those Stanford multitaskers, we carry the residue of one task into another, but instead of that being irrelevant or a problem, it is exactly what makes navigating complex new worlds possible. We can see the periphery as well as the center, especially if we work to remain connected to those who see the world in a different way than we do. Without that collaboration of different minds, we can insist on gorillas *or* basketballs and never understand that they, like everything else, are all part of the same experiment. So are we. We just need to be linked to one another to be able to see it all. When I talk to my students about the way we select the worlds we see in our everyday life, they often ask how they can possibly change the way they see. It's easy, I always answer. I'll *assign* you the task of seeing differently. And you will. That's what learning is.

IF I WERE TO DISTILL one simple lesson from all the science and all the stories in this book, it would be that with the right practice and the right tools, we can begin to see what we've been missing. With the right tools and the right people to share them with, we have new options.

From infancy on, we are learning what to pay attention to, what to value, what is important, what counts. Whether on the largest level of our institutions or the most immediate level of concentrating on the task before us, whether in the classroom or at work or in our sense of ourselves as human beings, what we value and what we pay attention to can blind us to everything else we could be seeing. *The fact that we don't see it doesn't mean it's not there.*

Why is this important? Because sometimes we can make ourselves miserable seeing only what we think we are supposed to see. When there is a difference between expectation and reality, we feel like failures. In one generation, our world has changed radically. Our habits and practices have been transformed seemingly overnight. But our key institutions of school and work have not kept up. We're often in a position of judging our new lives by old standards. We can feel loss and feel as if we are lost, failed, living in a condition of deficit.

The changes of the digital age are not going to go away, and they are not going to slow down in the future. They'll accelerate. It's time to reconsider the traditional standards and expectations handed down to us from the linear, assembly-line arrangements of the industrial age and to think about better ways to structure and to measure our interactive digital lives.

By retraining our focus, we can learn to take advantage of a contemporary world that may, at first, seem distracting or unproductive. By working with others who do not share our values, skills, and point of view, change becomes not only possible but productive—and sometimes, if we're lucky, exhilarating.

The brain is constantly learning, unlearning, and relearning, and its possibilities are as open as our attitudes toward them. Right now, our classrooms and workplaces are structured for success in the last century, not this one. We can change that. By maximizing opportunities for collaboration, by rethinking everything from our approach to work to how we measure progress, we can begin to see the things we've been missing and catch hold of what's passing us by.

If you change the context, if you change the questions you ask, if you change the structure, the test, and the task, then you stop gazing one way and begin to look in a different way and in a different direction. You know what happens next:

Now you see it.

Acknowledgments

||||||||||||||||||||||||

A book has many origins and an author as many debts. Mine go back to 1998, when I accepted an offer from Nannerl Keohane, then president of Duke University, and the late John Strohbehn, the provost, to become the first person at Duke or at any university to have the full-time position of vice provost for interdisciplinary studies. I was at a crossroads in my career, considering a move to another university to accept a distinguished chair in my original field of specialization, when a brief letter in my fax machine offered me a tantalizing alternative: to stay at Duke and work with faculty and students from any and every part of the university to create the most innovative programs in higher education. When I met with Nan and John later that week, I learned there was no real job description, no set of rules, no bureaucracy. Shaking things up was the mandate. Our first meeting was downright Socratic. "What kind of budget will I have?" I asked. "What budget would you like to have?" I paused, then said I didn't really want a budget, at least not at the beginning. If we were aspiring to a creative, entrepreneurial model of learning, shouldn't we identify worthy projects first, then figure out the way to fund them?

That turned out to be the right answer. Within days of moving into my new office, I was told I'd be working with one of our youngest trustees, Melinda French Gates. She was as new at philanthropy as I was at administration. With her two degrees from Duke and an incisively astute mind matched only by her caring, Melinda had a vision for an entirely new kind of scholarship program wherein innovative and original undergraduate students who happened to require financial aid would learn side by side with the best graduate and professional school students. We worked on the University Scholars Program together, finding a way through a thicket of national and local rules to create a bold and adventurous scholarship. I still have the plain, unmarked envelope that held the first endowment check, for $10 million, from the newly named Bill and Melinda Gates Foundation.

During a particularly tedious pass through regulatory red tape, Nan quipped that my job was really to "break things and make things." I thank her for entrusting me with the daring challenge that, many years later, infuses the message of this book. With everything new, something is altered. Sometimes you have to break things in order to make things. In times of tremendous change, such as our own, it can seem as if the wreckage is too great. That's when you seek out the right partners and the right tools, and, if you are lucky, you can find a better way.

That adage doesn't apply just to programs but also to brains. During my first year as vice provost, I worked with colleagues to help create our first Center for Cognitive Neuroscience. Here I thank my beloved colleague Lew Siegel, dean of the Graduate School, an extraordinary guide in every way. I began an intensive self-education in neuroscience as we read dossiers for potential directors and then continued reading after the distinguished neuroscientist Ron Mangun came to head our program. My lunches with Ron melded with the thinking I was doing about reforming the structures of higher education for the scholarship program I was helping to create with Melinda. Innovative uses of new digital technologies for research and thinking became an additional part of my job description, and soon I was working with another inspiring entrepreneur, Kimberly Jenkins, who had started the Education Division at Microsoft before moving on to NeXT, Apple, and Cisco. Tracy Futhey, chief informaton officer at Duke, was key to making our iPod experiment possible. I thank all of those who helped me rethink education, brain studies, innovation, and new technologies together, for that is the terrain of *Now You See It*.

I also thank Provost Peter Lange, my next boss, with whom I worked for seven years and who continues to be as honest, canny, and wise a friend and adviser as anyone could want. I am grateful for the unflagging support of President Dick Brodhead and Vice President Mike Schoenfeld. I cannot begin to list all the colleagues in the provost's office, all the trustees, deans, faculty members, staff, and students who contributed to the ideas in this book, so I simply acknowledge a debt to you all that is far greater than I can ever repay.

At the risk of leaving someone out, I do want to single out a few exemplary teachers, mentors, and guides, past and present, who have shaped this book and its author: E. M. Broner, Radonna Patterson, Terry Vance, and Linda Wagner-Martin, as well as three very dear colleagues who passed away in my last years of writing this book: Emory Elliott, Eve Kosofsky Sedgwick, and John Hope Franklin, whose name I carry, with humility and love, in the title of my professorship.

Hundreds of students influenced this book, and I thank them all for their courage to learn vigorously and new. I especially extend my gratitude to the students in This Is Your Brain on the Internet, a course that has pushed the boundaries of content, discipline, pedagogical practice, and peer-learning methods, and all under the unflattering gaze of a skeptical media. I owe my students a lot, not least appreciation for their sense of intellectual adventure. Even when it has felt more like a roller-coaster ride than the straight-and-narrow path of *A*'s that leads to law school or medical school, they have proved that, in the end, the cultivation of learning's lifelong exhilaration is well worth the risk.

And sometimes one wants not the wild ride but an absolutely meticulous, diligent, thoughtful, yet always imaginative brand of knowledge, and, in that, my students have served me equally well. I cannot begin to thank the research assistants—all doctoral students at Duke—who were smart, careful, flexible, thorough, and caring as this manuscript was coming together into a book. My deep thanks go to Lindsey Andrews, Erica Fretwell, Nathan Hensley, and Patrick Jagoda. My friend Priscilla Wald circled the Duke Forest trail with me over and over, discussing ideas and giving me insights from her vast knowledge across many disciplines. The insight and intellectual generosity of Anne Allison, Alice Kaplan, and Robin Kirk helped me get this book started. With a meticulous eye and perfect pitch, Sharon Holland read several chapters as they neared their final form. I wish I could name all of the dozens, maybe hundreds, of friends and colleagues whose conversations and insights (in person, on Facebook, on Twitter, on blogs) have made me smarter over the years. My colleagues at the incomparable Bogliasco Foundation, in Liguria, Italy, were interlocutors during a divine month of quiet and beauty. I thank my HASTAC colleagues at Duke—Fiona Barnett, Erin Ennis, Sheryl Grant, Erin Gentry Lamb, Mark Olson, Ruby Sinreich, Fred Stutzman, and Jonathan Tarr, and especially my two constant, daily colleagues, Mandy Dailey and Nancy Kimberly. This writing would not have happened without them and their remarkable humor, wit, and expertise at collaboration by difference. For energy and kindness at every turn, I am also indebted to my colleagues at the John Hope Franklin Center for Interdisciplinary and International Studies and at the John Hope Franklin Humanities Institute.

Beyond Duke, I thank David Theo Goldberg, my coauthor on *The Future of Thinking* and collaborator from the very beginning on HASTAC, the wonderful organization with the clunky (if inclusive) name: Humanities, Arts, Sciences, and Technology Advanced Collaboratory. Nor did we imagine HASTAC (pronounced "haystack") would grow to more than five thousand network members from hundreds of institutions, all dedicated to the future of learning in a digital age. At HASTAC from the beginning, and still, were dear friends and colleagues Ruzena Bajcsy, Anne Balsamo, Allison Clarke, Kevin Franklin, Tara McPherson, and Kathy Woodward. I thank John Seely Brown, Howard Rheingold, and Larry Smarr for "getting" what we were up to earlier than just about anyone, as did the visionaries heading the John D. and Catherine T. MacArthur Foundation's Digital Media and Learning Initiative, especially Julie Stasch and Connie Yowell. More recently, friendship with ethnographer and digital youth analyst danah boyd has been inspiring, as has been HASTAC's new partnership with the Mozilla Foundation. Several deans and department chairs at Duke since my return to the faculty, after eight years as a vice provost, have put up with my wanderings far beyond the bounds of normal departments or standard pedagogies. I thank them here, collectively, for their patience and support.

A lovely former editor, Rosemary Ahern, introduced me to Deirdre Mullane, in 2008. Writers dream of an agent like Deirdre. I didn't know anyone like her really existed. I thank her for her brilliance, humor, and practical clarity. My editor, Kevin Doughten, made this a far better book than the one he acquired, and he has worked tirelessly for its success. I thank you both, and the entire Viking team, for dedication and vision beyond compare.

I thank my stepson, Charles R. Davidson, and his wife, Susan Brown, and wonderful Gavin and Morag for their love. My parents, Marlene Fineman Notari and Paul Notari, continue to work and love with a joy and exuberance and idealism that inspires me always. In his eighties, my dad contributes to the future of the environment and civic life with a gusto, energy, and good humor that motivates my own life and work. Thanks, Dad. This book is for you, as well as for my sister, Sharon, and brother, Ken. My brother and his family—his wife, Mary Lou Shioji Notari, and their son, Kris—live many issues in this book, and together they have been nothing short of heroic through life's greatest challenges.

Every word and idea and passion in this book was written with and for my partner in life, Ken Wissoker, editorial director at Duke University Press. Love and work, Freud said, are the key ingredients of happiness, and I could not be luckier. Thank you, Ken.

Appendix

|||||||||||||||||||||||||

Twenty-first-Century
Literacies—a Checklist

Teachers at every level should be addressing the requirements, possibilities, and limitations of the digital media that now structure our lives. The new skills required to take advantage of the Internet have been called twenty-first-century literacies. Media theorist Howard Rheingold has named four of these—attention, participation, collaboration, and network awareness (see www.sfgate.com/cgi-bin/blogs/rheingold/category?blogid=108&cat=2538), and I have added many more. It's a fluid list to which anyone, at any age, could add other skills required for our digital age.

- Attention: What are the new ways that we pay attention in a digital era? How do we need to change our concepts and practices of attention for a new era? How do we learn and *practice* new forms of attention in a digital age?
- Participation: How do we encourage meaningful interaction and participation in a digital age? How can the Internet be useful on a cultural, social, or civic level?
- Collaboration: Collaboration can simply reconfirm consensus, acting more as peer pressure than a lever to truly original thinking. HASTAC has cultivated the methodology of "collaboration by difference" to inspire meaningful ways of working together.
- Network Awareness: How can we thrive as creative individuals and at the same time understand our contribution within a network of others? How do we gain a sense of what that extended network is and what it can do?

- Global Consciousness: How does the World Wide Web change our responsibilities in and to the world we live in?
- Design: How is information conveyed differently, effectively, and beautifully in diverse digital forms? Aesthetics form a key part of digital communication. How do we understand and practice the elements of good design as part of our communication and interactive practices?
- Affordance: How do we assess all of the particular networked features, limitations, and liabilities of a particular technology in order to know when and how to use it to our best advantage?
- Narrative, Storytelling: How do narrative elements shape the information we wish to convey, helping it to have force in a world of competing information?
- Procedural (Game) Literacy: What are the new tactics and strategies of interactive games, wherein the multimedia narrative form changes because of our success or failure? How can we use game mechanics for learning and for motivation in our lives?
- Critical Consumption of Information: Without a filter (editors, experts, and professionals), much information on the Internet can be inaccurate, deceptive, or inadequate. How do we learn to be critical? What are the standards of credibility?
- Digital Divides, Digital Participation: What divisions still remain in digital culture? Who is included and who excluded? How do basic aspects of economics and culture dictate not only who participates in the digital age but how they participate?
- Ethics: What are the new moral imperatives of our interconnected age?
- Assessment: What are the best, most fluid, most adaptive, and helpful ways to measure progress and productivity, not as fixed goals, but as a part of a productive process that also requires innovation and creativity?
- Data Mining: How can we better understand how everything we contribute to the Internet produces data? How do we become savvy users and interpreters of this data?
- Preservation: What are the requirements for preserving the digital world we are creating? Paper lasts. Platforms change.
- Sustainability: What are the metrics for sustainability in a world where we live on more kilowatts than ever before? How do we protect the environment in a plugged-in era?

- Learning, Unlearning, and Relearning: Alvin Toffler has said that in the rapidly changing world of the twenty-first century, the most important skill anyone can have is the ability to stop in one's tracks, see what isn't working, and then find ways to unlearn old patterns and relearn how to learn. How is this process especially important in our digital world?

Notes

||||||||||||||||||||||||

Introduction: I'll Count—You Take Care of the Gorilla

1. The researcher that day was my colleague Güven Güzeldere, coauthor, with Murat Aydede, of *Sensing, Perceiving, Introspecting: Cognitive Architecture and Phenomenal Consciousness* (Oxford: Oxford University Press, 2007).

2. The gorilla experiment by Daniel Simons and Christopher Chabris is the cleverest of dozens of experiments that go all the way back to the 1970s, beginning with the work of Ulric Neisser. His foundational work can be seen in "The Control of Information Pickup in Selective Looking," in A. D. Pick, ed., *Perception and Its Development: A Tribute to Eleanor J. Gibson* (Hillsdale, NJ: Lawrence Erlbaum, 1979): 201–19; and Ulric Neisser and Robert Becklen, "Selective Looking: Attending to Visually Specified Events." *Cognitive Psychology* 7, no. 4 (1975): 480–94.

3. Special thanks to Dan Simons for sharing this anecdote about the gorilla suit, in a conversation on Dec. 21, 2009.

4. The term "inattentional blindness" was coined by Arien Mack and Irvin Rock in the 1980s and 1990s and published in articles gathered together in the landmark study *Inattentional Blindness* (Cambridge, MA: MIT Press, 1998): ix.

5. Roger Highfield, "Did You See the Gorilla?" *Telegraph.co.uk*, May 5, 2004.

6. The man in this experiment, viewed millions of times on YouTube, is British psychologist Richard Wiseman. See "The Amazing Color-Changing Card Trick," Apr. 28, 2007, www.youtube.com/watch?v-voAntzB7EwE (accessed Jan. 10, 2010).

7. John Helleberg, Christopher D. Wickens, and Xidong Xu, *Pilot Maneuver Choice and Safety in a Simulated Free Flight Scenario* (Savoy, IL: Aviation Research Lab, Institute of Aviation, University of Illinois at Urbana-Champaign, 2000).

8. For a succinct discussion of why it is important to understand the difference between the Internet and the World Wide Web, see Thomas L. Friedman, *The World Is Flat: A Brief History of the Twenty-first Century* (New York: Farrar, Straus & Giroux, 2005): 60–66.

9. "Exactly How Much Are the Times A-Changin?" *Newsweek*, July 26, 2010, p. 56.

10. Julie Logan, "Failures in Education System Cause UK to Produce Less Dyslexic Entrepreneurs than the US," Nov. 5, 2007, www.cass.city.ac.uk/press/press_release_pdfs/Dyslexia%20-%20education.pdf (accessed Jan. 10, 2010).

11. "Survey Shows Widespread Enthusiasm for High Technology," NPR Online, www.npr.org/programs/specials/poll/technology (accessed Apr. 7, 2010).

12. Robert Darnton, "The Library in the New Age," *New York Review of Books*, June 12, 2008, www.nybooks.com/articles/archives/2008/jun/12/the-library-in-the-new-age (accessed June 15, 2009).

13. Clay Shirky, "The Collapse of Complex Business Systems," www.shirky.com/weblog/2010/04/the-collapse-of-complex-business-models (accessed Apr. 8, 2010).

14. Norman Doidge, *The Brain that Changes Itself: Stories of Personal Triumph from the Frontiers of Brain Science* (New York: Penguin Books, 2007).

15. "A History of Speeding Tickets," Wallstreet Fighter, www.mademan.com/2007/01/history-of-speeding-tickets (accessed Mar. 5, 2010).

16. U.S. Department of Labor study cited in "The Changing Workplace," FamilyEducation .com, http://life.familyeducation.com/work/personal-finance/47248.html (accessed Jan. 10, 2010). Futurist Jim Carroll cites a study of future career trends by the Australian Innovation Council in *Ready, Set, Done: How to Innovate When Faster Is the New Fast* (Mississauga, Ont.: Oblio Press, 2007): 61.

17. Alvin Toffler, et al., *Rethinking the Future: Rethinking Business Principles, Competition, Control and Complexity, Leadership, Markets and the World* (London: Nicholas Brealey, 1997).

1. Learning from the Distraction Experts

1. John J. Boren, Allan M. Leventhal, and H. Edmund Pigott, "Just How Effective Are Antidepressant Medications? Results of a Major New Study," *Journal of Contemporary Psychotherapy* 39, no. 2 (June 2009); see also Shankar Vedantam, "Drugs Cure Depression in Half of Patients: Doctors Have Mixed Reactions to Study Findings," *Washington Post*, Mar. 23, 2006, www.washingtonpost.com/

wp-dyn/content/article/2006/03/22/AR2006032202450.html (accessed Dec. 18, 2009).

2. As an industry magazine reports, "Cymbalta had $386 million in US first-quarter sales [in 2007], up 88% compared with the first three months of 2006." See "Lilly Launches New Cymbalta TV Spots," *Medical Marketing & Media*, June 1, 2007, www.mmm-online.com/lilly-launches-new-cymbalta-tv-spots/article/24337 (accessed Dec. 17, 2009).

3. According to the Draftfcb mission statement, "In delivering its clients a high Return on Ideas[SM] the agency is driven by The 6.5 Seconds That Matter[SM], a creative expression recognizing the brief period of time marketers have to capture consumers' attention and motivate them to act." See www.interpublic.com/companyfinder/companydetail?company=1241&cname=Draftfcb (accessed Nov. 25, 2009).

4. See, for example, Ruth S. Day, "Direct-to-Consumer Ads for Medical Devices: What Do People Understand and Remember?" Testimony to the U.S. Congress, Nov. 18, 2008, Senate Committee on Aging, www.duke.edu/~ruthday/medcog .html (accessed Dec. 15, 2009). Day's Medical Cognition Laboratory at Duke University evaluates "cognitive accessibility" to determine how much consumers of prescription and over-the-counter drugs, medical devices, and medical treatments are influenced by ads. Such research might be funded by the National Science Foundation—or Glaxo Welcome.

5. Amy Shaw, "Direct to Consumer Advertising of Pharmaceuticals," Proquest Discovery Guides, www.csa.com/discoveryguides/direct/review3.php (accessed Nov. 20, 2009).

6. For testimony of Draftfcb's success on the Cymbalta account, see "Best Branded TV Campaign," *Medical Marketing & Media*, Oct. 29, 2009, www.mmm-online.com/best-branded-tv-campaign/article/156510 (accessed Dec. 16, 2009). According to another account, "Industry watchers liked Lilly's 'Depression Hurts' campaign for Cymbalta; *Medical Marketing & Media* deemed it the best overall consumer campaign. And Cymbalta sales growth reached a whopping 60 percent, giving Lilly a $2.1 billion infusion to the top line, $1.8 billion of that in the U.S. alone." See "Top 13 Advertising Budgets," *Fierce Pharma*, Sept. 23, 2008, www.fiercepharma.com/special-reports/eli-lilly-top-13-advertising-budgets (accessed Dec. 16, 2009).

7. As the *Los Angeles Times* reported, "For every 10% increase in direct-to-consumer advertisements within a class of similar drugs, sales of drugs in that class (say,

antidepressants or erectile dysfunction drugs) went up 1%, Kaiser found in a 2003 study. In 2000, direct-to-consumer advertising alone boosted drug sales 12%, at an additional cost of $2.6 billion to consumers and insurers." Melissa Healy, "The Payoff: In Short, Marketing Works; by Targeting Consumers and Doctors—Directly and Indirectly—Drug Makers Are Driving Sales. Why Mess with Success?" *Los Angeles Times,* Aug. 6, 2007, F6. See also Miriam Shuchman, "Drug Risks and Free Speech—Can Congress Ban Consumer Drug Ads?" *New England Journal of Medicine,* 356 (May 31, 2007): 2236–39; and "How Consumers' Attitudes Toward Direct-to-Consumer Advertising of Prescription Drugs Influence Ad Effectiveness, and Consumer and Physician Behavior," *Marketing Letters* 15, no. 4 (Dec. 2004): 201–12.

8. Special thanks to my student Ian Ballard for relating this incident.

9. Ludwig Wittgenstein was an influence on Edward Tronick's important work on infant-mother scaffolding. See, especially, Edward Z. Tronick, "Dyadically Expanded States of Consciousness and the Process of Therapeutic Change," *Infant Mental Health Journal* 19, no. 3 (1998): 290–99.

10. The primers of the socio-interactionist view of language acquisition are Clare Gallaway and Brian J. Richards, *Input and Interaction in Language Acquisition* (New York: Cambridge University Press, 1994); Alison F. Garton, *Social Interaction and the Development of Language and Cognition* (New York: Psychology Press, 1995); and Brian MacWhinney, *Mechanisms of Language Acquisition* (Mahwah, NJ: Lawrence Erlbaum, 1987).

11. Birgit Mampe, Angela D. Friederici, Anne Christophe, and Kathleen Wermke, "Newborns' Cry Melody Is Shaped by Their Native Language," *Current Biology* 19, no. 23 (Nov. 5, 2009).

12. Cross-cultural studies by many investigators show that American parents tend to interact with their babies primarily by talking to, looking at, or smiling at them. See Berry Brazelton et al., "The Origins of Reciprocity: The Early Mother-Infant Interaction," in H. Rudolph Schaffer, ed. *Studies in Mother-Infant Interaction* (New York: Wiley, 1974); Robert A. LeVine et al., *Child Care and Culture: Lessons from Africa* (Cambridge, UK: Cambridge University Press, 1994). With newborns, American mothers tend to respond to hand movements, burps, and other physical responses as if these bodily functions were conversational.

13. Madeline Shakin et al., "Infant Clothing: Sex Labeling for Strangers," *Sex Roles* 12, no. 9–10 (1985): 55–64; Thomas Eckes and Hanns Martin Trautner, *The*

Developmental Social Psychology of Gender (Mahwah, NJ: Lawrence Erlbaum, 2000); and Katherine H. Karraker et al., "Parents' Gender-Stereotyped Perceptions of Newborn Infants: The Eye of the Beholder Revisited," *Sex Roles* 33, no. 9–10 (1995): 687–701.

14. Nancy Shand and Yorio Kosawa, "Japanese and American Behavior Types at Three Months: Infants and Infant-Mother Dyads," *Infant Behavior and Development* 8, no. 2 (1985): 225–40.

15. Richard E. Nisbett, *The Geography of Thought: How Asians and Westerners Think Differently—and Why* (New York: Free Press, 2003), 150.

16. My analysis of "concepts" and "categories" is much influenced by philosopher Elizabeth Grosz and the theoretical genealogies her work encompasses. See, for example, *The Nick of Time: Politics, Evolution, and the Untimely* (Durham, NC: Duke University Press, 2004).

17. For a comprehensive overview of Piaget's theories of concept development as well as a survey of many studies showing that children learn racial mappings long before they have a defined concept of race, see Debra Van Ausdale and Joe R. Feagin, *The First R: How Children Learn Race and Racism* (New York: Rowman & Littlefield, 2001).

18. See David J. Kelly, Paul C. Quinn, Alan M. Slater, Kang Lee, Liezhong Ge, and Olivier Pascalis, "The Other-Race Effect Develops During Infancy: Evidence of Perceptual Narrowing," *Psychological Science* 18, no. 12 (2007): 1084–89; Gizelle Anzures, Paul C. Quinn, Olivier Pascalis, Alan M. Slater, and Kang Lee, "Categorization of Faces in Infancy: A New Other-Race Effect," *Developmental Science* 11, no. 1 (2009): 78–83; and Charles Barkley, et al., *Who's Afraid of a Large Black Man?* (New York: Penguin, 2005). Barkley's collection of thirteen interviews on the issue of fear and black men is filled with insights.

2. Learning Ourselves

1. Gerald M. Edelman, *Second Nature: Brain Science and Human Knowledge* (New Haven, CT: Yale University Press, 2006), 18.

2. See Eric R. Kandel et al., eds., *Principles of Neural Science* (New York: McGraw-Hill, Health Professions Division, 2000), ch. 6.

3. Ideas about infant response to auditory stimulus began to change in the 1970s with the introduction of nonverbal mechanisms for measurement. See, for example,

W. Alan Eisele et al., "Infant Sucking Response Patterns as a Conjugate Function of Changes in the Sound Pressure Level of Auditory Stimuli," *Journal of Speech and Hearing Research* 18, no. 2 (June 1975): 296–307.

4. For a selection of some of the vast research on this topic, see Joseph L. Jacobson et al., "Paralinguistic Features of Adult Speech to Infants and Small Children," *Child Development* 54, no. 2 (1983): 436–42; and Christiane Dietrich et al., "Native Language Governs Interpretation of Salient Sounds Speech Differences at 18 Months," *Proceedings of the National Academy of Sciences of the USA* 104, no. 41 (2007): 16027–31.

5. An overview that includes the work of Anne and L. Dodge Fernald is in Robert S. Siegler, Judy S. DeLoache, and Nancy Eisenberg, *How Children Develop* (New York: Worth, 2002).

6. Whitney M. Weikum et al., "Visual Language Discrimination in Infancy," *Science* 316, no. 5828 (2007): 1159.

7. See Jean-Pierre Changeux, *Neuronal Man: The Biology of Mind* (New York: Pantheon, 1985); Michael S. Gazzaniga, *Nature's Mind: The Biological Roots of Thinking, Emotions, Sexuality, Language, and Intelligence* (New York: Basic Books, 1992); John C. Eccles, *Evolution of the Brain: Creation of the Self* (New York: Routledge, 1989); and a good historical discussion in the standard textbook, E. Bruce Goldstein, *Sensation and Perception* (Belmont, CA: Wadsworth, 1989).

8. George Lakoff, *The Political Mind: Why You Can't Understand 21st-Century American Politics with an 18th-Century Brain* (New York: Viking, 2008): 34.

9. Many brain discoveries are made by studying brains that don't function according to expectation. Recent research on autism, for example, is revealing more information about how mirror neurons work, not just in the cognitive part of the brain, but also in the most emotional parts of the brain, like the amygdala. See Vilayanur Ramachandran and Lindsay M. Oberman, "Broken Mirrors: A Theory of Autism," *Scientific American* 295, no. 5 (2006): 62–69.

10. Sandra Blakeslee, "Odd Disorder of Brain May Offer New Clues," *New York Times* Aug. 2, 1994, sec. C, p. 4. See also, Ursula Bellugi et al., "Nature and Nuture: Williams Syndrome Across Cultures," *Developmental Science* 10, no. 6 (2007): 755–62; and Ursula Bellugi and Albert M. Galaburda, "Multi-Level Analysis of Cortical Neuroanatomy in Williams Syndrome," *Journal of Cognitive Neuroscience* 12 (2000): S74–S88.

11. Even the numbers on this disease vary widely, with some studies suggesting it occurs in one in every 7,500 births worldwide, while others insist it occurs only

in one in every 20,000 births. While genetically the condition is the same, and while physical and personality syndromes manifest in similar ways, how those syndromes are evaluated differ drastically. Ursula Bellugi et al., "Nature and Nurture."

12. Martin J. Sommerville, et al., "Severe Expressive-Language Delay Related to Duplication of the Williams-Beuren Locus," *New England Journal of Medicine* 353, no. 16 (2005): 1694–1701; and Lucy R. Osborne and Carolyn B. Mervis, "Rearrangements of the Williams-Beuren Syndrome Locus: Molecular Basis and Implications for Speech and Language Development," *Expert Reviews in Molecular Medicine* 9, no. 15 (June 13, 2007): 1–16.

13. Ursula Bellugi, "Nature and Nurture."

14. Giacomo Rizzolatti, Luciano Fadiga, Leonardo Fogassi, and Vittorio Gallese began this work with macaque monkeys and have since extended their work to speculations about humans in the transmission of language, which is to say culture. The groundbreaking work on mirror neurons by the Parma research group has appeared in numerous articles over the last decade, including: Giacomo Rizzolatti et al., "Premotor Cortex and the Recognition of Motor Actions," *Cognitive Brain Research* 3 (1996): 131–41; Vittorio Gallese, et al., "Action Recognition in the Premotor Cortex, *Brain* 119, no. 2 (1996): 593–610; Leonardo Fogassi et al., "Parietal Lobe: From Action Organization to Intention Understanding," *Science* 308, no. 5722 (Apr. 25, 2005): 662–66; Giacomo Rizzolatti and Laila Craighero, "The Mirror-Neuron System," *Annual Review of Neuroscience* 27 (2004): 169–92; Giacomo Rizzolatti and Michael A. Arbib, "Language Within Our Grasp," *Trends in Neurosciences* 21, no. 5 (1998): 188–94; Marco Iacoboni et al., "Cortical Mechanisms of Human Imitation," *Science* 286, no. 5449 (Dec. 24, 1999): 2526–28; Leslie Brothers et al., "Response of Neurons in the Macaque Amygdala to Complex Social Stimuli," *Behavioral Brain Research* 41, no. 3 (1990): 199–213; and Alfred Walter Campbell, *Histological Studies on the Localisation of Cerebral Function* (Cambridge, UK: Cambridge University Press, 1905). Gallese recently began working with linguist George Lakoff to investigate how the structure and concept of "grasping" embodies the interconnection of mind and body prefigured by the discovery of mirror neurons, and that radically reverses the Enlightenment separation of mind and body, mind and brain, brain and world. All are interconnected in profound ways. See Vittorio Gallese and George Lakoff, "The Brain's Concepts: The Role of the Sensory-Motor System in Conceptual Knowledge," *Cognitive Neuropsychology* 22, no. 3–4 (2005): 455–79.

15. Valeria Gazzola and Christian Keysers, "The Observation and Execution of Actions Share Motor and Somatosensory Voxels in All Tested Subjects: Single-Subject Analyses of Unsmoothed fMRI Data," *Cerebral Cortex* 19, no. 6 (2009): 1239–55; Vilayanur S. Ramachandran, "Mirror Neurons and Imitation Learning as the Driving Force Behind 'The Great Leap Forward' in Human Evolution," available from Edge Foundation, www.edge.org/3rd_culture/ramachandran (accessed Nov. 16, 2006); and Norman Doidge, "The Brain That Changes Itself," *American Journal of Bioethics* 8, no. 1 (2008): 62–63.

16. Vittorio Gallese et al., "A Unifying View of the Basis of Social Cognition," *Trends in Cognitive Sciences* 8, no. 9 (2004): 396–403.

17. Frans de Waal, *Primates and Philosophers: How Morality Evolved*. (Princeton, NJ: Princeton University Press, 2009).

18. Dan Ariely, *Predictably Irrational: The Hidden Forces That Shape Our Decisions* (New York: HarperCollins, 2008).

19. The persistence or reappearance of the Babinski reflex in an older child or adult (except during sleep or after a long walk) can signal a serious neurological condition such as brain or spinal damage. See James S. Harrop et al., "Neurological Manifestations of Cervical Spondylosis: An Overview of Signs, Symptoms, and Pathophysiology," *Neurosurgery* 60, no. 1, suppl. 1 (2007): 14–20.

3. Project Classroom Makeover

1. These are all quoted and the experiment is discussed in James Todd, "The iPod Idea: Wired for Scholarship," *Duke Magazine* 91, no. 5, Sept.–Oct. 2005.

2. The iPod experiment would never have happened without approval and funding of this forward-looking initiative, for which credit goes to my colleagues Tracy Futhey, vice president for information technology, and Provost Peter Lange.

3. On values formation and the blind spots it leaves, see Barbara Herrnstein Smith, *Contingencies of Value: Alternate Perspectives for Critical Theory* (Cambridge, MA: Harvard University Press, 1988).

4. Jeff Howe, *Crowdsourcing: Why the Power of the Crowd Is Driving the Future of Business* (New York: Crown, 2008).

5. Professor Marie Lynn Miranda is a pioneer in using new technologies to help shape community activism on environmental policy. Her Web site is: www.nicholas.duke.edu/DEarchives/f02/f-mapping.html (accessed May 6, 2010).

6. Maria Magher, "iPod Gets Top Marks: Culbreth Middle School Is the First in the State to Require Device," *Chapel Hill News*, Mar. 14, 2010.

7. I have written about this at length with my HASTAC cofounder, David Theo Goldberg, in a research report that was first put up on the Web for comment from anyone who wished to offer it, then published in a research report based on colloquia we held all over the country, *The Future of Learning Institutions in a Digital Age* (Cambridge, MA: MIT Press, 2009). The expanded book form of this project is Cathy N. Davidson and David Theo Goldberg, *The Future of Thinking: Learning Institutions in a Digital Age* (Cambridge, MA: MIT Press, 2010).

8. "States Push for Nationalized Educational Standard," *U.S. and World News*, CBS Interactive, Mar. 10, 2010, http://cbs4.com/national/national.standardized .education.2.1552390.html (accessed Mar. 19, 2010).

9. The classic studies of American education include Lawrence Cremin, *American Education: The National Experience* (New York: HarperCollins, 1980); Carl Kaestle, *Pillars of the Republic: Common Schools and American Society, 1780–1860* (New York: Hill & Wang, 1983); and Michael Katz, *Reconstructing American Education* (Cambridge, MA: Harvard University Press, 1987).

10. See Alvin Toffler and Heidi Toffler, *Revolutionary Wealth: How It Will Be Created and How It Will Change Our Lives* (New York: Knopf, 2006): 357–62.

11. Robert Schwartz, "The American High School in Context," paper delivered to Sino-U.S. Seminar on Diversity in High School, Mar. 23, 2009, http://cnier .ac.cn/english/news/english_20090407094352_7842.html (accessed Mar. 19, 2010). This is the single most concise survey and set of statistics I have found anywhere, and one remarkably free of the polemics and politics (left or right) that confuse many of the statistics. This debt to him is not just for this wise assessment and sorting of the numbers but for his distinguished career of contribution to national educational policy since the Carter administration.

12. These are official figures from the Organisation for Economic Co-operation and Development (OECD), an international organization of thirty countries "committed to democracy and the market economy." "About OECD," www.oecd .org/pages/0,3417,en_36734052_36734103_1_1_1_1_1,00.html (accessed Mar. 16, 2010).

13. Ibid.

14. Special thanks to tweeter Michael Josefowicz, a retired printer who tweets as ToughLoveforX, for this clarifying distinction between *standards* and

standardization. Schwartz's essay discusses the surveys of why students drop out of school.

15. On class size and academic performance, see foundational work in, for example, Glen E. Robinson and James H. Wittebols, *Class Size Research: A Related Cluster Analysis for Decision-Making.* (Arlington, VA: Education Research Service, 1986); Jeremy D. Finn, *Class Size and Students at Risk: What Is Known? What Is Next?* (Washington, DC: U.S. Department of Education, Office of Educational Research and Improvement, National Institute on the Education of At-Risk Students, 1998); and Gene V. Glass, Leonard S. Cahen, Mary L. Smith, and Nikola N. Filby, *School Class Size: Research and Policy* (Beverly Hills, CA: Sage, 1982). This work is reviewed at www2.ed.gov/pubs/ReducingClass/Class_size .html (accessed May 2, 2010).

16. See William R. Dagged and Paul David Nussbaum, "How Brain Research Relates to Rigor, Relevance and Relationships," www.leadered.com/pdf/Brain%20 Research%20White%20Paper.pdf (accessed Apr. 29, 2010). See also Peter S. Eriksson, Ekaterina Perfilieva, Thomas Bjork-Eriksson, Ann-Marie Alborn, Claes Nordborg, Daniel A. Peterson, and Fred H. Gage, "Neurogenesis in the Adult Human Hippocampus," *Nature Medicine* 4 (1998): 1313–17. There is a prolific body of research and policy statements on the "new three *Rs*"—including other *Rs*, such as responsibility, representing, relating, and so forth. See, for example, Clifford Adelman, *Principal Indicators of Student Academic Histories in Postsecondary Education, 1972–2000* (Washington, DC: U.S. Department of Education, Institute of Education Sciences, 2004); and Anthony P. Carnevale and Donna M. Desrochers, *Connecting Education Standards and Employment: Course-Taking Patterns of Young Workers* (Washington, DC: Achieve Inc., 2002).

17. Seth Godin, "What You Can Learn from a Lousy Teacher," http://sethgodin .typepad.com/seths_blog/2010/03/what-you-can-learn-from-a-lousy-teacher .html (accessed Mar. 20, 2010).

18. Matthew B. Crawford, *Shop Class as Soulcraft: An Inquiry into the Value of Work* (New York: Penguin, 2009).

19. All of this comes from weekly faculty blog posts at Quest 2 Learn, available online for parents and all the interested world to see: http://pta.q2l.org/relay-week-of-december-14-december-18-2009 (accessed Jan. 10, 2009).

20. Katie Salen, ed., *The Ecology of Games: Connecting Youth, Games, and Learning,* John D. and Catherine T. MacArthur Foundation Series on Digital Media and Learning (Cambridge, MA: MIT Press, 2008), 2.

21. Ibid., 9.

22. Quoted in David Kushner, "Can Video Games Teach Kids?" *Parade*, Dec. 20, 2009.

23. FAQ on the Quest 2 Learn/Institute of Play Web site, www.instituteofplay.com/node/71#8 (accessed May 6, 2010).

24. Quoted in Kushner, "Can Video Games Teach Kids?"

25. Daniel J. Levitin, *This Is Your Brain on Music: The Science of a Human Obsession* (New York: Dutton, 2006), 106.

26. *Cognitive surplus* is a term coined by Clay Shirky in *Here Comes Everybody: The Power of Organizing Without Organizations* (New York: Penguin, 2008).

27. "Stanford Study of Writing," results available at http://ssw.stanford.edu (accessed Jan 4, 2010).

28. Quoted in "A Vision of Students Today," YouTube video by Kansas State University professor Michael Wesch and the students in Wesch's Cultural Anthropology 101, www.youtube.com/watch?v=dGCJ46vyR9o (accessed Jan. 4, 2010).

29. Jeff Hawkins and Sandra Blakeslee, *On Intelligence* (New York: Holt, 2005).

4. How We Measure

1. Cathy N. Davidson, "How to Crowdsource Grading," www.hastac.org/blogs/cathy-davidson/how-crowdsource-grading (accessed Jan. 10, 2010).

2. One of the most widely discussed papers of 2008 debunks the field of "social neuroscience" by showing the specious interpretation, application, and circular research design of studies using fMRIs to test human behavior. It uses the controversial term "voodoo" science. See Edward Vul, Christine Harris, Piotr Winkelman, and Harold Pashler, "Puzzlingly High Correlations in fMRI Studies of Emotion, Personality, and Social Cognition," *Perspectives on Psychological Science* 4, no. 3 (2009): 274–90.

3. Special thanks to graduate student William Hunt for his insights into grading. For an analysis of the evolution of grading, see Christopher Stray, "From Oral to Written Examinations: Cambridge, Oxford and Dublin 1700–1914," *History of Universities* 20, no. 2 (2005): 94–95.

4. Mark W. Durm, "An A Is Not an A Is Not an A: A History of Grading," *Educational Forum* 57 (Spring 1993). See also Mary Lovett Smallwood, *Examinations and Grading Systems in Early American Universities* (Cambridge, MA: Harvard University Press, 1935).

5. Jay Mathews, "A to F Scale Gets Poor Marks but Is Likely to Stay," *Washington Post*, Oct. 18, 2005, www.washingtonpost.com/wp-dyn/content/article/2005/10/17/AR2005101701565_2.html (accessed May 4, 2010).

6. Joseph J. Harris, H. Russell Cross, and Jeff W. Savell, "History of Meat Grading in the United States," Meat Science at Texas A&M University, 1990, rev. 1996, http://meat.tamu.edu/history.html (accessed Mar. 22, 2010); and Herbert Windsor Mumford, "Market Classes and Grades of Cattle with Suggestions for Interpreting Market Quotations," *Bulletin of the University of Illinois (Urbana-Champaign Campus), Agricultural Experiment Station*, no. 78 (1902), www.ideals.illinois.edu/handle/2142/3059 (accessed Mar. 22, 2010). Special thanks to William R. Hunt for this work on grading students and meat.

7. Franz Samelson, "Was Early Mental Testing: (a) Racist Inspired, (b) Objective Science, (c) A Technology for Democracy, (d) The Origin of the Multiple-Choice Exams, (e) None of the Above? (Mark the RIGHT Answer)," pp. 113–27, in *Psychological Testing and American Society, 1890–1930*, ed. Michael M. Sokal (New Brunswick, NJ: Rutgers University Press, 1987), 122–23.

8. Peter Sacks, *Standardized Minds: The High Price of America's Testing Culture and What We Can Do to Change It* (New York, Da Capo, 2001), 221–22.

9. Frederick J. Kelly, "The Kansas Silent Reading Tests," *Journal of Educational Psychology* 8, no. 2 (Feb. 1916).

10. Ibid.

11. Theodore M. Porter, *Trust in Numbers: The Pursuit of Objectivity in Science and Public Life* (Princeton, NJ: Princeton University Press, 1995): ix.

12. Samelson, "Early Mental Testing," 122–23.

13. Porter, *Trust in Numbers*, 208–10.

14. Samelson, "Early Mental Testing," 122–23.

15. The debates about classical test theory, item response theory (IRT), latent trait theory, modern mental test theory, and so forth are extremely complicated. All attempt to justify the fact that a larger body of knowledge is tested according to filling in one "bubble" about one topic. Critics of IRT compare assessing overall knowledge based on response to one item as equivalent to assessing a Beethoven symphony based on one note. Relationships among ideas and sets of facts are not tested in IRT. For a readable overview, see Susan E. Embretson and Steven Paul Reise, *Item Response Theory for Psychologists* (Mahwah, NJ: Lawrence Erlbaum, 2000).

16. Frederick James Kelly, *The University in Prospect: On the Occasion of His Inauguration as President of the University of Idaho* (Moscow, ID, 1928): 21.

17. My special thanks to Julie Munroe, Reference Library, Special Collections, University of Idaho, for finding materials about Frederick Kelly that were thought not to exist.

18. Stephen J. Ceci, *On Intelligence: A Bioecological Treatise on Intellectual Development* (Cambridge, MA: Harvard University Press, 1996); and Michael M. Sokal, ed., *Psychological Testing and American Society, 1890–1930* (New Brunswick, NJ: Rutgers University Press, 1987).

19. See Jay Mathews, "Just Whose Idea Was All This Testing?" *Washington Post*, Nov. 14, 2006; and Richard Phelps, "Are US Students the Most Heavily Tested on Earth?" *Educational Measurement: Issues and Practice* 15, no. 3 (Fall, 1996): 19–27.

20. Quoted in Mark J. Garrison, *A Measure of Failure: The Political Origins of Standardized Testing* (Albany, NY: SUNY Press, 2009): 84.

21. Garrison, *A Measure of Failure*, 29–30. See also Howard Gardner, *Multiple Intelligences: The Theory in Practice* (New York: Basic Books, 1993).

22. William Stern, *The Psychological Methods of Intelligence Testing* (Baltimore: Warwick & York, 1912).

23. It is possible that psychologists themselves have overestimated the harmful role of IQ tests in the immigration and eugenics controversy, a point made in a revisionist collection of essays, edited by Michael M. Sokal, *Psychological Testing and American Society, 1890–1930* (New Brunswick, NJ: Rutgers University Press, 1987).

24. Stephen Jay Gould reproduces the confusing early pictorial Army Beta IQ tests and challenges readers to score higher than the "moron" level, *The Mismeasure of Man*, rev. ed. (New York: Norton, 1996).

25. Frederick McGuire, "Army Alpha and Beta Tests of Intelligence," in Robert J. Sternberg, ed., *Encyclopedia of Intelligence*, vol. 1 (New York: Macmillan, 1994): 125–29; and Leon J. Kamin, "Behind the Curve," review of *The Bell Curve: Intelligence and Class Structure in American Life*, Richard J. Herrnstein and Charles Murray (New York: Free Press, 1994), *Scientific American*, Feb. 1995, pp. 99–103.

26. Larry V. Hedges and Amy Nowell, "Sex Differences in Mental Test Scores, Variability, and Numbers of High-Scoring Individuals," *Science* 269 (1995): 41–45; see also Cyril L. Burt and Robert C. Moore, "The Mental Differences Between the Sexes," *Journal of Experimental Pedagogy* 1 (1912): 273–84, 355–88.

27. See Michael Bulmer, *Francis Galton: Pioneer of Heredity and Biometry* (Baltimore: Johns Hopkins University Press, 2003); and Nicholas Wright Gillham, *A Life of Sir Francis Galton: From African Exploration to the Birth of Eugenics* (New York: Oxford University Press, 1996).

28. Gould, *Mismeasure of Man,* 177.

29. Richard E. Nisbett, *Intelligence and How to Get It: Why Schools and Cultures Count* (New York: Norton, 2009).

30. David Gibson, "Assessment & Digital Media Learning," presented at the Peer to Peer Pedagogy (P3) HASTAC Conference, John Hope Franklin Center for Interdisciplinary and International Studies, Duke University, Sept. 10, 2010. Gibson is a computer scientist and educational visionary working on digital media assessment theory, a combination of evidence-centered test design and machine-readable methods for creating and scoring assembly-style (adaptive) testing, such as found in game mechanics. See, for example, David Gibson et al., *Games and Simulations in Online Learning: Research and Development Frameworks* (Hershey, PA: Information Science Publishing, 2007); see also Robert J. Mislevy et al., *Computerized Adaptive Testing: A Primer* (Hillsdale, NJ: Lawrence Erlbaum, 2000).

31. Timothy Salthouse, "Implications of Within-Person Variability in Cognitive and Neuropsychological Functioning for the Interpretation of Change,"*Neuropsychology* 21, no. 6 (2007): 401–11. For additional studies by Salthouse of cognition, aging, and testing procedures, see also "Groups Versus Individuals as the Comparison Unit in Cognitive Aging Research," *Developmental Neuropsychology 2* (1986): 363–72; and Timothy A. Salthouse, John R. Nesselroade, and Diane E. Berish, "Short-Term Variability and the Calibration of Change," *Journal of Gerontology: Psychological Sciences* 61 (2006): 144–51.

32. Salthouse, "Implications of Within-Person Variability."

33. "Cognitive Scores Vary as Much Within Test Takers as Between Age Groups," *UVA Today,* July 2, 2007, available at www.virginia.edu/uvatoday/newsRelease .php?id=2325 (accessed Jan. 10, 2010).

34. Diane Ravitch, *The Death and Life of the Great American School System: How Testing and Choice Are Undermining Education* (New York: Basic Books, 2010).

35. Daniel J. Cohen and Roy Rosenzweig, "No Computer Left Behind," *Chronicle Review, Chronicle of Higher Education,* Feb. 24, 2006.

36. This is not simply a pipe dream. In November 2010, the Mozilla Foundation sponsored its first international festival in Barcelona—Learning, Freedom, and the

Open Web—at which programmers and educators codeveloped learning portfolios and learning badges for this purpose. HASTAC members worked on some of these for higher education.

37. Helen Barrett, "Authentic Assessment with Electronic Portfolios Using Common Software and Web 2.0 Tools," www.electronicportfolios.org/web20.html (accessed Sept. 12, 2010); and, for an overview of new digital measurement, David Weinberger, *Everything Is Miscellaneous: The Power of the New Digital Disorder* (New York: Times Books, 2007).

38. Danah Boyd, "Taken Out of Context: American Teen Sociality in Networked Publics," PhD dissertation, University of California Berkeley, Jan. 2009, www.zephoria.org/thoughts/archives/2009/01/18/taken_out_of_co.html (accessed Sept. 12, 2010).

39. See "Framework for Twenty-First-Century Learning," an invaluable set of resources compiled by the Partnership for Twenty-First-Century Skills, a partnership among the U.S. Department of Education and many private and nonprofit agencies, foundations, and corporations. For more information, see www.p21.org/index.php (accessed Sept. 12, 2010).

40. Historian Todd Pressner is the leader of the Hypercities project, which was a winner of a 2008 Digital Media and Learning Competition award, http://hypercities.com (accessed Oct. 6, 2010).

5. The Epic Win

1. "Epic Win," *Urban Dictionary,* www.urbandictionary.com/define.php?term= epic%20win (accessed Mar. 30, 2010).

2. Jane McGonigal's Web site, www.avantgame.com (accessed Mar. 20, 2010).

3. "About World Without Oil," www.worldwithoutoil.org/metaabout.htm (accessed Mar. 30, 2010).

4. Evoke, http://blog.urgentevoke.net/2010/01/27/about-the-evoke-game (accessed Apr. 5, 2010).

5. Joseph Kahne, "Major New Study Shatters Stereotypes About Teens and Video Games," John D. and Catherine T. MacArthur Foundation, Nov. 16, 2008. All statistics are taken from this study.

6. For the best overview of this early research of the 1980s and 1990s, see C. Shawn Green and Daphne Bavelier, "The Cognitive Neuroscience of Video Games" (Dec. 2004), repr. Lee Humphreys and Paul Messaris, *Digital Media: Transformations in*

Human Communications (New York: Peter Lang, 2006). Green and Bavelier use the results from this research on arcade games to retest these skills on contemporary digital games. They are replicating, in several new experiments, many of the older kinds of cognitive research and finding similar results with contemporary online and mobile games. Although the Green and Bavelier paper does not make the kinds of historical-critical points I am raising here, its excellent survey of this older research informs my discussion.

7. D. Drew and J. Walters, "Video Games: Utilization of a Novel Strategy to Improve Perceptual Motor Skills and Cognitive Functioning in the Non-institutionalized Elderly," *Cognitive Rehabilitation* 4 (1986): 26–31.

8. Tim Lenoir has astutely documented what he calls the "military-entertainment complex." See Tim Lenoir and Henry Lowood, "All but War Is Simulation: The Military-Entertainment Complex," *Configurations* 8, part 3 (2000): 289–36. Donkey Kong was used to train soldiers, and excellence at the game made one more attractive to recruiters.

9. This 1994 study by Patricia M. Greenfield, Patricia deWinstanley, Heidi Kilpatrick, and Daniel Kaye is especially relevant to this discussion of attention and needs to be replicated with current techniques using contemporary video games and contemporary kids. See "Action Video Games and Informal Education: Effects and Strategies for Dividing Visual Attention," *Journal of Applied Developmental Psychology* 15, no. 1 (1994): 105–123.

10. C. Shawn Green and Daphne Bavelier, "Effect of Action Video Games on the Spatial Distribution of Spatiovisual Attention," *Journal of Experimental Psychology: Human Perception and Performance* 32, no. 6 (2009): 1465.

11. Richard J. Haier et al., "Regional Glucose Metabolic Changes After Learning a Complex Visuospatial/Motor Task: A Positron Emission Tomographic Study," *Brain Research* 570 (1992): 134–43.

12. Kebritch Mansureh, "Factors Affecting Teachers' Adoption of Educational Computer Games: A Case Study," *British Journal of Educational Technology* 41, no. 2 (2010): 256–70.

13. Early Nintendo Gameboy Commercial, www.dailymotion.com/video/x67spw_early-nintendo-gameboy-commercial-1_tech (accessed Mar. 25, 2010).

14. Bryan Vossekuil et al., "The Final Report and Findings of the Safe School Initiative: Implications for the Prevention of School Attacks in the United States," United States Secret Service and United States Department of Education (Washington, DC, 2002).

15. Mark Ward, "Columbine Families Sue Computer Game Makers," *BBC News*, May 1, 2001, http://news.bbc.co.uk/2/hi/science/nature/1295920.stm (accessed Mar. 25, 2010).

16. Maggie Jackson, *Distracted: The Erosion of Attention and the Coming Dark Age* (New York: Prometheus Books, 2009); Mark Bauerlein, *The Dumbest Generation: How the Digital Age Stupefies Young Americans and Jeopardizes Our Future (or, Don't Trust Anyone Under 30)* (New York: Tarcher/Penguin, 2008), Nicholas Carr, *The Shallows: What the Internet Is Doing to Our Brains* (New York: Norton, 2010); and Nicholas Carr, "Is Google Making Us Stupid? Why You Can't Read the Way You Used To," *Atlantic Monthly*, July 2008.

17. Psychiatrists and psychologists, including Peter Breggin, Thomas Hartmann, and Sami Timimi, propose that ADHD is not a developmental, neurobehavioral disorder but a "socially constructed" disorder that varies widely from society to society, with diagnostic rates ranging from less than 1 percent to over 26 percent. While some extreme cases now diagnosed with ADHD may well need medical intervention (and possibly a more accurate diagnosis), they believe the pathologizing is harmful, especially as we have so little research on the long-term neuroanatomical effects of many of the drugs being prescribed as "treatment" for ADHD. See Sami Timimi and Maitra Begum, *Critical Voices in Child and Adolescent Mental Health* (London: Free Association Books, 2006); and Thomas Hartmann, *The Edison Gene: ADHD and the Gift of the Hunter Child* (Rochester, VT: Park Street Press, 2003).

18. "Adolescent Violence and Unintentional Injury in the United States: Facts for Policy Makers," National Center for Children in Poverty, Oct. 2009, shows that teen years are the most susceptible to self-perpetrated violence or violence perpetrated by others. In addition, both are highly correlated with poverty as well. See www .nccp.org/publications/pub_890.html (accessed May 2, 2010). However, these numbers have gone down, rather than up, in recent years, especially when poverty is factored in or out proportionally. See also Alexandra S. Beatty and Rosemary A. Chalk, *A Study of Interactions: Emerging Issues in the Science of Adolescence* (Washington, DC: National Research Council and Institute of Medicine, 2007); and "Criminal Victimization in the United States: Victimization Rates for Persons Age 12 and Over, by Type of Crime and Age of Victims," www.ojp.usdoj.gov/bjs/pub/pdf/cvus/current/cv0603.pdf (accessed Mar. 15, 2009). For a sampling of data on declining rates for suicide, murders, depression, and other factors, see Mike Males, "The Kids Are (Mostly) All Right," *Los Angeles Times*, May 27, 2007; and

Lawrence Grossberg, *Caught in the Crossfire* (Boulder, CO: Paradigm Publishers, 2005).

19. Joseph Kahne quoted in "Major New Study Shatters Stereotypes About Teens and Video Games," PEW Research Center, www.civicsurvey.org/Pew%20DML%20 press%20release%20FINAL.pdf (accessed Jan. 10, 2010).

20. See John Seely Brown and Douglas Thomas, "The Play of Imagination: Beyond the Literary Mind," working paper, Aug. 22, 2006, www.hastac.org/resources/ papers (accessed Jan. 10, 2010).

21. To learn more about Virtual Peace, a game designed under the leadership of Timothy Lenoir, the Kimberly Jenkins Professor of New Technologies and Society at Duke University, visit www.virtualpeace.org.

22. Mihaly Csikszentmihalyi, *Creativity: Flow and the Psychology of Discovery and Invention* (New York: HarperCollins, 1996).

23. John Seely Brown and Douglas Thomas have teamed up for a number of articles on the importance of games and on the personal and social qualities games reward, encourage, support, and nourish. See "The Gamer Disposition," *Harvard Business Review* 86, no. 5 (2008); "You Play World of Warcraft? You're Hired! Why Multiplayer Games May Be the Best Kind of Job Training," *Wired* 14, no. 4 (2006); "The Power of Dispositions," forthcoming, *International Journal of Media and Learning;* and "Why Virtual Worlds Can Matter," working paper, Oct. 21, 2007. I am very grateful to John Seely Brown for sharing these papers with me, including the two unpublished ones.

24. In 2009, the MacArthur Foundation released a second important study based on a three-year collaborative qualitative team effort by scholars, graduate students, and other researchers at the University of Southern California and the University of California–Berkeley. See Mizuko Ito et al., *Hanging Out, Messing Around, and Geeking Out: Kids Living and Learning with New Media*, John D. and Catherine T. MacArthur Foundation Series on Digital Media and Learning (Cambridge, MA: MIT Press, 2009).

25. The research-based, six-volume MacArthur Foundation Digital Media and Learning Series, for which I serve as a series adviser, was published by MIT Press in February 2008. It consists of W. Lance Bennett, ed., *Civic Life Online: Learning How Digital Media Can Engage Youth;* Miriam J. Metzger and Andrew J. Flanagin, ed., *Digital Media, Youth, and Credibility;* Tara McPherson, ed., *Digital Youth, Innovation, and the Unexpected;* Katie Salen, ed., *The Ecology of Games: Connecting Youth, Games, and*

Learning; Anna Everett, ed., *Learning Race and Ethnicity: Youth and Digital Media;* and David Buckingham, ed., *Youth, Identity, and Digital Media.* See also John Palfrey and Urs Gasser, *Born Digital: Understanding the First Generation of Digital Natives* (New York: Basic Books, 2008).

26. James Daly, "Reshaping Learning from the Ground Up: Alvin Toffler Tells Us What's Wrong—and Right—with Public Education," *Edutopia*, www.edutopia .org/future-school (accessed Oct. 6, 2010).

27. Benedict Carey, "Studying Young Minds, and How to Teach Them," *New York Times*, Dec. 21, 2009.

6. The Changing Workplace

1. OECD Manual: "Measuring Productivity: Measurement of Aggregate and Industry-Level Productivity Growth" (Paris: Organisation for Economic Co-operation and Development, 2001), www.oecd.org/dataoecd/59/29/2352458 .pdf (accessed Oct. 6, 2010).

2. See especially Carr, *The Shallows;* and for a debate on the topic of how our brain is surviving the Internet onslaught, Steven Pinker, "Mind Over Mass Media," *New York Times*, June 1, 2010. See also Jackson, *Distracted;* James Surowiecki, *The Wisdom of Crowds* (New York: Random House, 2005); and Clay Shirky *Here Comes Everybody: The Power of Organizing Without Organizations* (New York: Penguin, 2008).

3. See, for example, Sven Birkerts, "Reading in a Digital Age," *American Scholar*, Spring 2010, www.theamericanscholar.org/reading-in-a-digital-age (accessed July 23, 2010). For a survey of ideas of reading and attention, see Maryanne Wolf, *Proust and the Squid: The Story and Science of the Reading Brain* (New York: HarperCollins, 2007).

4. For a concise compendium of the numbers, comparing information sources and flows between 2000 and 2010, see "Exactly How Much Are the Times A-Changin'?" *Newsweek*, July 26, 2010, p. 56.

5. Gloria Mark, www.ics.uci.edu/~gmark (accessed Jan. 7, 2010). See also Gloria Mark, Victor M. Gonzalez, and Justin Harris, "No Task Left Behind? Examining the Nature of Fragmented Work," *Take a Number, Stand in Line (Interruptions & Attention 1): Proceedings of the SIGCHI Conference on Human Factors in Computing Systems, Portland, OR*, Apr. 2–7, 2005 (New York: Association for Computing

Machinery, 2005), 321–30; and Gloria Mark, Daniela Gudith, and Ulrich Klocke, "The Cost of Interrupted Work: More Speed and Stress," *Proceedings of the CHI 2008*, n.p. A video presentation of Mark's findings is available at http://videolectures .net/chi08_mark_tci (accessed Jan. 7, 2010).

6. Mark et al., "No Task Left Behind?" See also A. J. Bernheim Brush, et al., "Understanding Memory Triggers for Task Tracking," *Proceedings of ACM CHI 2007 Conference on Human Factors in Computing Systems* (2007), 947–50; and Mary Czerwinski, Eric Horvitz, and Susan Wilhite: "A Diary Study of Task Switching and Interruptions," in Elizabeth Dykstra-Erickson and Manfred Tscheligi, eds., *Proceedings of ACM CHI 2004 Conference on Human Factors in Computing Systems* (2004): 175–82.

7. "So Where Was I? A Conversation on Workplace Interruptions with Gloria Mark," Information Overload Research Group, http://iorgforum.org/articles/ June08Article.htm (accessed Jan. 10, 2010).

8. Excellent overviews of the research on attention for the last five decades are: Hal Pashler, *The Psychology of Attention* (Cambridge, MA: MIT Press, 1998): and Richard D. Wright and Lawrence M. Ward, *Orienting of Attention* (Oxford, UK: Oxford University Press, 2008).

9. Peter Borsay, *A History of Leisure: The British Experience since 1500* (London: Palgrave Macmillan, 2006).

10. For a survey and an analysis of crosscultural adaptation, see Andrew Molinsky, "Cross-Cultural Code-Switching: The Psychological Challenges of Adapting Behavior in Foreign Cultural Interactions," *Academy of Management Review* 32, no. 2 (2007), 622–40.

11. "Web Browser Market Share," *StatOwl,* http://statowl.com/web_browser_ market_share.php (accessed Jan. 3, 2010); and "StatCounter Global Stats," *StatCounter,* http://gs.statcounter.com (accessed Jan. 3, 2010). Special thanks to Aza Raskin for allowing me to interview him in March 2010. In December 2010, Raskin announced his resignation from his position as creative lead for Firefox to start his own company, Massive Health, a venture dedicated to applying design principles to health.

12. Jef Raskin, "Computers by the Millions," 1979, *DigiBarn Computer Museum,* www .digibarn.com/friends/jef-raskin/writings/millions.html (accessed July 28, 2010).

13. For a full biography and analysis of Taylor's contribution, see Robert Kanigel, *The One Best Way: Frederick Winslow Taylor and the Enigma of Efficiency* (New York: Viking, 1997). See also Charles D. Wrege and Ronald G. Greenwood, *Frederick*

W. Taylor, the Father of Scientific Management: Myth and Reality (Homewood, IL: Business One Irwin, 1991).

14. Frederick Winslow Taylor, *Shop Management* (New York: Harper & Bros., 1911); and *The Principles of Scientific Management* (New York: Norton, 1967), 7.

15. Robert Kanigel, *The One Best Way*, 202.

16. Peter F. Drucker, *Management,* rev. ed. (New York: HarperCollins Business, 2008). This quotation comes from the Amazon Web site advertisement for Kanigel's study of Taylor, www.amazon.com/One-Best-Way-Efficiency-Technology/dp/0140260803 (accessed Jan. 10, 2010).

17. Historian Jennifer Karns Alexander has shown that human worker efficiency has been compared (usually unfavorably) to the machine at least since the invention of the waterwheel in the Middle Ages. See *The Mantra of Efficiency: From Waterwheel to Social Control* (Baltimore, MD: Johns Hopkins University Press, 2008). Alexander analyzes the intellectual concept of efficiency in Enlightenment ideas of rationality, especially the bifurcation of mind and body. Bruno Latour discusses the idea of "constitution" arising from English natural philosophy in the mid-1750s as significant for later studies of will, thought, and attention, in *We Have Never Been Modern* (Cambridge, MA: Harvard University Press, 1993), 33. See also Robert L. Martensen, *The Brain Takes Shape: An Early History* (Oxford, UK: Oxford University Press, 2004), for an analysis of the history of mind and brain and the gradual extrication of "brain" from "mind."

18. Harry Braverman's *Labor and Monopoly Capital: The Degradation of Work in the Twentieth Century* (New York: Monthly Review Press, 1974) is a classic critique of Taylorism. In *The New Ruthless Economy: Work and Power in the Digital Age* (New York: Oxford University Press, 2005), Simon Head argues that the digital workplace is more Taylorized than ever and extends the operations of work into every aspect of private life and leisure time.

19. For excellent overviews of women and the modern workplace, see Barbara J. Harris, *Beyond Her Sphere: Women and the Professions in American History* (New York: Greenwood Press, 1978); Elizabeth Smyth, Sandra Acker, Paula Bourne, and Alison Prentice, *Challenging Professions: Historical and Contemporary Perspectives on Women's Professional Work* (Toronto: University of Toronto Press, 1999); and Natalie J. Sokoloff, *Black Women and White Women in the Professions: Occupational Segregation by Race and Gender, 1960–1980* (New York: Routledge, 1992).

20. Theodore M. Porter, *Trust in Numbers: The Pursuit of Objectivity in Science and Public Life* (Princeton, NJ: Princeton University Press, 1995): 76, 210.

21. To be exact, 904,900. U.S. Department of Labor, Bureau of Labor Statistics, "Human Resources, Training, and Labor Relations Managers and Specialists," www.bls.gov/oco/ocos021.htm (accessed July 18, 2010).

22. Clive Thompson's 2005 article in the *New York Times Magazine* drew on the language of sociology to coin the phrase "interrupt-driven." See Clive Thompson, "Meet the Life Hackers," *New York Times Magazine,* Oct. 16, 2005, www.nytimes.com/2005/10/16/magazine/16guru.html?_r=1&pagewanted=1 (accessed Jan. 10, 2010). For a contrast in the ideology of two very different technologies, see Tom Standage, *The Victorian Internet: The Remarkable Story of the Telegraph and the Nineteenth Century's On-line Pioneers* (New York: Walker, 1998). See also Katie Hafner and Matthew Lyon, *Where Wizards Stay Up Late: The Origins of the Internet* (New York: Simon & Schuster, 1996); and Tim Berners-Lee and Mark Fischetti, *Weaving the Web: The Original Design and Ultimate Destiny of the World Wide Web by Its Inventor* (San Francisco: Harper San Francisco, 1999).

23. Paul Sutter, *Driven Wild: How the Fight Against Automobiles Launched the Modern Wilderness Movement* (Seattle: University of Washington Press, 2002), 104.

24. U.S. Department of Labor, Bureau of Statistics, "American Time Use Survey Summary" for 2009, released June 22, 2010, USDL-10-0855, www.bls.gov/news/release/atus.nr0.html (accessed July 15, 2010).

25. Paul Carroll, *Big Blues: The Unmaking of IBM* (New York: Crown, 1994); Richard Thomas De Lamarter, *Big Blue: IBM's Use and Abuse of Power* (New York: Dodd, Mead, 1986); Emerson W. Pugh. *Building IBM: Shaping an Industry* (Cambridge, MA: MIT Press, 1995); and Louis V. Gerstner, *Who Says Elephants Can't Dance? Inside IBM's Historic Turnaround* (New York: HarperCollins, 2004).

26. Gerstner, *Who Says Elephants Can't Dance?*

27. Chuck Hamilton, e-mail to author, Mar. 30, 2010.

28. Special thanks to Charles (Chuck) Hamilton for talking with me on March 31 and April 6, 2010, and for taking the time to send me materials from his job and for answering my questions about his work life.

29. See Bill Bamberger and Cathy N. Davidson, *Closing: The Life and Death of an American Factory* (New York: Norton, 1998).

30. Henning Boecker et al., "The Runner's High: Opioidergic Mechanisms in the Human Brain," *Cerebral Cortex* 18, no. 11 (1991): 2523, http://cercor.oxfordjournals.org/content/18/11/2523.full (accessed Oct. 18, 2010).

31. IBM Jam Events, https://www.collaborationjam.com (accessed Apr. 8, 2010).

32. Ibid.; and Dean Takahashi, "IBM's Innovation Jam 2008 Shows How Far Crowdsourcing Has Come," *Venture Beat,* Oct. 9, 2008, http://venturebeat .com/2008/10/09/ibms-innovation-jam-2008-shows-how-far-crowdsourcing-has-come (accessed Apr. 8, 2010).

33. Nicole Kobie, "The Chief Executive of Second Life Thinks Virtual Worlds Will Be the Future of Work," *IT Pro,* Mar. 5, 2010, www.itpro.co.uk/621181/q-a-mark-kingdon-on-second-life-for-business/2 (accessed Mar. 30, 2010).

34. Quoted in Karl M. Kapp and Tony O'Driscoll, *Learning in 3D: Adding a New Dimension to Enterprise Learning and Collaboration* (San Francisco: Wiley, 2010), 330.

7. The Changing Worker

1. Karl M. Kapp and Tony O'Driscoll, *Learning in 3D: Adding a New Dimension to Enterprise Learning and Collaboration* (San Francisco: Wiley, 2010), 86–87.

2. For a compendium of the effects of obesity on health-care treatment, see Ginny Graves, "The Surprising Reason Why Being Overweight Isn't Healthy," Jan. 21, 2010, Health.com, www.cnn.com/2010/HEALTH/01/21/obesity.discrimination/ index.html (accessed Mar. 25, 2010).

3. For a more detailed discussion of women's struggles to obtain equal treatment in the workplace and of women's advancement in the corporate hierarchy, see Douglas Branson, *No Seat at the Table: How Corporate Governance and Law Keep Women Out of the Boardroom* (New York University Press, 2007); and Douglas S. Massey, *Categorically Unequal: The American Stratification System* (New York: Russel Sage Foundation, 2008). See especially research on women and the law in Karen Maschke, *The Employment Context* (New York: Taylor & Francis, 1997); and Evelyn Murphy and E. J. Graff, *Getting Even: Why Women Don't Get Paid Like Men—and What to Do About It* (New York: Touchstone, 2005).

4. Marianne Bertrand and Sendhil Mullainathan, "Are Emily and Greg More Employable than Lakisha and Jamal? A Field Experiment on Labor Market Discrimination," *American Economic Review* 94, no. 4 (2004): 991–1013.

5. Special thanks to my colleague and friend Tony O'Driscoll for many conversations over the years about his techniques in Second Life and other virtual environments. I am also grateful that he introduced me to two of the other visionaries profiled in

this book, Chuck Hamilton and Margaret Regan (who facilitated the Second Life session with "Lawanda" described here).

6. For a further discussion of the idea of "theory" and "rapid feedback," see Kapp and O'Driscoll, *Learning in 3D*, 86–87.

7. See Kapp and O'Driscoll, *Learning in 3D*, p. 85, for further discussion of neurological responses to changing environments, and p. 87 for the idea of integration, not just application.

8. Dan Ariely's experiments about the way emotions influence future decision making are detailed in *The Upside of Irrationality: The Unexpected Benefits of Defying Logic at Work and at Home* (New York: HarperCollins, 2010).

9. "Brave New Thinkers," *Atlantic Monthly*, Nov. 2009, www.theatlantic.com/doc/200911/brave-thinkers2 (accessed Jan. 13, 2010).

10. Special thanks to Thorkil Sonne for his correspondence of December 2009. Additional quotations come from Chris Tachibana, "Autism Seen as Asset, Not Liability, in Some Jobs: A New Movement Helps Hone Unique Traits of Disorder into Valuable Skills," MSNBC.com, Dec. 8, 2009, www.msnbc.msn.com/id/34047713/ns/health-mental_health (accessed Dec. 8, 2009); see also "Entrepreneur Thorkil Sonne on What You Can Learn from Employees with Autism," *Harvard Business Review*, Sept. 2009, http://hbr.org/2008/09/entrepreneur-thorkil-sonne-on-what-you-can-learn-from-employees-with-autism/ar/1 (accessed Jan. 12, 2010); and "Brave New Thinkers," *Atlantic Monthly*, Nov. 2009, www.theatlantic.com/doc/200911/brave-thinkers2 (accessed Jan 13, 2010).

11. "Entrepreneur Thorkil Sonne."

12. Quoted in Tachibana, "Autism Seen as Asset."

13. Malcolm Gladwell writes about the analyses of Gretzky's "field sense" on the ice in "The Physical Genius: What Do Wayne Gretzky, Yo-Yo Ma, and a Brain Surgeon Named Charlie Wilson Have in Common?" *New Yorker*, Aug. 2, 1999, www.gladwell.com/1999/1999_08_02_a_genius.htm (accessed May 7, 2010).

14. My thanks to Margaret Regan for her time and her candor during our conversation of Apr. 2, 2010. Quotations are from this interview and the FutureWork Institute Web site.

15. "The Future of Work in America," FutureWork Institute slide presentation, http://futureworkinstitute.com/futureofwork/futurework/Future_of_Work.html (accessed Apr. 15, 2010).

16. Michael Lewis, "The No-Stats All Star," *New York Times*, Feb. 13, 2009.

17. Katherine Mangu-Ward, "Wikipedia and Beyond: Jimmy Wales' Sprawling Vision," *Reason* 39, no. 2 (June 2007): 21, www.reason.com/news/show/119689.html (accessed Mar. 31, 2010).

18. See "Jimmy Donal 'Jimbo' Wales," Wikipedia, http://en.wikipedia.org/wiki/Jimmy_Wales (accessed Dec. 20, 2009).

19. Wikipedia Community Portal, http://en.wikipedia.org/wiki/Wikipedia:Community_portal (accessed Jan. 12, 2010).

20. Ibid.

21. Richard Pérez-Peña, "Keeping News of Kidnapping off Wikipedia," *New York Times,* June 28, 2009, www.nytimes.com/2009/06/29/technology/internet/29wiki.html, (accessed July 20, 2010).

22. See especially Jonathon Cummings and Sara Kiesler, "Who Collaborates Successfully? Prior Experience Reduces Collaboration Barriers in Distributed Interdisciplinary Research," *Proceedings of the ACM Conference on Computer-Supported Cooperative Work,* San Diego, CA, 2008. These results are highlighted in *Nature,* 455 (2008): 720–23. I first began developing "collaboration by difference" as a method in 1999, and HASTAC adopted this as our official method sometime after. For a more detailed discussion, see Cathy N. Davidson and David Theo Goldberg, *The Future of Thinking: Learning Institutions in a Digital Age* (Cambridge, MA: MIT Press, 2010).

8. You, Too, Can Program Your VCR (and Probably Should)

1. Vilayanur S. Ramachandran and Sandra Blakeslee, *Phantoms in the Brain: Probing the Mysteries of the Human Mind* (New York: Morrow, 1998), 43.

2. R. Douglas Fields, "Myelination: An Overlooked Mechanism of Synaptic Plasticity?" *Neuroscientist* 11, no. 6 (2005): 528–31. The literature on adolescent myelination is extensive, contradictory, and hypothetical. It's interesting that a number of recent adolescent-onset psychosocial diseases, including schizophrenia, have been attributed to irregularities in myelination. See, for example, Frances M. Benes, "Myelination of Cortical-Hippocampal Relays During Late Adolescence," *Schizophrenia Bulletin* 15, no. 4 (1989): 585–93. See also "Myelination," Centre for Educational Research and Innovation, Organisation for Economic Co-operation and Development, *Understanding the Brain: The Birth of a Learning Science* (Paris: OECD, 2007), 187.

3. Wendy Todaro, "Want to Be a CEO? Stay Put," Forbes.com, Mar. 31, 2003, www.forbes.com/2003/03/31/cx_wt_0401exec.html (accessed May 7, 2010).

4. Special thanks to James R. Gaines for taking the time to correspond in December 2009. See James R. Gaines, "Over 60, and Proud to Join the Digerati," *New York Times,* Nov. 28, 2009.

5. Brandeis University, "Confidence in Memory Performance Helps Older Adults Remember," *Science Daily,* Mar. 8, 2006. Lachman is one of the finest and most prolific researchers on cognition in middle age. See also Carrie Andreoletti, Bridget W. Veratti, and Margie E. Lachman, "Age Differences in the Relationship Between Anxiety and Recall," *Aging and Mental Health* 10, no. 3 (2006): 265–71; Margie E. Lachman and Carrie Andreoletti, "Strategy Use Mediates the Relationship Between Control Beliefs and Memory Performance for Middle-Aged and Older Adults," *Journals of Gerontology,* series B, *Psychological Sciences and Social Sciences,* 61, no. 2 (2006): 88–94; Margie E. Lachman, Carrie Andreoletti, and Ann Pearman, "Memory Control Beliefs: How Are They Related to Age, Strategy Use and Memory Improvement?" *Social Cognition* 24, no. 3 (2006): 359–85.

6. Peter S. Eriksson, "Neurogenesis in the Adult Human Hippocampus," *Nature Medicine* 4, no. 11 (1998): 1313–17.

7. Florin Dolcos, Kevin S. Labar, and Roberto Cabeza, "Interaction Between the Amygdala and the Medial Temporal Lobe Memory System Predicts Better Memory for Emotional Events," *Neuron* 42, no. 5 (2004): 855–63; Sander M. Daselaar et al., "Effects of Healthy Aging on Hippocampal and Rhinal Memory Functions: An Event-Related fMRI Study," *Cerebral Cortex* 16, no. 12 (2006): 1771–82; and Roberto Cabeza et al., "Brain Activity During Episodic Retrieval of Autobiographical and Laboratory Events: An fMRI Study using a Novel Photo Paradigm," *Journal of Cognitive Neuroscience* 16, no. 9 (2004): 1583–94.

8. Howard Gardner, "Creating the Future: Intelligence in Seven Steps," www.newhorizons.org/future/Creating_the_Future/crfut_gardner.html (accessed May 7, 2010).

9. The ongoing National Survey of Midlife Development in the United States (MIDUS) is attempting to gather survey data on midlife patterns that address a greater range of social, mental, and health issues. For early results see, Orville G. Brim et al., *National Survey of Midlife Development in the United States (MIDUS), 1995–1996* (Ann Arbor, MI: Inter-university Consortium for Political and Social Research, 2007).

10. Yaakov Stern, "What Is Cognitive Reserve? Theory and Research Application of the Reserve Concept," *Journal of International Neuropsychological Society* 8 (2002): 448–60; Nikolaos Scarmeas et al., "Association Between the APOE Genotype and Psychopathologic Symptoms in Alzheimer's Disease," *Neurology* 58, no. 8 (2002): 1182–88; Yaakov Stern et al., "Exploring the Neural Basis of Cognitive Reserve," *Journal of Clinical and Experimental Neuropsychology* 25 (2003): 691–701; and Yaakov Stern et al., "Brain Networks Associated with Cognitive Reserve in Healthy Young and Old Adults," *Cerebral Cortex* 15, no. 4 (2005): 394–402.

11. Sydney Jones and Susannah Fox, "Generations Online in 2009," Pew Internet & American Life Project (Pew Research Center Publications: 2009), http://pewresearch.org/pubs/1093/generations-online (accessed May 7, 2010).

12. Sam Nurmi, "Study: Ages of Social Network Users," Royal Pingdom, Feb. 16, 2010, http://royal.pingdom.com/2010/02/16/study-ages-of-social-network-users (accessed Apr. 15, 2010).

13. David Meyer, "Women and Elderly Lead Internet Charge," ZDNet.co.uk (2007), http://news.zdnet.co.uk/communications/0,1000000085,39288730,00.htm (accessed May 7, 2010).

14. "New Data on Twitter's Users and Engagement," Metric System, http://themetricsystem.rjmetrics.com/2010/01/26/new-data-on-twitters-users-and-engagement (accessed Apr. 18, 2010).

15. Joan M. Kiel, "The Digital Divide: Internet and E-mail Use by the Elderly," *Informatics for Health and Social Care* 30, no. 1 (2005): 19–24. See also "Internet a Long-Term Benefit for Depression," *Science Daily*, Oct. 12, 2006.

16. Robert J. Campbell and James Wabby, "The Elderly and the Internet: A Case Study," *Internet Journal of Health* 3, no. 1 (2003); and Robert J. Campbell, Kimberly D. Harris, and James Wabby, "The Internet and Locus of Control in Older Adults," *Proceedings of the American Informatics Association* (2002): 96–100. Much excellent geriatric Internet research, not surprisingly, comes from the Netherlands. See, for example, Laurence L. Alpay et al., "Easing Internet Access of Health Information for Elderly Users," *Health Informatics Journal* 10, no. 3 (2004): 185–94.

17. My special thanks to David Theo Goldberg, my HASTAC cofounder and codirector of the HASTAC/MacArthur Foundation Digital Media and Learning Competitions, for introducing me to the technology visionaries at the Waag Society in Amsterdam, and especially to Marleen Stikker, founder and president of the Waag Society, for her hospitality during my trip to Amsterdam. I'm grateful

to the researchers and designers at the Waag Society for generously sharing their research and this anecdote about Holland's original work pairing elderly citizens with youth volunteers via the Internet. For the Netherlands' "Bill of Technological Rights" for senior citizens, see www.seniorwatch.de/country/netherlands/Policyenvironment(Netherl).htm (accessed Jan. 9, 2010). Similar findings have been reported by Gary Small and his team of researchers; see Gary Small, Teena Moody, Prabha Siddarth, and Susan Bookheimer, "Your Brain on Google: Patterns of Cerebral Activation During Internet Searching," *Journal of Geriatric Psychiatry* 17, no. 2 (Feb. 2009): 116–26.

18. Netherlands' "Bill of Technological Rights" for senior citizens.

19. Jenna Wortham, "MySpace Turns Over 90,000 Names of Registered Sex Offenders," *New York Times,* Feb. 3, 2009.

20. Matt Granfield, "Average Age of People Using Social Media," Zakazsukha Zoo, www.dpdialogue.com.au/zakazukhazoo/average-age-of-people-using-social-media (accessed May 6, 2010).

21. "'Oldest' Blogger Dies, Aged 108," BBC News, July 14, 2008, http://news.bbc.co.uk/2/hi/asia-pacific/7505029.stm (accessed May 6, 2010); and Saeed Ahmed, "'World's Oldest Blogger' Dies at 108," CNN.com, July 14, 2008, www.cnn.com/2008/WORLD/europe/07/14/oldest.blogger/?iref=hpmostpop (accessed May 6, 2010).

22. At 108, Olive Riley wasn't actually uploading her own material. Filmmaker Mike Rubbo met Olive in the course of doing research on centenarians and was so charmed that he made a movie, *All About Olive* (2005), and then subsequently transcribed what she dictated to him. He taped her vlogs and helped her post her e-mails to various virtual friends around the globe.

Conclusion: Now You See It

1. Howard Rheingold, *Smart Mobs: The Next Social Revolution* (Cambridge, MA: Perseus Publications, 2003); and *Virtual Reality* (New York: Summit Books, 1991). See also Howard Rheingold's vlog on this subject, which can be found on his Web site for his new Social Networking Classroom Co-Laboratory, (http://vlog.rheingold.com/index.php/site/video/social-media-classroom-co-laboratory). See also the forum on "Participatory Learning" he conducted with the HASTAC Scholars beginning on August 26, 2008, www.hastac.org/content/hastac-welcomes-howard-rheingold-discussion-participatory-learning. Here, too, Rheingold talks about the importance of students deciding their intellectual goals first, then deciding what

digital technology facilitates those goals. He advocates against beginning with the technology before those goals are defined.

2. Theresa Tamkins, "Drop that BlackBerry! Multitasking may be Harmful," *CNN Health,* Aug. 29, 2005, http://articles.cnn.com/2009-08-25/health/multitasking .harmful_1_instant-messaging-multitaskers-online-video?_s=PM:HEALTH (accessed Sept. 14, 2010).

3. See, for example, the excellent work by Debra A. Gusnard and Marcus E. Raichle, "Searching for a Baseline: Functional Imaging and the Resting Human Brain," *Nature Reviews Neuroscience* 2, no. 10 (2001): 685–94. See also, Harold Burton, Abraham Z. Snyder, and Marcus E. Raichle, "Default Brain Functionality in Blind People," *Proceedings of the National Academy of Sciences USA* 101, no. 43 (2004): 15500–505. This article looks at default functionality of the human brain, as revealed by "task-independent decreases in activity occurring during goal-directed behaviors" and the way such activity is functionally reorganized by blindness. See also Marcus E. Raichle, "The Neural Correlates of Consciousness: An Analysis of Cognitive Skill Learning," *Philosophical Transactions of the Royal Society B: Biological Sciences* 1377 (1998): 1889–1901, which uses fMRIs to isolate neural correlates of consciousness in the human brain. Raichle tests neural activity during skill mastery. For a summary of the latest work, see Marcus E. Raichle, "The Brain's Dark Energy," *Scientific American,* Mar. 2010, pp. 44–47.

4. Malia F. Mason et al., "Wandering Minds: The Default Network and Stimulus-Independent Thought," *Science* 315, no. 5810 (2007): 393–95.

5. Jonathan Smallwood et al., "Going AWOL in the Brain: Mind Wandering Reduces Cortical Analysis of External Events," *Journal of Cognitive Neuroscience* 20, no. 3 (2008): 458–69.

6. Mason, "Wandering Minds."

7. The great literary theorist Roland Barthes argues that our tendency toward distraction is precisely what makes reading so pleasurable. See *The Pleasure of the Text,* tr. Richard Miller (New York: Hill & Wang, 1975).

8. Studies conducted by Jonathan Schooler of the University of California–Santa Barbara show not only that our mind wanders but that we don't know it does until it's called to our attention. Jonathan Smallwood, Merrill McSpadden, Jonathan W. Schooler, "The Lights Are On but No One's Home: Meta-awareness and the Decoupling of Attention When the Mind Wanders," *Psychonomic Bulletin & Review* 14, no. 3 (2007): 527–33; and Jonathan Smallwood and Jonathan W. Schooler, "The Restless Mind," *Psychological Bulletin* 132, no. 6 (2006): 946–58. See also

Michael J. Kane et al.,"Working Memory, Attention Control, and the N-Back Task: A Question of Construct Validity," *Journal of Experimental Psychology: Learning, Memory, and Cognition* 33, no. 3 (2007): 615–22; and Carmi Schooler, "Use It and Keep It, Longer, Probably: A Reply to Salthouse," *Perspectives on Psychological Science* 2, no. 1 (2007): 24–29. As a writer, I am also intrigued by work on what is called prospective memory, that is, the ability to remember one's own planning in order to execute it. I'm convinced much writer's block comes from the tendency to be able to clearly envision one's argument and then to find the "transcription" of those ideas on paper to be a disappointingly faulty, laborious, and inconsistent rendition that has almost nothing to do with the prospective memory that allows one to envision the perfect, beautifully executed piece as a whole. See Peter Graf and Bob Uttl, "Prospective Memory: A New Focus for Research," *Consciousness and Cognition* 10, part 4 (2001): 437–50.

9. Dan Ariely, *Predictably Irrational: The Hidden Forces That Shape Our Decisions* (New York: HarperCollins, 2008); and Daniel Gilbert, *Stumbling on Happiness* (New York: Knopf, 2006). In different arenas, Ariely and Gilbert emphasize the myriad ways that we misestimate ourselves and others. Daniel Goleman, in *Vital Lies, Simple Truths: The Psychology of Self-Deception* (New York: Simon & Schuster, 1985), suggests that our brain actually reinforces us with various pleasant chemicals (typically dopamine) when we avoid thinking about that which makes us anxious, a biochemical explanation that supports the responses that Ariely and Gilbert test experimentally. See also Daniel T. Levin, *Thinking and Seeing: Visual Metacognition in Adults and Children* (Cambridge, MA: MIT Press, 2004).

10. Eyal Ophira, Clifford Nass, and Anthony D. Wagner, "Cognitive Control in Media Multitaskers," *Proceedings of the National Academy of Sciences of the United States of America (PNAS)* 10, no. 1073 (Aug. 24, 2009), www.pnas.org/content/early/2009/08/21/0903620106 (accessed May 7, 2010).

11. Alexa M. Morcom and Paul C. Fletcher, "Does the Brain Have a Baseline? Why We Should Be Resisting a Rest," *Neuroimage* 37, no. 4 (2007): 1073–82.

12. Daniel Levitin, quoted with permission from Facebook, Dec. 30, 2009.

13. Linda Stone first coined the term "continuous partial attention" in 1998 and writes about it on her blog at http://continuouspartialattention.jot.com/WikiHome (accessed Jan. 10, 2010).

14. KAIST stands for Korea Advanced Institute for Science and Technology.

Index

|||||||||||||||||||||||||

ACT, 9–10
Adams, Martha, 67
Adler, Dankmar, 182
advertising, *see* television commercials
affordance, 204
Africa, 145
aging, 254–70, 276
  brain and, 256, 259–60, 264–66,
    269–70
  rejuvenation and, 264–65, 276
airplane pilots, 4
Alzheimer's disease, 260, 264, 267–68
Amen, Daniel G., 265
America's Army, 155, 156
animals:
  humans as different from, 54
  teaching among, 54–55
Apple, 180, 192, 212
Apple Digital Campuses, 61, 66
  Duke University iPod experiment,
    61–70, 71, 77, 78, 85, 90, 99,
    140, 161
Ariely, Dan, 58
art, 79, 144
Asperger's syndrome, 214, 215
attention, 243, 279, 281, 284
  brain and, *see* brain
  continuous partial, 287

Dadaist approach to, 286
difference and, 49
distraction and, *see* distraction
learning of, 5, 39, 56
multitasking and, *see* multitasking
Rheingold experiment and, 277–79,
  280
stimulus-dependent, 283–84
to task, 6, 7, 73, 74
Taylorized model of, 279–80
attention blindness, 1–8, 10, 12, 39,
  140, 276, 281, 287
biology of, 49
standardized education and, 77
attention-blindness experiments, 19,
  176, 209
  card trick, 4
  flight training, 4
  gorilla, 1–5, 20, 282, 290
attention disorders, 9, 10, 79–80, 89,
  139, 152–53
Australian National University, 274
autism, 48, 214–19

Babinski reflex, 58
Baker, Beauchamp, 89
Baker, Lesli, 89
basketball, 224–25, 241

Battier, Shane, 224–25, 226, 227, 233, 241, 263
Baumann, Frederick, 182
Bavelier, Daphne, 149
Berners-Lee, Tim, 6, 17, 62, 242
Binet, Alfred, 117–19, 123
blogs, 100–101, 102, 108, 110, 169, 177, 188, 270
Boost program, 97
BP oil spill, 5
brain, 14–17, 20, 30, 31, 42, 76, 109, 171, 278, 284, 291
    aging and, 256, 259–60, 264–66, 269–70
    Alzheimer's disease and, 268
    brain chatter and mind wandering, 171, 172, 278, 281, 282, 285, 286
    demyelination in, 259, 260
    disabilities in, 265–66, 267–70
    division of, 258
    flow activities and, 157
    of infants, 44, 45–48, 69
    interactive nature of, 69–70
    learning and, 30
    memory and, 103, 259, 264
    metaphors for, 14–15
    midlife bilaterality in, 266
    multitasking and, 280–81
    music and, 99
    neurons in, *see* neurons and neural networks
    nineteenth-century science of, 15
    plasticity of, 15–16, 264, 265–66
    Rheingold experiment and, 277–79, 280
    self-exploration in, 285
    technology and, 16
Brain-Age, 156
Brain Science Research Center, 287–88
Brazil, 197–98
Bren School of Information and Computer Sciences, 170

Brix, Ron, 217
Broca, Paul Pierre, 15, 16
Brodmann, Korbinian, 15, 16
Brown, John Seely, 158, 196, 273, 290
Bruzga, Robert, 249, 255

Cabeza, Roberto, 266
calculators, 57, 125
Card, Orson Scott, 132
card trick experiment, 4
categories, 31, 32–33, 34, 37–38, 42, 47–48, 53
"Cathedral and the Bazaar, The" (Raymond), 229
Chabris, Christopher, 3, 4
change, 6, 7, 10–14, 16, 17, 19, 20, 50, 51, 90, 127, 172, 243–44, 258, 281, 287, 291
    confidence and, 86
    endism and, 290
    resentment of, 57
Chicago fire, 182
children:
    complexity of thought in, 160
    education of, *see* education and schools
    games and, *see* games, gaming
    Internet's effect on, 56–57, 101, 102, 101, 102
    see also infants
China, 160–61, 189, 198, 200, 203, 204
Chronicle of Higher Education, 61, 107
cognitive reserves, 268–70
Cohen, Daniel H., 124–25
collaboration, 90, 128, 135, 161, 194, 200, 212, 242, 263, 291
    Battier effect and, 225
    crowdsourcing, *see* crowdsourcing
    by difference, 2, 20, 100, 161, 191–92, 208, 213, 217, 225, 233, 252, 257, 266, 281
    games and, 154, 158–59, 161
    hive mind and, 143–44

mass, 225–34, 241
tools, 136, 137, 138, 145
wikinomics, 200
Columbia University, 267–68, 281
Columbine, 147–48, 150–51, 152, 155
commercials, *see* television commercials
computers, 10–11, 13–14, 167, 168,
    173, 175–79, 180
  cognitive reserves and, 269
  see also Internet
"Computers by the Millions" (Raskin),
    180
Concentrate, 174
Constitution, U.S., 71, 72
context, 212, 225
control, sense of, 258–59, 260, 276
corporate jamming, 198–200, 204
Cosby, Bill, 224
Council of Chief State School Officers, 71
Crawford, Matthew B., 81
Creative Commons, 232
Creative Productions, 132–40, 161
Cronkite, Walter, 27
cross-functional complementarity, 266
crowdsourcing, 64–65, 69, 70, 78, 109,
    175, 192, 197, 241
  grading and, 105–10
Csikszentmihalyi, Mihaly, 157–58
Culbreth Middle School, 69
cultural values and scripts, 29 30, 33,
    34, 35, 41, 58, 69, 171, 279
  Cymbalta ad and, 28–29
  learning of, 37–39
  and "natural" traits and forms, 33, 35,
    39, 41, 47, 50, 206
  new experiences and, 41–43, 50–51,
    206
  Williams syndrome and, 51–53
Cymbalta, 23–29, 30, 32, 33, 42, 50, 279

Darnton, Robert, 11
Darwin, Charles, 120

Davidson, Cathy N.:
  accident of, 247–56, 268
  brother of, 265–66
  childhood school experiences of,
    8–9, 80
  dyslexia of, 8–9
  grading philosophy of, 105–10
  iPod experiment and, 63–64, 161
  ISIS and, 68
  in Japan, 29–30, 36, 39
  schools visited by, 78–79, 80–81,
    92–95, 97–98, 110
  South Korea visited by, 287–90
  This Is Your Brain on the Internet
    course of, 99–104, 105 9, 111,
    161, 260–61, 264
Davidson, Charles, 90
Davidson, Inez, 82–86, 90, 104, 138, 161
Davidson, Ted, 82, 84
daydreams, 171–72, 280, 281
depression, 274
developeWorks, 188
De Waal, Frans, 54–55
Dewey, John, 186
*Diagnostic and Statistical Manual of
    Mental Disorders* (DSM), 79–80
Digital Media and Learning Initiative, 88
distraction, 31–32, 55–56, 58, 206, 278,
    280, 281, 283, 285, 290
  in advertising, 24, 25, 28
  infants and, 31, 32
  Rheingold experiment and, 277–79,
    280
  in workplace, 165, 167, 169–70, 171,
    180, 185, 195, 283
Doidge, Norman, 15
Dongdaemun, 289–90
Draftfcb Healthcare, 24, 25 29, 30,
    32, 33
Drucker, Peter, 181
Duke Sports Medicine Center, 247, 249,
    251, 252–55

Duke University, 92, 97, 224, 234,
    261, 266
  Boost program, 97
  Fuqua School of Business, 18–19, 212
  Information Science + Information
    Studies program, 68, 99
  iPod experiment, 61–70, 71, 77, 78,
    85, 90, 99, 140, 161
  podcasting conference, 68
dyslexia, 8–9, 218

education and schools, 6, 10, 11–13, 20,
    51, 57, 58, 61–104, 159–60, 161,
    191, 194–95, 206, 213, 243, 291
  art programs in, 79
  attention to task in, 73, 74,
    165–66, 181
  cognitive reserves and, 269
  college attendance and completion,
    74–75, 76
  college preparation, 64, 75, 76, 81, 90
  community-related projects in,
    130–31
  creative thinking and, 77, 125
  Davidson's visits to schools, 78–79,
    80–81, 92–95, 97–98, 110
  end-of-grade challenge proposed for,
    129–30
  expansion of, 74, 115
  expertise, specialization, and
    hierarchies in, 64, 65, 70, 73, 110,
    111, 116, 165
  Forest View Elementary, 97–98, 273
  games and, *see* games, gaming
  grading in, *see* grading
  high school, 64, 74–76, 186
  history of, in U.S., 71–74
  industrial work and, 6, 12, 13, 17,
    57, 72–74, 76, 77–78, 160,
    165–66, 206
  Inez Davidson and, 82–86, 90, 104,
    138, 161
  iPod programs in, 61–70, 71, 77, 78,
    85, 90, 99, 140, 161
  Mann and, 72, 120
  Middleton Elementary, 92–95, 96, 97
  No Child Left Behind and, 13, 93,
    96, 123–24, 152
  and pacing of school year, 129
  preschools, 46, 64
  Quest 2 Learn, 87–91, 133, 156, 161
  relationships with teachers and
    counselors in, 76, 154
  relevance in, 76, 130, 154
  rigor in, 75–76, 126, 154
  space program and, 13, 74, 81
  spending on, 75
  standardization in, 13, 57, 71–72, 75,
    76, 77–78, 79, 81, 86, 93, 96–97,
    111, 113, 114, 116, 153–54, 160;
    see also grading; tests
  student writing in, 100–102
  teachers in, 76, 81–86, 126, 128,
    131, 141
  tests in, *see* tests
  This Is Your Brain on the Internet
    course, 99–104, 105–9, 111, 161,
    260–61, 264
  time and timeliness valued in, 73, 173
  Voyager Academy, 132–40, 141,
    147, 197
  white-collar work and, 76–77, 81
  worldwide statistics on, 75
Einstein, Albert, 16, 143
elderly, 155–56, 264, 270–71, 274–76
Electronic Arts, 91
Eli Lilly and Company, 23
e-mail, 7, 14, 173–74, 177, 179, 282
encyclopedias, 62, 169
  Wikipedia, 70, 100, 169, 192,
    224–33, 241, 242
Ender's Game, 132, 136, 141
endism, 290
energy use, 144

epic wins, 141–42, 143
eugenics, 120
Evoke, 144–45
expertise, 64, 65, 70, 257–58

Facebook, 70, 171, 174, 192, 232, 270, 275
Farmer, Paul, 156 57
Firefox, 70, 174, 175, 176, 180
Fletcher, Paul C., 285–86
flight training, 4
flow, 157–58
FLYP, 262–63
focus, 2–3, 5, 10, 19, 100, 171, 276, 277, 280, 281
   games and, 154
Food and Drug Administration (FDA), 23–24, 25, 27
Ford, Henry, 181
Forest View Elementary, 97–98, 273
Franklin, John Hope, 224
Freedom, 174
freerunning (parkour), 138–39, 140, 285
Freud, Sigmund, 90, 180, 205
From Slavery to Freedom (Franklin), 224
Fuqua School of Business, 18–19, 212
FutureWork Institute, 219–23, 241

Gaines, James R., 262–63
Galton, Francis, 120, 147
games, gaming, 50, 63, 86, 87–92, 95–96, 102, 131, 132–61, 177, 188
   America's Army, 155, 156
   Brain-Age, 156
   Columbine and, 147 48, 150 51, 152, 155
   and complexity of thought, 160
   Creative Productions and, 132–40, 161
   dismissal of, 145–46
   elderly and, 155–56
   epic wins in, 141–42, 143
   Evoke, 144–45

flow and, 157–58
Game Boy, 149–50
gamer mentality and, 158, 196
Halo 2, 143, 144
in history, 146
learning simulators and, 155
lifelong learning and, 155
LittleBigPlanet, 87, 91, 95 96, 138, 160
magic circle and, 159
McGonigal and, 141–45, 146, 159, 209
multiplayer, 142, 144
Nim, 160–61
number of people playing, 145
Quest 2 Learn and, 87 91, 133, 156, 161
research on, 147–50, 152, 155, 157
social activism and, 143, 144–45, 154, 159
social life and, 154
time spent playing, 146
video, 88, 89, 145–56, 158–59, 195, 269
video, as bad for children, 146–48, 150–54
violent, 147, 150–52, 153, 154
Virtual Peace, 156–57
World of Warcraft, 70, 102, 142, 146, 158
World Without Oil, 144
Gardner, Howard, 118, 266–67
Garrison, Mark, 118
Gates, Bill, 159
genetic counseling, 18
Germain, Duncan, 132–40, 141, 143, 147, 161, 197
globalization, 12–13, 128, 167, 190, 191, 287
global teaming, 190, 197–98
Godin, Seth, 78–79
Goldberg, David Theo, 100

Google, 12, 17, 125, 152, 169, 170, 192, 212
gorilla experiment, 1–5, 20, 282, 290
Gould, Stephen Jay, 121
grading, 53, 64, 78, 101, 107–13, 119, 159
  crowdsourcing and, 105–10
  Davidson and, 105–10
  history of, 111–12
  see also tests
Green, C. Shawn, 149
Gretzky, Wayne, 218–19
Gulf of Mexico oil spill, 5

H1N1 flu epidemic, 202–3
Habitat Jam, 198–99
habits, 17, 19, 35, 42, 49, 51, 172, 206, 243, 281, 290, 291
Haiti, 69, 156–57
Halo 2, 143, 144
Hamilton, Chuck, 187–98, 200–207
Harris, Eric, 150–51
Harvard Business Review, 158
Harvard Medical School, 281
HASTAC (Humanities, Arts, Sciences, and Technology Advanced Collaboratory), 91, 100, 105, 106, 161, 185, 187
HASTAC/MacArthur Foundation Digital Media and Learning Competition, 91, 130, 156, 175
Hawkins, Jeff, 103, 104
Hayek, Friedrich von, 229
H-Bot, 124–25, 267
heart arrhythmias, 67
Hebb, Donald O., 45
Hebbian principle, 45, 56, 259
Higher Education Research Institute, 153
hive mind, 143–44
hockey, 218–19
Home Insurance Building, 182

Howe, Jeff, 64–65
Hypercities, 130–31

IBM, 109, 188–94, 196–200, 202–5, 208, 284
I Love Bees, 143, 144
immigrants, 94, 110, 119
India, 160–61, 203
infants, 5, 31–43, 58, 279, 291
  brain of, 44, 45–48, 69
  crying of, 34–35, 37
  gender differences in treatment of, 37, 40
  language acquisition in, 33, 39, 47–48
  Motherese and, 36, 38
  music and, 36–37
  object naming and, 37–38
  race and ethnicity discerned by, 40–41
  touching and holding of, 37, 40
information age, 11, 12, 13
Information Science + Information Studies (ISIS), 68, 99
Inside Higher Ed, 61, 107
inspiration, 179, 285
instinct, 58
Institute for Play, 88
intelligence, 117
  forms of, 118, 266–67
  IQ tests, 51, 114, 117–19, 120, 121, 149, 183, 186, 254, 266
Internet, 6–7, 11, 13, 16, 62, 67, 85, 188, 243, 258, 270–71, 282
  crowdsourcing and, 64–65, 109, 175
  depression and, 274
  effects on children, 56–57, 101, 102, 101, 102
  elderly and, 270–71, 274–76
  hive mind and, 143–44
  open-source production and, 65, 191–92, 229, 242
  software for managing time spent on, 174

surfing/searching experience on, 70, 161, 283
work and, 165–69, 172, 174–78, 186
World Wide Web, 6, 7, 14, 17, 62, 63, 66, 70, 141, 165–67, 169, 177–80, 191–92, 205, 242, 258, 270
iPhone, 14–15
iPod, 62, 69
Duke University experiment with, 61–70, 71, 77, 78, 85, 90, 99, 140, 161
IQ tests, 51, 114, 117–19, 120, 121, 183, 186, 254, 266
mental rotations, 149
Irving, Washington, 71
iTunes U, 68

Jacobsen, Thomas, 214
Japan, 36, 37, 38
Davidson in, 29–30, 36, 39
electronics and, 150
Williams syndrome in, 52–53
Japanese language, 47
Japanese proverb, 287
Jefferson, Thomas, 72
Jennings, Peter, 61
Jobs, Steve, 66
John D. and Catherine T. MacArthur Foundation, 88, 91, 103, 130, 156, 175
Johnson, Steve, 238–40, 241
Josefowicz, Michael, 271–74

Kahne, Joseph, 154
KAIST University, 287–88
Kansas Silent Reading Test, 114, 115, 116
Keller, Bill, 231
Kelly, Frederick J., 113–17, 123, 131
Keohane, Nannerl, 261–62
kinship, 32, 39
Klebold, Dylan, 150–51
Korea, 37, 75, 287–90

Lachman, Margie, 258–59, 260, 264
Lakoff, George, 50
language, 32–33, 39
cognitive reserves and, 269
crying and, 37
learning of, 33, 39, 47–48
Motherese and, 36, 38
object naming in, 37–38
sounds in, 33, 47, 48
tests and, 94, 110, 119
Leadership in Energy and Environmental Design (LEED), 234, 238, 239, 240
lead paint project, 66
learning, 19, 20, 30, 32, 38, 41, 43, 47, 53, 54, 56, 58, 85, 86, 110, 161, 243, 258, 276, 291
of attention, 5, 39, 56
cognitive reserves and, 269–70
of culture, 37–39
excitement in, 50
games and, *see* games, gaming
impulse for, 204–7
in infants, 33
of language, 33, 39, 47–48
multitasking and, 285
neurons and, 45
programmed cell death and, 48
relearning, 19, 20, 49, 53, 57, 110, 161, 244, 258, 291
styles of, 141
unlearning, 19, 20, 42, 47, 49, 51, 53, 57, 63, 86, 106, 110, 111, 161, 179, 236, 244, 258, 281, 291
see also education and schools
learning disabilities and disorders, 9, 10, 77, 79–80, 83, 89, 90, 133, 139, 140–41, 152–53
autism, 48, 214–19
dyslexia, 8–9, 218
Learning in 3D (O'Driscoll), 209
learning simulations, 155

"Legend of Sleepy Hollow, The" (Irving), 71
leisure, 170, 186–87, 269, 279
Lenoir, Timothy, 156
Lenovo, 189, 192
Levitin, Daniel, 99, 286
Lewis, Michael, 225
Linden Lab, 201
Linux, 109, 192
LittleBigPlanet (LBP), 87, 91, 95–96, 138, 160
long tail, 141, 218
Lunsford, Andrea, 101

MacArthur Foundation, 88, 91, 103, 130, 156, 175
Malone, Michael, 261
Mann, Horace, 72, 120
Mark, Gloria, 170–71, 174, 181
Mason, Malia, 281
mathematics, 160–61, 267
MBA programs, 182–84, 185, 191
    Weekend Executive, 18–19
McGonigal, Jane, 141–45, 146, 159, 209
meat, grading of, 112
memory, 103, 259, 260–62, 264, 266
    Alzheimer's disease and, 268
meta-awareness, 282
Mexico, 75
Microsoft, 143, 175, 192, 222, 229
Middleton Elementary, 92–95, 96, 97
military, 155
Miranda, Marie Lynn, 66
mirror neurons, 53–54, 55
Mismeasure of Man, The (Gould), 121
monkeys, 53–54
Monopoly, 146
Moorman, Claude T., III, 248–49
Morcom, Alexa M., 285–86
Motherese, 36, 38
movable type, 11

Mozilla, 65, 70, 175, 179–80, 192, 229
multitasking, 6, 7, 12, 14, 31, 45, 46, 175, 177, 180, 194, 280–87
    flow activities and, 157
    Stanford study on, 283–84, 290
    video games and, 149, 158–59
    worries about, 56, 204, 282–83
Museum of Contemporary Art, Los Angeles, 144
music, 195–96, 268
    in advertising, 27, 29
    brain and, 99
    infants and, 36–37
    iPod experiment and, 67
    Williams syndrome and, 51
Myers, Harry, 16
MySpace, 270, 275

names, remembering, 262
Nass, Clifford, 283
National Assessment of Educational Progress (NAEP), 125
National Governors Association, 71
National Institute of Aging (NIA), 259
National Park Service, 186
nervous system, 46, 49
neurons and neural networks, 14, 44–55, 57, 69, 278
    cognitive reserves and, 268–70
    cross-training and, 281
    Hebbian principle and, 45, 56, 259
    midlife bilaterality and, 266
    mirror neurons, 53–54, 55
    multitasking and, 285–86
    neurotransmitters, 195
    shearing and, 45, 47, 49, 51, 57, 69, 285, 286
Newsweek, 61, 64
New Visions for Public Schools, 88
New York Times, 231
Nim, 160–61
Nintendo, 149–50, 151, 156, 188

Nisbett, Richard, 37, 38
No Child Left Behind, 13, 93, 96,
123–24, 152
NTs (neurotypicals), 214–17
nursing homes, 264

O'Driscoll, Tony, 208–13, 214, 217,
218, 224, 225, 233
On Intelligence (Hawkins), 103
open-source production, 65, 191–92,
229, 242
see also crowdsourcing
O'Reilly, Tim, 219
Organisation for Economic
Co-operation and Development
(OECD), 75, 165, 169, 173

Pac-Man, 148
PalmPilot, 103
parkour (freerunning), 138–39, 140, 285
Parsons The New School for Design, 88
Partners in Health, 156–57
Pascalis, Olivier, 40–41
Patterson, Radonna, 255–56, 257, 262
Pew Internet and American Life Project,
145, 147, 154
phantom limbs, 250–51, 264
Phantoms in the Brain (Ramachandran),
250
phones, 57
iPhone, 14–15
physical therapy, 247–56, 268, 270
Piaget, Jean, 40
platoons (toons), 136, 137, 138, 145
play, 176–78, 195, 200, 204, 205
podcasts, 68
Pokémon, 62–63
Principles of Scientific Management, The
(Taylor), 181
printing, mass, 11, 71–72
programmed cell death, 48
Proximity Hotel, 234–41

Quaintance, Dennis, 235–41, 244, 272
Quaintance, Nancy, 235–36
Quest 2 Learn (Q2L), 87–91, 133, 156,
161

race and ethnicity, 40–41, 52
Raichle, Marcus, 280
Ramachandran, Vilayanur S., 250–51
Rand, Ayn, 228
Raskin, Aza, 174–80, 195, 204
Raskin, Jef, 180
rational choice theory, 232
Ravitch, Diane, 123
Raymond, Eric S., 229
reflexive behaviors, 45, 46
Regan, Margaret, 219–23, 224, 225, 233
Rheingold, Howard, 277–79, 280, 283
Richardson, Henry Hobson, 182
Riley, Olive, 275, 276
Rizzolatti, Giacomo, 53
Rohde, David, 231
Rosenzweig, Roy, 124–25

Salen, Katie, 87–88, 89, 92, 104, 161
Salthouse, Timothy, 121–23, 254
Samsung, 288
Sang, Hang, 84–85
Sanger, Larry, 227
Scholastic Aptitude Test (SAT), 115–16,
122, 125
Schooler, Jonathan, 282
schools, *see* education and schools
Schwartz, Robert, 74
Second Life, 201–2, 203, 207, 209–13,
218, 233
self-awareness, 50
Seoul, 288–90
Shirky, Clay, 11
Shop Class as Soulcraft (Crawford), 81
Shop Management (Taylor), 181
Simon, Théodore, 117–18
Simons, Daniel, 3, 4

Sitkin, Sim, 18
skills:
  cognitive reserves and, 269
  complementary, 263
  literacy, 19
  and seeing talents where others see
    limits, 214–19
  twenty-first-century, 125–26, 127
skyscrapers, 182, 184
Smart Mobs (Rheingold), 277
social Darwinism, 120
social networks, 70, 174, 185, 232,
    269, 270
  Facebook, 70, 171, 174, 192, 232,
    270, 275
  MySpace, 270, 275
  Twitter, 12, 65, 70, 107, 127, 174,
    177, 185, 192, 242, 271–73
software performance testers, 214–19
Sonne, Thorkil, 214–19, 225, 233, 241
Sony, 66, 91
South Korea, 75
  Davidson's visit to, 287–90
Space Invaders, 148
space program, 13, 74, 81
Specialisterne, 214–19
specialization, 70, 73–74
speech patterns, 27
standards, 279
  in education, 13, 57, 71–72, 75, 76,
    77–78, 79, 81, 86, 93, 96–97, 111,
    113, 114, 116, 153–54, 160;
    see also grading; tests
Stanford-Binet Intelligence Scale,
    118–19
Stanford University, 277, 283–84, 290
statistical methods, 119–20, 183, 186
Stein, David, 92, 93, 97, 98
Stern, William, 118
Stern, Yaakov, 267–69
Stone, Linda, 287
Stutzman, Fred, 174

Sullivan, Louis, 182
surprise, 286
Switzerland, 75

talents, 223, 241, 257–58
  limitations as, 214–19
Taliban, 231
Tapscott, Don, 200, 231
Taylor, Frederick Winslow, 181–83,
    186–87, 190, 199, 205, 233, 241,
    242, 279–80, 282, 286
teachers, 126
  inspiring, 81–86, 128, 131, 141
teaching, 54–55
technology, 13, 14, 16, 152, 172, 281, 287
television, 7, 17, 70, 157, 271
  studies on, 148
television commercials, 24–25, 28, 29, 58
  cultural differences and, 30
  for Cymbalta, 23–29, 30, 32, 33, 42,
    50, 279
  voice-overs in, 27
Terman, Lewis, 118
term papers, 100–102
tests, 13, 17, 75, 77, 79, 85, 107–10,
    113–31, 154, 183
  adaptive or progressive, 126
  end-of-grade, 79, 80, 89, 93, 94, 117,
    121, 124, 125, 129–31, 134,
    160, 191
  essay, 115
  H-Bot and, 124–25, 267
  IQ, 51, 114, 117–19, 120, 121, 149,
    183, 186, 254, 266
  languages and, 94, 110, 119
  multiple-choice, 57, 81, 110, 112,
    113–15, 117, 120, 121, 124, 125,
    128, 139, 152, 160, 169, 183, 186
  Salthouse and, 121–23, 254
  SAT, 115–16, 122, 125
  silent reading, 114, 115, 116
  standard of living and, 121

statistical methods and, 119–20
for twenty-first-century skills, 125–26,
127
within-person deviations in, 122–23
see also grading
text messaging, 7, 101, 169
This Is Your Brain on Music (Levitin), 99
This Is Your Brain on the Internet,
99–104, 105–9, 111, 161,
260–61, 264
Thomas, Douglas, 158
Tichnor, Sarah, 97–98
time:
school and, 73, 173
work and, 172–73, 181, 184, 223, 279
Toffler, Alvin, 19, 159, 269
toons (platoons), 136, 137, 138, 145
Towers Perrin, 219–20, 221, 223
travel, 202
Turkey, 75
Twitter, 12, 65, 70, 107, 127, 174, 177,
185, 192, 242, 271–73

uniquely qualified minds, 231, 263
University of California–Irvine, 170
University of California–Los Angeles,
153
University of California–Santa Barbara,
282
University of Idaho, 116–17
University of North Carolina, 103
University of Pennsylvania, 182–83, 187
United Nations World Urban Forum, 198
"Use of Knowledge in Society, The"
(Hayek), 229

values, *see* cultural values and scripts
VCRs, 256–57, 258, 259, 260, 262,
263, 270
violence, 153
video games and, 147, 150–52,
153, 154

virtual environments, 201–3, 204, 206,
207, 209–13, 218, 233
Virtual Peace, 156–57
Virtual Reality (Rheingold), 277
voice characteristics, 27
of Motherese, 36, 38
voice-overs, 27
Voyager Academy, 132–40, 141, 147, 197

Waag Society, 274–75
Wales, Jimmy, 225–33, 242
walking, 46, 48–49
Walkman, 66
Washington University, 280
Weekend Executive MBA Program,
18–19
whales, 54–55
Wharton School of Business, 182–83, 187
Whitney Museum of American Art, 144
wikinomics, 200
Wikipedia, 70, 100, 169, 192, 225–33,
241, 242
Williams syndrome, 51–53
Wilson, E. O., 90–91
Wilson, Woodrow, 186
Wired, 64–65
work, workplace, 6, 10, 11, 12–14, 20,
46, 51, 165–207, 208–44, 257, 291
Battier and, 224–25, 226, 241, 263
business letters and, 173–74
business schools and, 182–84, 187
careers and occupations, 18–19
cognitive reserves and, 269
computers in, 167, 168, 173, 175–79
corporate jamming and, 198–200, 204
creative thinking and, 77
customizing workflow in, 178–79
decentralization in, 191, 218
distraction and, 165, 167, 169–70,
171, 180, 185, 195, 283
e-mail and, 14, 173–74, 177, 179, 282
endeavor-based, 196–97, 200

work (*cont.*)
  human resources departments and,
    183, 184, 185, 189
  IBM and, 109, 188–94, 196–200,
    202–5, 208, 284
  industrial, 6, 12, 13, 17, 57, 72–74, 76,
    77–78, 160, 165–66, 172, 180–82,
    187, 190, 204, 206, 243, 264
  Internet and, 165–69, 172, 174–78,
    186
  labor force growth rate, 221
  leisure and, 170, 186–87, 279
  management theory and, 181–86
  Mark and, 170–71, 174, 181
  mass collaboration in, 225–34, 241
  multitasking and, 175, 177, 180, 204
  number of hours worked, 165
  O'Driscoll and, 208–13, 214, 217,
    218, 224, 225, 233
  office hierarchy and, 183
  as part of life, 219–23
  physical spaces for work, 172,
    182–83, 184
  pink-collar, 183–84, 186, 257
  play and, 176–78, 195, 200, 204, 205
  procrastination and, 178
  Proximity Hotel and, 234–41
  Raskin and, 174–80, 195, 204
  Regan and, 219–23, 224, 225, 233
  Second Life and, 201–2, 203, 207,
    209–13, 218, 233
  seeing future of, by refusing its past,
    234–41
  and seeing how and when to best
    work together, 224–25
  seeing ourselves as others see us in,
    208–13
  and seeing talents where others see
    limits, 214–19
  self-motivation and, 184
  Sonne and, 214–19, 225, 233, 241
  stress and, 282
  task-specific buildings and, 184
  task switching in, 170–71
  Taylorism and, 181–83, 186–87, 190,
    199, 205, 233, 241, 242, 279–80,
    282, 286
  time and, 172–73, 181, 184, 223,
    279
  virtual meeting spaces and, 201–3,
    204, 206, 207, 209–13, 218, 233
  Wales and, 225–33, 242
  white-collar, 76–77, 81, 183, 186,
    257
World Bank Institute, 144
World of Warcraft, 70, 102, 142, 146,
    158
World Wide Web, 6, 7, 14, 17, 62, 63,
    66, 70, 141, 165–67, 169, 177–80,
    191–92, 205, 242, 258, 270
  see also Internet
World Without Oil, 144
Wrigley Corporation, 217
writing, invention of, 11

Xerox, 273

Yerkes, Robert, 119

Zuckerberg, Mark, 232